Cambridge Studies in Biological and Evolutionary Anthropology 59

Reproduction and Adaptation

In the space of one generation major changes have begun to take place in the field of human reproduction. A rapid increase in the control of fertility, and the understanding and treatment of sexual health issues, has been accompanied by an emerging threat to reproductive function linked to increasing environmental pollution and dramatic changes in lifestyle.

Organised around four key themes, this book provides a valuable review of some of the most important recent findings in human reproductive ecology. Major topics include the impact of the environment on reproduction; the role of physical activity and energetics in regulating reproduction; sexual maturation and ovulation assessment; and demographic, health and family planning issues. Both theoretical and practical issues are covered, including the evolution and importance of the menopause and the various statistical methods by which researchers can analyse characteristics of the menstrual cycle in field studies.

C. G. NICHOLAS MASCIE-TAYLOR is Professor of Human Population Biology and Health at the University of Cambridge. His main fields of research are the inter-relationships between nutrition, growth, disease and reproductive ecology.

LYLIANE ROSETTA is a recently retired senior scientist from the National Centre for Scientific Research (CNRS), France. Her research is devoted to the study of the regulation of fertility in relation to nutrition, energetics and the resulting energy availability.

Cambridge Studies in Biological and Evolutionary Anthropology

Series editors

Also available in the series

42 *Simulating Human Origins and Evolution* Ken Wessen 0 521 84399 5
43 *Bioarchaeology of Southeast Asia* Marc Oxenham & Nancy Tayles (eds.) 0 521 82580 6
44 *Seasonality in Primates* Diane K. Brockman & Carel P. van Schaik 0 521 82069 3
45 *Human Biology of Afro-Caribbean Populations* Lorena Madrigal 0 521 81931 8
46 *Primate and Human Evolution* Susan Cachel 0 521 82942 9
47 *The First Boat People* Steve Webb 0 521 85656 6
48 *Feeding Ecology in Apes and Other Primates* Gottfried Hohmann, Martha Robbins & Christophe Boesch (eds.) 0 521 85837 2
49 *Measuring Stress in Humans: A Practical Guide for the Field* Gillian Ice & Gary James (eds.) 0 521 84479 7
50 *The Bioarchaeology of Children: Perspectives from Biological and Forensic Anthropology* Mary Lewis 0 521 83602 6
51 *Monkeys of the Taï Forest* W. Scott McGraw, Klaus Zuberbühler & Ronald Noe (eds.) 0 521 81633 5
52 *Health Change in the Asia-Pacific Region: Biocultural and Epidemiological Approaches* Ryutaro Ohtsuka & Stanley I. Ulijaszek (eds.) 978 0 521 83792 7
53 *Technique and Application in Dental Anthropology* Joel D. Irish & Greg C. Nelson (eds.) 978 0 521 870 610
54 *Western Diseases: An Evolutionary Perspective* Tessa M. Pollard 978 0 521 61737 6
55 *Spider Monkeys: The Biology, Behavior and Ecology of the Genus Ateles* Christina J. Campbell 978 0 521 86750 4
56 *Between Biology and Culture* Holger Schutkowski (ed.) 978 0 521 85936 3
57 *Primate Parasite Ecology: The Dynamics and Study of Host-Parasite Relationships* Michael A. Huffman & Colin A. Chapman (eds.) 978 0 521 87246 1
58 *The Evolutionary Biology of Human Body Fatness: Thrift and Control* Jonathan C. K. Wells 978 0 521 88420 4

Reproduction and Adaptation

Edited by

C. G. Nicholas Mascie-Taylor

Department of Biological Anthropology, University of Cambridge, UK

Lyliane Rosetta

CNRS (National Centre for Scientific Research), Paris, France

The Second Parkes Foundation Workshop

CAMBRIDGE
UNIVERSITY PRESS

University Printing House, Cambridge CB2 8BS, United Kingdom

One Liberty Plaza, 20th Floor, New York, NY 10006, USA

477 Williamstown Road, Port Melbourne, VIC 3207, Australia

314-321, 3rd Floor, Plot 3, Splendor Forum, Jasola District Centre, New Delhi - 110025, India

103 Penang Road, #05-06/07, Visioncrest Commercial, Singapore 238467

Cambridge University Press is part of the University of Cambridge.

It furthers the University's mission by disseminating knowledge in the pursuit of education, learning and research at the highest international levels of excellence.

www.cambridge.org
Information on this title: www.cambridge.org/9780521509633

First published 2011

A catalogue record for this publication is available from the British Library

Library of Congress Cataloging in Publication data
Reproduction and adaptation : topics in human reproductive ecology / [edited by] C. G. Nicholas Mascie-Taylor, Lyliane Rosetta.
p. cm. – (Cambridge studies in biological and evolutionary anthropology)
ISBN 978-0-521-50963-3 (hardback)
1. Human reproduction – Environmental aspects. 2. Fertility, Human – Environmental aspects. 3. Human ecology. I. Mascie-Taylor, C. G. N. II. Rosetta, Lyliane. III. Title. IV. Series.
QP251.R4432 2010
612.6 – dc22 2010038764

ISBN 978-0-521-50963-3 Hardback

Contents

Contributors

LINDA H. BEARINGER Center for Adolescent Nursing, University of Minnesota, Minneapolis, USA

DARNA L. DUFOUR Department of Anthropology, University of Colorado, Boulder, Colorado, USA

GEOFF P. GARNETT Department of Infectious Disease Epidemiology, Imperial College London, UK

ANNA GLASIER University of Edinburgh, UK

ANDREW HINDE Southampton Statistical Sciences Research Institute, University of Southampton, Southampton, UK

MICHAEL JOFFE Division of Epidemiology and Biostatistics, Imperial College London, UK

LAURENCE JOUBIN Unité de Recherche Clinique (URC) Hôpital Necker-Enfants Malades, Paris, France.

PHYLLIS C. LEE Behaviour and Evolution Research Group, Department of Psychology, University of Stirling, UK

C. G. NICHOLAS MASCIE-TAYLOR Department of Biological Anthropology, University of Cambridge, Cambridge, UK

ALAN S. McNEILLY Medical Research Council Human Reproductive Sciences Unit, Queens Medical Research Institute, Edinburgh, UK

LYLIANE ROSETTA National Centre for Scientific Research (CNRS), GDR 2655, Paris, France

MOLLY SECOR-TURNER Center for Adolescent Nursing, University of Minnesota, Minneapolis, USA

PRAKASH SHETTY School of Medicine, University of Southampton, Southampton, UK

LYNNETTE LEIDY SIEVERT Department of Anthropology, UMass Amherst, Amherst, Massachusetts, USA

RENEE E. SIEVING Center for Adolescent Nursing, University of Minnesota, Minneapolis, USA

JEAN-CHRISTOPHE THALABARD University Paris Descartes, MAP5 UMR CNRS 8145 and Endocrine–Gynaecology Unit, Paris, France

JACQUELINE M. WALLACE Rowett Research Institute, Bucksburn, Aberdeen, UK

JOHN P. WIEBE Hormonal Regulatory Mechanisms, Department of Biology, The University of Western Ontario, London, Ontario, Canada

Preface

The Second Parkes Foundation Workshop was held in Cambridge in September 2007. The Foundation is a charity formed to promote the study of the biosocial sciences, i.e. the interdisciplinary area between biology and sociology, the publication of the results of such studies and to support related charitable activities of an educational or scientific nature. Each year it supports six to eight Masters and PhD students in their field studies.

The theme of the First Parkes Foundation Workshop was Human Adaptability. Here in the second volume the focus is on Reproduction and Adaptation in which four themes are explored: reproduction and the environment, energetics and adaptation, sexual maturation and ovulation assessment, and demography, health and fertility.

The editors would like to thank The Parkes Foundation for their generosity in funding the two-day meeting, and the speakers and authors of the chapters for their interest and commitment. We would like to express special thanks to Dr Rie Goto, who was responsible for most of the conference logistics, and also helped in preparing the manuscript for publication.

1 *Reproduction and environment*

ALAN S. McNEILLY

Introduction: when is the timing of birth a priority for a species?

Two factors play an important part in determining the optimum number of offspring that any species should produce each pregnancy, the optimum number for the fetuses to develop normally in the uterus, and the ability of the mother to adequately suckle the newborn to the point where they are able to survive the rigors of the environment. The first is of prime importance and is determined by a relatively tight control over the number of pre-ovulatory follicles that develop and ovulate, while the second is determined by when the offspring are born, and how frequently births occur in relation to the development of the young. In some species time of year of birth is not necessarily a major consideration, e.g. where seasonal variations in food supply are not great, but in more extreme climates timing of birth is of crucial importance as there must be sufficient time for the young to be able to withstand harsh climates and variations in food supply. In these situations the time of breeding needs to be controlled to limit the periods of fertility, so that if conception occurs birth can still take place at an optimum time of year. In most species reproduction is very efficient and usually ovulations will result in pregnancy. Only when we humans try to influence, or interfere with, either the timing of events or the numbers of ovulations that take place in normal-weight women on adequate diets, can problems arise.

It is also important to remember that of all the oocytes laid down in fetal life, and of all the oocytes that activate and induce follicle development, the number of ovulations that actually take place and lead to offspring are infinitesimally small. Thus the development and selection of the pre-ovulatory dominant follicle is a very rare event. For instance in women the maximum recorded numbers of babies born to any one woman is around thirty in her reproductive life span, and if the woman had breastfed this would naturally have been reduced to ten

Reproduction and Adaptation, eds. C. G. Nicholas Mascie-Taylor and Lyliane Rosetta.
Published by Cambridge University Press.

or eleven, owing to the suppressive effects of breastfeeding on the return of ovulations after pregnancy (McNeilly, 2006). Even in highly fecund species such as mice, the total numbers of offspring are quite small compared with the undoubted continued availability of oocytes and the capacity of the mouse ovary to develop viable pre-ovulatory follicles.

Because activation of oocytes in the ovary and the subsequent development of the activated oocyte/follicle complex to the antral stage is independent of regulators outside the ovary, this is not normally the major controller of the ovulation rate. The crucial element in the timing of successful reproduction thus lies in the regulation of the final stages of follicle development through selection and final maturation of the dominant pre-ovulatory follicle(s), and this is regulated from outside the ovary by the pituitary gonadotrophins LH (luteinizing hormone) and FSH (follicle stimulating hormone) acting in concert and sequentially on the follicle(s). The fundamental controller of the timing of these events is the GnRH (gonadotrophin-releasing hormone) pulse generator in the hypothalamus which determines the pulse frequency of release of GnRH. This determines the pattern of gonadotrophin secretion, which ultimately regulates the final stages of follicle development and timing of ovulation, leading to potential conception and pregnancy.

Of course exactly when ovulation takes place is dependent for timing on when parturition takes place, since this determines when the young are born and must start their survival strategy. Thus the length of pregnancy is also a timing controller. Where pregnancy is short, such as in rodents, then all these events can take place in the breeding season, while in others – e.g. sheep and red deer, where birth should ideally take place at a time in the spring when weather conditions are not too adverse – the length of pregnancy determines that fertility resumes in the autumn to control when conception occurs. In humans the timing of birth normally has no consequences, provided there is an adequate supply of food for the mother to produce milk. There is evidence that there is a seasonal regulation over the time of year when births and thus conceptions occur (Lerchl *et al.*, 1993; Roenneberg & Aschoff, 1990a, b), and it has been suggested that this occurs through changes in light intensity, which impacts on when conceptions occur (Cummins, 2007). However, understanding the mechanisms whereby this may occur is still lacking, and the effect is only seen in a minority of women, and thus might be viewed as a weak physiological controller of the timing of conception. Other physiological factors which are certainly known to regulate the timing of conception in humans, and hence adaptation of reproduction, will be discussed later. This chapter will discuss the fundamental physiological mechanisms regulating exactly when ovulation will take place, and thus conception and ultimately birth of a normal healthy baby.

Regulating the timing of ovulation

All oocytes in the ovary are produced during fetal life (see Krysko *et al.*, 2008). While there has been a suggestion that there may be some renewal of oocytes after birth (Johnson *et al.*, 2005) this remains very controversial (Bristol-Gould *et al.*, 2006; Eggan *et al.*, 2006; Liu *et al.*, 2007; Telfer *et al.*, 2005). It is clear that the menopause in women is caused by a complete loss of oocytes and follicles in the ovary and so if any renewal was taking place this would only occur for a minimal time period. Approximately 20 oocytes are activated within the ovary every day. The mechanisms are not known but it is now clear that oocyte activation is associated with growth of follicles, which continue to develop under the control of local factors within the ovary and are independent of gonadotrophins from the pituitary (Scaramuzzi *et al.*, 1993). In the absence of gonadotrophins the growth of follicles is arrested at sizes specific to different species (mice: 180 um; sheep: 2.5 mm; cattle: 5 mm; women: 5 mm). From these sizes onwards growth is dependent on stimulation by FSH that drives follicle growth to the pre-ovulatory size (Figure 1.1).

In the menstrual cycle, during the luteal phase, follicle growth remains at the gonadotrophin-independent arrested stage because FSH levels in plasma are suppressed below the threshold for stimulation of follicle growth (Figure 1.1). A rise in plasma concentrations of bio- FSH and immunoreactive FSH occurs around menses when secretion of estradiol and inhibin A from the corpus luteum, which are inhibitory to FSH secretion, decline as the corpus luteum fails in the absence of conception (Groome *et al.*, 1996; Reddi *et al.*, 1990). This rise in FSH above the threshold stimulates the final stages of follicle development over the next 14 days. However, as the follicles grow each begins to secrete estradiol (Backstrom *et al.*, 1982) and inhibin B (Groome *et al.*, 1996) which feed back directly on to the pituitary to suppress FSH secretion below the threshold for all but one or two follicles, which continue to grow to a pre-ovulatory size (McNeilly *et al.*, 2003).

The mechanisms whereby a single follicle(s) is selected to continue to grow relates to the development of LH receptors in the granulosa cell compartment of the follicle, which allows LH to act as a surrogate FSH and support the final stages of follicle development. In addition the follicle may become more sensitive to the low levels of FSH. Throughout this time follicle growth remains determined by the levels of FSH. On its own, LH cannot stimulate follicle growth, but is essential for steroid secretion by the developing follicle. It stimulates androgen production by the thecal layer surrounding the follicle, and this is converted to estradiol by aromatase induced in granulosa cells of the selected follicle(s). As the selected follicle develops through to a pre-ovulatory follicle it secretes increasing amounts of estradiol, which eventually

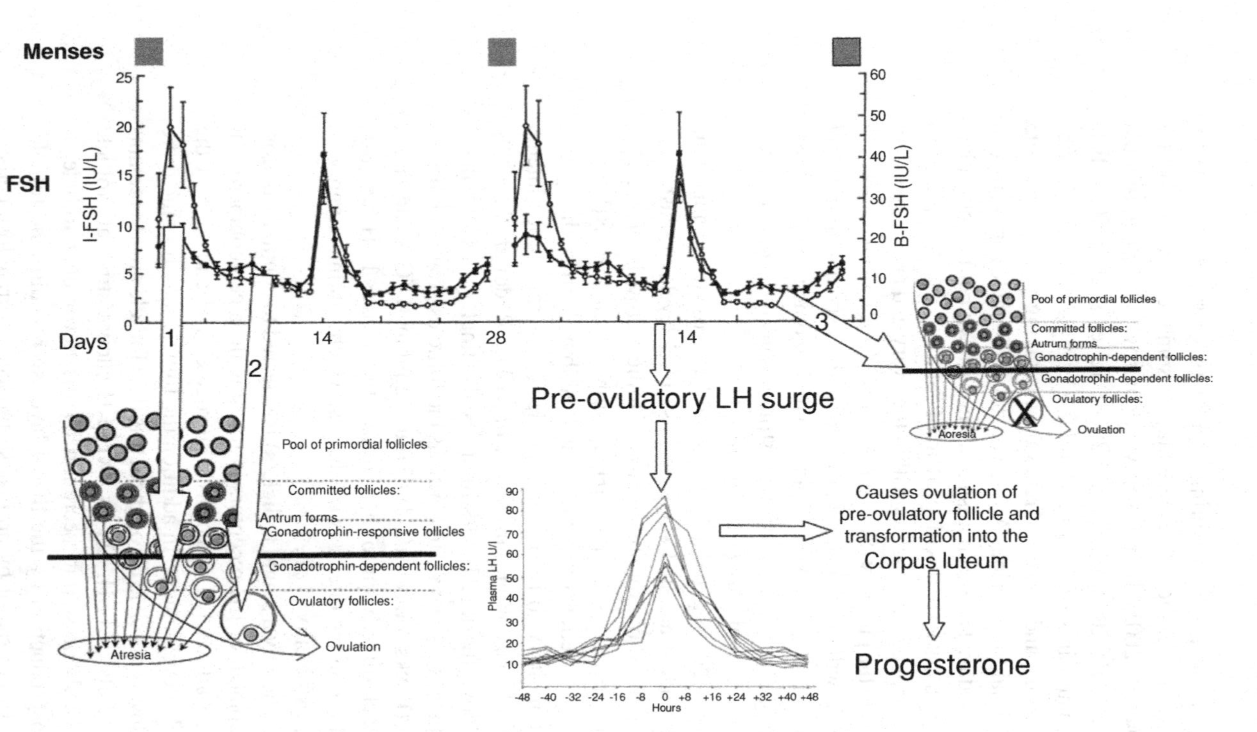

Figure 1.1. Changes in plasma concentrations of bioactive FSH (B-FSH: open circles), and immunoreactive FSH (I-FSH: closed circles) across the menstrual cycle and the relationship with follicle development, and selection of the dominant pre-ovulatory follicle. At menses the corpus luteum stops functioning, thus removing the source of estradiol and inhibin A that directly suppress secretion of FSH from the pituitary. As a consequence, plasma levels of FSH increase above the threshold necessary to stimulate the final stages of follicle development (Arrow 1). As follicles increase in size they secrete increasing amounts of estradiol (see Figure 1.2) and inhibin B that suppress plasma FSH below this threshold, and all but a single follicle that now has increased sensitivity to FSH and can respond to LH can survive (Arrow 2). After ovulation induced by the midcycle pre-ovulatory LH surge, the combined effects of estradiol and inhibin A secreted by the corpus luteum again suppress FSH secretion below the threshold required for follicle growth (Arrow 3). This sequence of events is maintained until conception occurs. (Redrawn from Djahanbakhch *et al.*, 1981; Reddi *et al.*, 1990, and Scaramuzzi *et al.*, 1993).

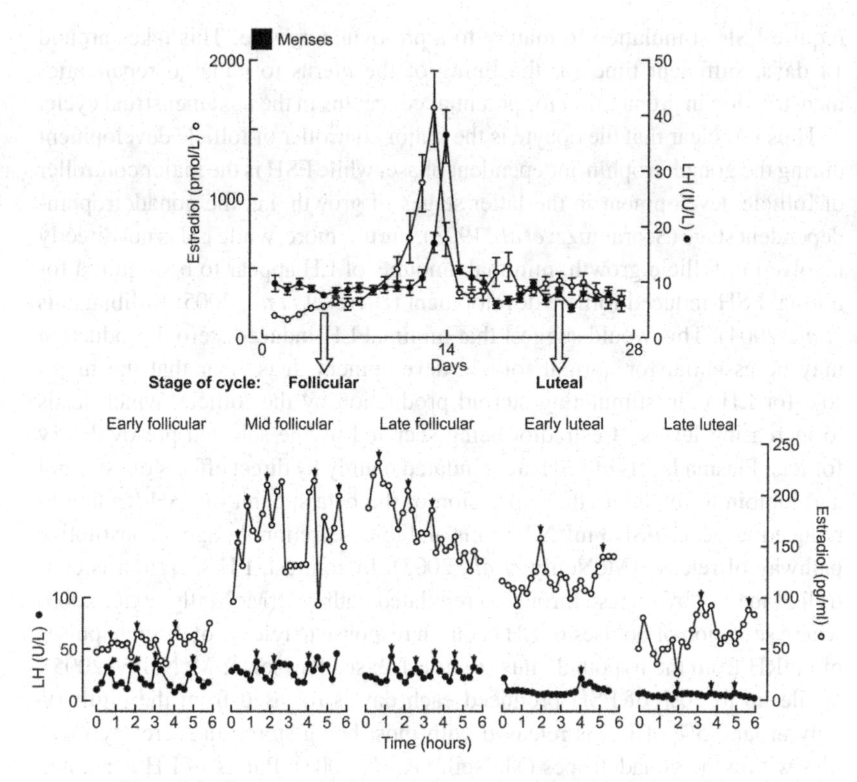

Figure 1.2. Changes in A) the mean plasma concentrations of estradiol (open circles) and LH (closed circles) and B) the pulsatile pattern of LH and estradiol secretion (indicated by arrows) over 6-hour periods during the menstrual cycle. The decline in progesterone secretion as the corpus luteum stops functioning at menses removes the inhibition on the GnRH pulse generator. This results in an increase in the frequency of LH/GnRH pulsatile secretion to around one pulse per hour throughout the follicular phase of the cycle, stimulating increased secretion and plasma concentrations of estradiol as follicles develop. Note that in most cases each LH pulse is followed by a pulse of estradiol. When the plasma concentrations of estradiol secreted by the maturing pre-ovulatory follicle reach a threshold this results in a surge release of GnRH, causing the release of the pre-ovulatory LH surge that induces ovulation of the pre-ovulatory follicle. The increase in progesterone secretion by the corpus luteum in the luteal phase then supresses the LH/GnRH pulse frequency to around one pulse per 4 hours which, if conception occurs, will remained suppressed through pregnancy and lactation. (Redrawn from Backstrom *et al.*, 1982 and Reddi *et al.*, 1990).

reach a threshold necessary to stimulate the release of the pre-ovulatory LH surge, leading to ovulation of the follicle and transformation into the corpus luteum (Backstrom *et al.*, 1982; Figure 1.2). During the luteal phase the corpus luteum secretes progesterone, estradiol and inhibin A which feedback on to the pituitary to suppress FSH secretion below the threshold for follicle growth beyond 5 mm (Figure 1.1). In this way, at the time of menstruation, follicles require FSH stimulation to mature to a pre-ovulatory size. This takes around 14 days, sufficient time for the lining of the uterus to undergo repair after menstruation in preparation for potential conceptus in the next menstrual cycle.

Thus it is clear that the oocyte is the major controller of follicle development during the gonadotrophin-independent phase, while FSH is the major controller of follicle development in the latter stages of growth, i.e. the gonadotrophin-dependent stage (Scaramuzzi *et al.*, 1993). Furthermore, while LH is not directly involved in follicle growth, minimal amounts of LH appear to be required for normal FSH-induced follicle development (Bjerckel *et al.*, 2005; Kolibianakis *et al.*, 2004). This would suggest that minimal LH-induced steroid production may be essential for normal follicle development. It is clear that the major role for LH is in stimulating steroid production by the follicle, which leads to increasing levels of estradiol being secreted by the selected pre-ovulatory follicle. Plasma levels of FSH are regulated mainly by direct effects of estradiol and inhibin to modulate the expression of the beta subunit of FSH leading to reduced levels of FSHß mRNA levels and thus secretion though a constitutive pathway of release (McNeilly *et al.*, 2003). In contrast, LH secretion is controlled mainly by release through a regulated pathway (McNeilly *et al.*, 2003) where secretion of boluses of LH occur in response to release of discrete pulses of GnRH from the hypothalamus (Figure 1.2: see Pawson & McNeilly, 2005). While up to 50% of FSH produced each day is released from the pituitary, only around 5% of LH is released with most being stored in secretory granules within the gonadotropes (McNeilly *et al.*, 2003). Pulses of LH generated by GnRH occur as a result of release of stored LH containing granules over a few minutes, and GnRH-induced pulsatile secretion of LH only occurs if LH-containing granules are present in gonadotropes (Crawford *et al.*, 2000). The secretion as a series of pulses is not essential for the follicles to respond. Single daily injections of LH are sufficient to maintain near normal secretion of estradiol, while infusion of the same total amount of LH can induce greater secretion of estradiol from follicles as the same amount given as a series of pulses over the same time period (Campbell *et al.*, 2007). However, biologically LH is released in pulses, and this gives an excellent means of regulating LH output.

Estradiol is crucial in two ways. Firstly it is essential for the repair of the uterus during the proliferative phase after menstruation and also induces progesterone receptor expression to allow progesterone, together with estradiol to prime the uterus for implantation during the secretory luteal phase of the cycle after menstruation (Jabbour *et al.*, 2006). The estradiol secreted by the follicle during the follicular phase of the cycle is dependent on the pulsatile secretion of LH, and thus pulses of GnRH. Secondly estradiol is the signal to the hypothalamus to stimulate the onset of the GnRH surge which is essential for the generation of the pre-ovulatory LH surge to allow ovulation and formation of the corpus luteum (Duncan 2000). If LH pulses are stopped by treatments with GnRH antagonists estradiol secretion declines rapidly and even though a pre-ovulatory follicle may be present there is no signal to the hypothalamus to relay this information. Furthermore the GnRH pulse generator in the hypothalamus fires approximately every hour during the follicular phase of the cycle (Figure 1.2) but is slowed during the luteal phase through the effects of progesterone. It is now clearly established through many studies that slowing the GnRH pulse generator and thus the pulsatile section of LH is sufficient to inhibit the menstrual cycle by removing the drive to estradiol secretion by the pre-ovulatory follicle, and the ability of the hypothalamus to monitor the stages of follicle development in the ovary through sensing these changes in estradiol. Thus modulation of the GnRH pulse generator is key to controlling the timing of reproduction in all species that have been adequately studied.

Physiological controllers of reproduction

Puberty: Before puberty secretion of both LH and FSH are suppressed and this is related to an absence of GnRH secretion. In both girls and boys, as puberty progresses pulsatile secretion of LH occurs during the night associated with sleep and the amplitude of secretion increases throughout puberty until an adult pattern of pulses throughout the day and night occurs at the end of puberty (Plant & Barker-Gibb, 2004; Wu *et al.*, 1996).

Nutrition: While severe undernutrition or obesity is associated with an absence of fertility and menstrual cycles it appears that the effects are related to the amount of available energy and not necessarily to the amount of nutrition per se (Loucks, 2003; The ESHRE Capri Workshop Group, 2006). In a series of elegant studies it was shown that decreasing the level of available energy resulted in a rapid decrease in the frequency of pulsatile LH secretion implying that the effect was mediated through reduced frequency of GnRH pulsatile secretion (Loucks, 2003; Loucks & Thuma, 2003). The effects were rapidly reversible and there was no effect on FSH secretion. Exactly how the effects of energy are relayed to the hypothalamus are still unclear, but initial rapid effects occur directly at the hypothalamus. Longer-term effects of undernutrition leading to chronic energy deficiency also have direct effects on the ability of the ovary to respond to gonadotrophins, an effect caused in some part by altered growth hormone secretion affecting liver production of insulin-like growth-factor 1 (IGF1) which normally maintains ovarian gonadotrophin responsiveness (Webb *et al.*, 2004). Obesity, which is becoming a major pathology, also affects reproductive function but the several mechanisms – including interference with the normal activity of the GnRH pulse generator – are only currently being unraveled (The ESHRE Capri Workshop Group, 2006).

Lactation: There is no doubt that in adult life the major physiological controller of reproduction in most species is lactation (see McNeilly, 2006). Depending on the species, newborns can be barely developed as in marsupials, or be almost able to fend for themselves without much, if any, parental care, e.g. guinea pigs. However, almost all rely on maternal milk for a period until they are capable of taking and utilizing solid nutrients alone. As well as the requirement for nutrition, milk also supplies immunological protection from infections through immunoglobulins secreted into the milk (see McNeilly, 2001). Furthermore in most species the suckling stimulus suppresses fertility delaying conception and the arrival of the next infant until the present incumbent is ready to progress to a life independent of milk. In women the interbirth interval can have dramatic effects on the survival of siblings, with infant death rates as high as 50% with short interbirth intervals (Hobcraft *et al.*, 1985). It is now clear that the principal mechanism whereby suckling suppresses fertility in all species for which there are adequate data is initially by abolishing, and then maintaining, a reduced frequency of GnRH pulsatile secretion (McNeilly, 2006).

The degree and duration of suppression is dependent on the pattern of sucking, both in frequency and duration. Long durations of infertility are associated with both frequent short duration of suckling and less frequent but longer durations of over a day. Each mother–infant pair is individual and so the patterns of breastfeeding are different. Nevertheless guidelines have been developed which identify that a mother who is fully breastfeeding with minimal supplements will remain infertile for at least 6 months and this may be extended for prolonged periods even in well-nourished societies (Kennedy *et al.*, 1989, 1996; Labbok *et al.*, 1994; WHO 1998a, b). During the periods of infertility, pulsatile GnRH/LH remains below the required hourly frequency (Tay *et al.*, 1992), and a normal ovulatory cycle and menstruation can be induced by treatment with hourly GnRH pulses (Glasier *et al.*, 1986; Zinaman *et al.*, 1995). Furthermore, in the later stages of infertility, FSH secretion returns to within normal limits and normal follicle development to pre-ovulatory follicle sizes can occur (Peerheentupa *et al.*, 2000). Nevertheless, because of lack of normal pulsatile LH/GnRH secretion, estradiol production by these follicles is low and thus the hypothalamus has no readout of the presence of these potentially ovulatory follicles, a pre-ovulatory LH surge does not occur, and infertility is maintained. The timing of when an infant weans and allows the mother to return to fertility is dependent on the infant. If the baby is supplied with supplementary food then it will adjust its nutrient intake and reduce the amount of sucking and breast milk intake. This reduced suckling reduces the suppression on the GnRH pulse generator, which will return to normal frequency and the resulting return of ovulatory cycles. It is clear that suckling controls reproduction through affecting the frequency of pulsatile GnRH secretion.

Photoperiod: Photoperiod is a major regulator of fertility in a number of species. As with the other situations discussed above, the period of infertility is directly associated with a suppression of the GnRH pulse generator, and the mechanisms involved have been described in detail elsewhere (Hastings *et al.*, 2007; Hazlerigg & Loudon, 2008). Whether there is any influence from photoperiod, or similar seasonal effects on reproduction, in humans is much less clear (Roenneberg & Aschoff, 1990a, b). As discussed earlier, changes in the timings of birth during the year have been recorded but these reports do not discuss the important issues of how the timing of conception might have been altered – the duration of pregnancy is not a factor – or any serious mechanisms that might be involved. It has been suggested that alterations in the intensity of light might be involved since there is a correlation with these changes and occurrences of birth. However, this effect is very limited, and minimal in comparison with the effects of photoperiod seen in true seasonal species.

The fertile period and maintenance of pregnancy: Even if normal follicle development occurs and a fertilizable oocyte(s) is released in the menstrual cycle, both conception and then maintenance of pregnancy must be secured. In animals there is a clear indication of the fertile period when mating will normally result in pregnancies at a high frequency. This estrous behavior, induced by the same high levels of estradiol secreted from the pre-ovulatory follicles that induce the pre-ovulatory GnRH/LH surge, does not last long, but is the only time that a female will normally allow a male to mate. Furthermore it occurs within a few hours before the time of ovulation in both spontaneous and induced ovulating species. In women there is no similar overt sign of impending fertility, but the fertile window during which pregnancies occur after a single intercourse last up to six days before the onset of the pre-ovulatory LH surge, giving more flexibility in the timing of intercourse leading to a successful conception (Keulers *et al.*, 2007; Wilcox *et al.*, 1995, 2000). Nevertheless, the most fertile period is over the 2 days before and the day of the LH surge. However, while the duration of the menstrual cycle is often quoted as 28 days, this is a median; normal, potentially fertile menstrual cycles can last for shorter or longer periods. Most of the variation in cycle length is owing to variations in the duration of the follicular phase since the luteal phase of the cycle is relatively fixed at around 14 days (Duncan, 2000). This means that the time of ovulation and the fertile period could actually be timed accurately only from potential menses (14 to, say, 16 days), and is almost impossible to determine from the last menses. Observation of the changes in cervical mucus can predict the onset of the fertile period with a reasonable degree of success (Bigelow *et al.*, 2004). However, even if conception occurs this will not lead inevitably to a successful pregnancy. While miscarriages in the first trimester tend to occur because of fetal abnormalities, there are also failures of implantation in a number of women. Careful analysis of "normal" menstrual cycles for the appearance of human chorionic gonadotrophin (hCG), the pregnancy hormone released by a conceptus, has shown that as many as 22% of women lose a conceptus before implantation, while a further 31% loss occurs soon after implantation (Wilcox *et al.*, 1995). Thus fertility in humans is not high even if intercourse is timed perfectly.

Males: All the discussion so far has centered on the female since she is the one producing the egg, carrying any conceptus through to term, and then feeding the infant after birth. In fact the mechanisms initiating and then maintaining fertility in males are basically the same at the hypothalamic–pituitary level. During fetal life the testes produce testosterone independently of the pituitary, and this testosterone masculinizes the fetus. Follicle stimulating hormone stimulates the production of Sertoli cells, the cells that reside within the tubules in the testes and nurture sperm maturation. The number of Sertoli cells determines the final capacity of the testes to produce sperm since each Sertoli cell can only maintain a finite number of sperm at all stages of development (Sharpe, 2004). Before puberty the GnRH pulse generator is silent in males and as puberty is initiated pulsatile secretion of GnRH/LH occurs at night during sleep (Plant & Barker-Gibb, 2004; Wu *et al.*, 1996). The amplitude of secretion increases as puberty progresses until a normal 24-hour pattern of pulsatile secretion of LH every 1–4 hours begins (Bergendahl *et al.*, 1998). Once initiated, the GnRH/LH pulse generator continues throughout life with minimal changes in frequency apart from those imposed by energy deficits. In adult men, FSH seems to play a limited role, if any, in maintaining spermatogenesis. This is dependent on intra-testicular levels of testosterone stimulated by LH acting on the Leydig cells in the testes (Walton & Anderson, 2004; Zirkin, 1998). As in females, the frequency of the GnRH pulse generator can be altered by the effects of testosterone, estradiol and progesterone in men (Walton *et al.*, 2006). Thus human males remain fertile at all times unless there is some major pathology involved (Sharpe & Skakkebaek, 2008). While declines in sperm count have been noted in a number of countries, associated with an increase in hypospadias and cryptorchidism, male Testicular Dysgenisis Syndrome and a small increase in infertility, in the majority of cases males remain fertile (Sharpe & Skakkebaek, 2008). In seasonal breeding species such as rams, fertility is turned off as it is in females, by a photoperiodically induced suppression of the GnRH pulse generator (Lincoln *et al.*, 2006).

A final common pathway

While it has been clear for some time that steroids, particularly estradiol, can affect GnRH pulsatile secretion, either directly or through induction of other mediators, it was unclear how this effect was generated. Estradiol normally acts through estrogen receptors alpha or beta, which affect gene transcription. However, GnRH neurons have minimal or no estrogen receptors. While unraveling the causes of infertility in patients, a new controlling system was uncovered (see Gianetti & Seminara, 2008). The protein kisspeptin produced in the hypothalamus acts through the g-protein-coupled GPR 54 receptor to regulate neuronal activity and was shown to stimulate GnRH secretion (Castellano *et al.*, 2008; Dungan *et al.*, 2006; Gianetti & Seminara, 2008; Maeda *et al.*, 2007; Roa *et al.*, 2008). The GPR54 receptors are present on GnRH neurons, while estrogen receptors are present on kisspeptin neurons. In rodents there appear to be two sets of kisspeptin neurons with different responses to estrogen. The kisspeptin neurones in the arcuate (ARC) nucleus respond to estrogen by suppressing kisspeptin secretion leading to reduced GnRH secretion, while those in the anteroventral periventricular (AVPV) nucleus respond by increasing GnRH secretion. It has been suggested therefore that the regulation of pulsatile GnRH secretion is via the arcuate nucleus (ARC) kisspeptin neurons, while the GnRH surge is regulated via the AVPV nuclear kisspeptin neurones (Dungan *et al.*, 2006; Gianetti & Seminara, 2008; Maeda *et al.*, 2007). This emerging controlling system may answer how many factors regulate GnRH pulsatile secretion since the kisspeptin neurons appear to respond to changes in nutritional peptide modulating signals, and, in lactation at least, levels of kisspeptin are reduced in response to suckling (Yamada *et al.*, 2007).

Conclusions

Adapting reproduction to the prevailing external situations requires a simple all-or-none system. This is accomplished by the GnRH pulse generator that controls the secretion of the gonadotrophins, principally LH from the pituitary. In turn both FSH and LH regulate the final stages of ovarian follicle development to the point of ovulation. All the major factors which determine the reproductive cycle in both females and males appear to feed directly through changes in frequency of the GnRH pulse generator. It now appears that the kisspeptin neuronal system may play a pivotal role in the overall integration of the factors regulating GnRH neurons. The final common pathway appears to be universal and ensures a simple frequency regulator of reproduction.

Acknowledgements

It is a pleasure to thank my colleagues who have contributed to the research of my own group reported here, to Judy McNeilly for comments on this paper, and Ted Pinner for help with the graphics. My work is funded by the Medical Research Council (Unit grant G7 002.00007.01).

References

Backstrom, C.T., McNeilly, A.S., Leask, R.M. and Baird, D.T. (1982). Pulsatile secretion of LH, FSH, prolactin, oestradiol and progesterone during the human menstrual cycle. *Clinical Endocrinology*, 17, 29–42.

Bergendahl, M., Aloi, J.A., Iranmanesh, A., Mulligan, T.M. and Veldhuis, J.D. (1998). Fasting suppresses pulsatile luteinizing hormone (LH) secretion and enhances orderliness of LH release in young but not older men. *Journal of Clinical Endocrinology and Metabolism*, 83, 1967–75.

Bigelow, J.L., Dunson, D.B., Stanford, J.B., Ecochard, R. *et al.* (2004). Mucus observations in the fertile window: a better predictor of conception timing of intercourse. *Human Reproduction*, 19, 889–92.

Bjercke1, S., Fedorcsak, P., Åbyholm, T. *et al.* (2005). IVF/ICSI outcome and serum LH concentration on day 1 of ovarian stimulation with recombinant FSH under pituitary suppression. *Human Reproduction*, 20, 2441–7.

Bristol-Gould, S.K., Kreeger, P.K., Selkirk, C.G. *et al.* (2006). Fate of the initial follicle pool: empirical and mathematical evidence supporting its sufficiency for adult fertility. *Developmental Biology*, 298, 149–54.

Campbell, B.K., Kendall, N.R. and Baird, D.T. (2007). The effect of the presence and pattern of luteinizing hormone stimulation on ovulatory follicle development in sheep. *Biology of Reproduction*, 76, 719–27.

Castellano, J.M., Roa, J., Luque, R.M. *et al.* (2008). KiSS-1/kisspeptins and the metabolic control of reproduction: Physiologic roles and putative physiopathological implications. *Peptides*, Jun 21. [Epub ahead of print].

Crawford, J., Currie, R. and McNeilly, A.S. (2000). Replenishment of LH stores of gonadotrophs in relation to gene expression, synthesis and secretion of LH after the preovulatory phase of the sheep oestrous cycle. *Journal of Endocrinology*, 167, 453–63.

Cummins, D.R. (2007). Additional confirmation for the effect of environmental light intensity on the seasonality of human conceptions. *Journal of Biosocial Science*, 39, 383–96.

Djahanbakhch, O., McNeilly, A.S., Hobson, B.M. and Templeton, A.A. (1981). A rapid luteinising hormone radioimmunoassay for the prediction of ovulation. *British Journal of Obstetrics and Gynaecology*, 88, 1016–20.

Duncan, W.C. (2000). The human corpus luteum: remodelling during luteolysis and maternal recognition of pregnancy. *Reviews of Reproduction*, 5, 12–17.

Dungan, H.M., Clifton, D.K. and Steiner, R.A. (2006). Minireview: kisspeptin neurons as central processors in the regulation of gonadotropin-releasing hormone secretion. *Endocrinology*, 147, 1154–8.

Eggan, K., Jurga, S., Gosden, R., Min, I.M. and Wagers, A.J. (2006). Ovulated oocytes in adult mice derive from non-circulating germ cells. *Nature*, 441, 1109–14.

Gianetti, E. and Seminara, S. (2008). Kisspeptin and KISS1R: a critical pathway in the reproductive system. *Reproduction*, 136, 295–301.

Glasier, A., McNeilly, A.S. and Baird, D.T. (1986). Induction of ovarian activity by pulsatile infusion of LHRH in women with lactational amenorrhoea. *Clinical Endocrinology*, 24, 243–52.

Groome, N.P., Illingworth, P.J., O'Brien, M. *et al.* (1996). Measurement of dimeric inhibin-B throughout the human menstrual cycle. *Journal of Clinical Endocrinology and Metabolism*, 81, 1401–5.

Hastings, M., O'Neill, J.S. and Maywood, E.S. (2007). Circadian clocks: regulators of endocrine and metabolic rhythms. *Journal of Endocrinology*, 195, 187–98.

Hazlerigg, D. and Loudon, A. (2008). New insights into ancient seasonal life timers. *Current Biology*, 18, R795–804.

Hobcraft, J., McDonald, J.W. and Rutstein, S. (1985). Demographic determinants of infant and early child mortality: A comparative analysis. *Population Studies*, 39, 363–85.

Jabbour, H.N., Kelly, R.W., Fraser, H.M. and Critchley, H.O.D. (2006). Endocrine regulation of menstruation. *Endocrine Reviews*, 27, 17–46.

Johnson, J., Skaznik-Wikiel, M., Lee, H.J. *et al.* (2005). Setting the record straight on data supporting post-natal oogenesis in female mammals. *Cell Cycle*, 4, 1471–7.

Kennedy, K.I., Rivero, R. and McNeilly, A.S. (1989). Consensus statement on the use of breastfeeding as a family planning method. *Contraception*, 39, 477–96.

Kennedy, K.I., Labbok, M.H. and Van Look, P.F.A. (1996). Consensus statement: lactational amenorrhoea method for family planning. *International Journal of Gynaecology*, 54, 55–7.

Keulers, M.J., Hamilton, C.J.C.M., Franx, A., Ewvers, J.L.H. and Bots, R.S.G.M. (2007). The length of the fertile window is associated with the chance of spontaneously conceiving an ongoing pregnancy in subfertile couples. *Human Reproduction*, 22, 1652–6.

Kolibianakis, E.M., Zikopoulos, K., Schiettecatte, J. *et al.* (2004). Profound LH suppression after GnRH antagonist administration is associated with a significantly higher ongoing pregnancy rate in IVF. *Human Reproduction*, 19, 2490–6.

Krysko, D.V., Diez-Fraile, A., Criel, G. and Svistunov, A.A. (2008). Life and death of female gametes during oogenesis and folliculogenesis. *Apoptosis*, 13, 1065–87.

Labbok, M.H., Pérez, A., Valdés, V. *et al.* (1994). The lactational amenorrhea method (LAM): a postpartum introductory family planning method with policy and program implications. *Advances in Contraception*, 10, 93–109.
Lerchl, A., Simoni, M. and Nieschlag, E. (1993). Changes in seasonality of birth rates in Germany from 1951 to 1990. *Naturwissenschaften*, 80, 516–18.
Lincoln, G.A., Clarke, I.J., Hut, R.A. and Hazlerigg, D.G. (2006). Characterizing a mammalian circannual pacemaker. *Science*, 314, 1941–44.
Liu, Y., Wu, C., Lyu, Q. *et al.* (2007). Germline stem cells and neo-oogenesis in adult human ovary. *Developmental Biology*, 306, 112–20.
Loucks, A.B. (2003). Energy availability, not body fatness, regulates reproductive function in women. *Exercise and Sport Sciences Review*, 31, 144–8.
Loucks, A.B. and Thuma, J.R. (2003). Luteinizing hormone pulsatility is disrupted at a threshold of energy availability in regularly menstruating women. *Journal of Clinical Endocrinology and Metabolism*, 88, 297–311.
Maeda, K.-I., Adachi, S., Inoue, K., Ohkura, S. and Tsukamura, H. (2007). Metastin/kisspetin and control of estrous cycle in rats. *Reviews in Endocrinology and Metabolic Disorders*, 8, 21–9.
McNeilly, A.S. (2001). Maternal reproductive and lactational physiology in relation to the duration of exclusive breastfeeding. In: *Developmental Readiness of Normal Full-Term Infants to the Progress from Exclusive Breastfeeding to the Introduction of Complementary Foods: Reviews of the Relevant Literature Concerning Infant Immunologic, Gastrointestinal, Oral, Motor and Maternal Reproductive and Lactational Development*, ed. A.J. Naylor and A. Morrow. Wellstart International and the LINKAGES Project/Academy for Educational Development, (USAID), Washington, DC., pp. 27–34.
McNeilly, A.S. (2006). Suckling and the control of gonadotropin secretion. In *Physiology of Reproduction*, 3rd edition, ed. J.D. Neill and A. Plant, pp. 2511–51.
McNeilly, A.S., Crawford, J.L., Taragnat, C., Nicol, L. and McNeilly, J.R. (2003). The differential secretion of FSH and LH: regulation through genes, feedback and packaging. *Reproduction Supplement*, 61, 463–76.
Pawson, A.J. and McNeilly, A.S. (2005). The pituitary effects of GnRH. Animal Reproduction. *Science*, 88, 75–94.
Peerhentupa, A., Chritchley, H.O.D., Illingworth, P.J. and McNeilly, A.S. (2000). Enhanced sensitivity to steroid negative feedback during breast-feeding: Low-dose estradiol (Transdermal Estradiol Supplementation) suppresses gonadotrophins and ovarian activity assessed by inhibin B. *Journal of Clinical Endocrinology and Metabolism*, 85, 4280–6.
Plant, T.M. and Barker-Gibb, M.L. (2004). Neurobiological mechanisms of puberty in higher primates. *Human Reproduction Update*, 10, 67–77.
Reddi, K., Wickings, E.J., McNeilly, A.S., Baird, D.T. and Hillier, S.G. (1990). Circulating bioactive follicle stimulating hormone and immunoreactive inhibin during the normal menstrual cycle. *Clinical Endocrinology*, 33, 547–57.
Roa, J., Aguilar, E., Dieguez, C., Pinilla, L. and Tena-Sempere, M. (2008). New frontiers in kisspeptin/GPR54 physiology as fundamental gatekeepers of reproductive function. *Frontiers in Neuroendocrinology*, 29, 48–69.

Roenneberg, T. and Aschoff, J. (1990a). Annual rhythm of human reproduction: I. Biology, sociology, or both? *Journal of Biological Rhythms*, 5, 195–216.

Roenneberg, T. and Aschoff, J. (1990b). Annual rhythm of human reproduction: II. Environmental correlations. *Journal of Biological Rhythms*, 5, 217–39.

Scaramuzzi, R.J., Adams, N.R., Baird, D.T. *et al.* (1993). A model for follicle selection and the determination of ovulation rate in the ewe. *Reproduction, Fertility and Development*, 5, 459–78.

Sharpe, R.M. (2004). Sertoli cell endocrinology and signal transduction: androgen regulation. In: *Sertoli Cell Biology*, ed. M.K. Skinner and M.D. Griswold. Elsevier Science (USA), Vol. 12, pp. 199–216.

Sharpe, R.M. and Skakkebaek, N.E. (2008). Testicular dysgenesis syndrome: mechanistic insights and potential new downstream effects. *Fertility and Sterility*, 89, e33–8.

Tay, C.C.K., Glasier, A.F. and McNeilly, A.S. (1992). The twenty-four hour pattern of pulsatile luteinizing hormone, follicle stimulating hormone and prolactin release during the first eight weeks of lactational amenorhoea in breastfeeding women. *Human Reproduction*, 7, 951–8.

Telfer, E.E., Gosden, R.G., Bsykov, A.G. *et al.* (2005). On regenerating the ovary and generating controversy. *Cell*, 122, 821–22.

The ESHRE Capri Workshop Group (2006). Nutrition and reproduction in women. *Human Reproduction Update*, 12, 193–207.

Walton, M. and Anderson, R.A. (2004). Update on male hormonal contraceptive agents. *Expert Opinion on Investigational Drugs*, 13, 1123–33.

Walton, M.J., Bayne, R.A., Wallace, I., Baird, D.T. and Anderson, R.A. (2006). Direct effect of progestogen on gene expression in the testis during gonadotropin withdrawal and early suppression of spermatogenesis. *Journal of Clinical Endocrinology and Metabolism*, 91, 2526–33.

Webb, R., Garnsworthy, P.C., Gong, J.-G. and Armstrong, D.G. (2004). Control of follicular growth: Local interactions and nutritional influences. *Journal of Animal Science*, 82, E63–74.

WHO (1998a). The World Health Organization multinational study of breastfeeding and lactational amenorrhoea. I. Description of infant feeding patterns and of the return of menses. *Fertility and Sterility*, 70, 448–60.

WHO (1998b). The World Health Organization multinational study of breastfeeding and lactational amenorrhoea. II. Factors associated with the length of amenorrhoea. *Fertility and Sterility*, 70, 461–71.

Wilcox, A.J., Weinberg, C.R. and Baird, D.D. (1995). Timing of sexual intercourse in relation to ovulation. *New England Journal of Medicine*, 333, 1517–21.

Wilcox, A.J., Dunson, D. and Baird, D.D. (2000). The timing of the "fertile window" in the menstrual cycle: day-specific estimates from a prospective study. *British Medical Journal*, 321, 1259–62.

Wu, F.C.W, Butler, G.E., Kelnar, C.J.H, Huhtaniemi, I. and Veldhuis, J.D. (1996). Ontogeny of pulsatile gonadotropin-releasing hormone secretion from mid-childhood, through puberty, to adulthood in the human male: a study using deconvolution analysis and an ultrasensitive immunofluorometric assay. *Journal of Clinical Endocrinology and Metabolism*, 81, 1798–805.

Yamada, S., Uenoyama, Y., Kinoshta, M. *et al.* (2007). Inhibition of metastin (kisspeptin-54)-GPR54 signaling in the arcuate nucleus-median eminence region during lactation in rats. *Endocrinology*, 148, 2226–32.

Zinaman, M.J., Cartledge, T., Tomai, T., Tippett, P. and Merriam, G.R. (1995). Pulsatile GnRH stimulates normal-cycle ovarian function in amenorrhoeic lactating postpartum women. *Journal of Clinical Endocrinology and Metabolism*, 80, 2088–93.

Zirkin, B.R. (1998). Spermatogenesis: its regulation by testosterone and FSH. *Seminars in Cell and Developmental Biology*, 9, 417–21.

2 *Genetic damage and male reproduction*

MICHAEL JOFFE

What has happened to the male reproductive system?

The health of the male reproductive system deteriorated sharply during the twentieth century. Testicular cancer increased four-fold or more in the space of several decades, throughout the world in populations of European ancestry, and in certain others, e.g. Maoris in New Zealand (Adami *et al.*, 1994; Joffe, 2001; Parkin, 2005). There is pathological evidence that the disease process starts in early life (Skakkebaek *et al.*, 1987), and in accordance with this the time trends show the clearest patterns if looked at by birth cohort, e.g. the incidence stopped increasing for ten years around 1940 in Denmark, Norway and Sweden (but not Finland, East Germany or Poland), then resumed its rapid rise (Figure 2.1)(Bergström *et al.*, 1996). The increase started in men born in the late nineteenth century in England and Wales (Davies, 1981), and in the first decade of the twentieth century in the Nordic countries, Germany and Poland (Bergström *et al.*, 1996). Many features of the epidemiological data are consistent with risk being associated with increasing prosperity.

Paradoxically, although testis biology is extremely well conserved through evolution, there is no satisfactory animal model for this disease, so the evidence is limited to epidemiological studies, genetic studies and clinical research (Skakkebaek, 2007). These are complicated by the existence of two main types of testicular cancer: seminoma and non-seminoma (divided into embryonal cell carcinoma, teratoma, choriocarcinoma and mixed-cell type). They share epidemiological characteristics, and therefore these different outcomes probably share one or more risk factors.

The evidence on male fertility is less certain. A report that sperm density declined 50% in 50 years worldwide (Carlsen *et al.*, 1992) can be criticised on methodological grounds, and for its implicit assumption that not only trends but also levels of sperm density are the same everywhere (Joffe, 2001). However, more reliable single-centre studies have shown a decline of approximately 2% a year in density, as well as 0.5% deterioration in sperm motility and morphology,

Reproduction and Adaptation, eds. C. G. Nicholas Mascie-Taylor and Lyliane Rosetta.

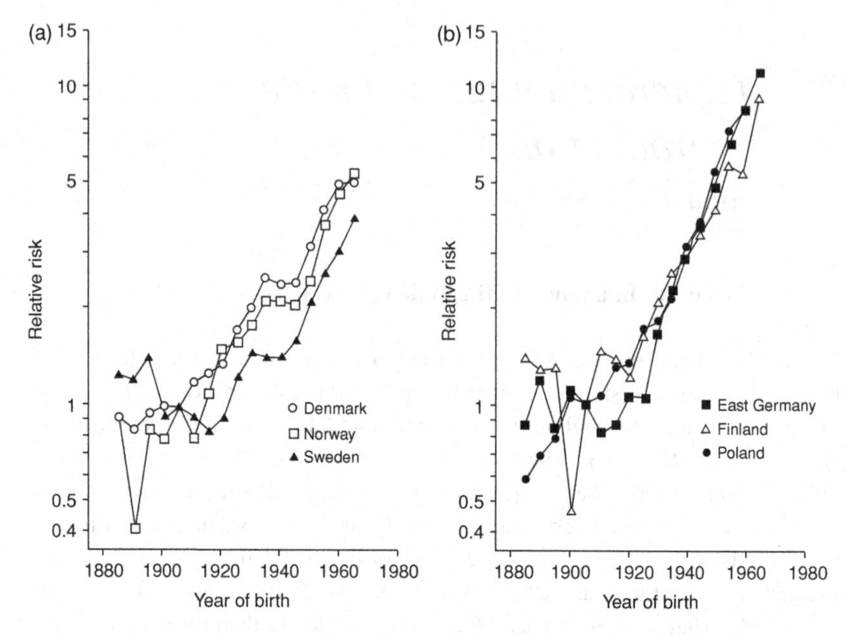

Figure 2.1. Trends in testicular cancer incidence in six European countries, age-adjusted, relative to 1905=100, by year of birth. Reproduced with kind permission from Oxford University Press (Bergström *et al.*, 1996).

in parts of northwest Europe during the 1970s to 1990s (men born in the 1950s to 1970s)(Auger *et al.*, 1995; Irvine *et al.*, 1996; Van Waeleghem *et al.*, 1996), although not in certain other places, notably the USA (Wittmaack & Shapiro, 1992; Fisch *et al.*, 1996; Paulsen *et al.*, 1996) or Finland (Vierula *et al.*, 1996). It is unclear when such deterioration began, or what the pre-decline levels were. As human daily sperm production is several-fold lower than in other mammals (Sharpe, 1994; França *et al.*, 2005), it is possible that there is a long but invisible history to this decline.

In addition, male subfertility shows the unusual feature that the precise manifestation, e.g. the particular combination of impairment of sperm density, motility and morphology, and other pathological characteristics, varies between affected men, but is a characteristic "fingerprint" for that person that endures over time (Sloter *et al.*, 2000; Marchetti & Wyrobek, 2005; Sergerie *et al.*, 2005), albeit with some fluctuation in severity (Erenpreiss *et al.*, 2006).

Testicular cancer and (some cases of) male subfertility are sometimes grouped together as the "testicular dysgenesis syndrome" (TDS) along with cryptorchidism and hypospadias, as they share histological characteristics such as microlithiasis and Sertoli cell-only tubules (Skakkebaek *et al.*, 2001). They also tend to occur together, although rarely all four in one individual. However, it may be better to see this in terms of one or more common risk factors rather than a syndrome.

Time trends in cryptorchidism and hypospadias are less clear, as they are difficult to study (Toppari *et al.*, 2001). Cryptorchidism has apparently increased in southern England (John Radcliffe Hospital Cryptorchidism Study Group, 1992) and Denmark (Boisen *et al.*, 2004), but the evidence on trends in hypospadias is unclear (Toppari *et al.*, 2001).

Hypospadias and cryptorchidism clearly arise in utero, and it is probable that the same is true for predisposition to the other two endpoints as well. Exposures in later life can also affect semen quality (Setchell, 1998), and possibly the risk of testicular cancer (Davies *et al.*, 1996), but here the focus is on shared risk factors affecting male reproductive development. In particular, carcinoma-in-situ (CIS) is the forerunner of testicular cancer and is present at birth (Skakkebaek *et al.*, 2001), and impairment of semen quality is already present in Danish conscripts in their late teens (Jensen *et al.*, 2004a).

Spatial/ethnic variation is marked. The risk of all four endpoints is considerably higher in Denmark than in Finland (Møller Jensen *et al.*, 1988; Adami *et al.*, 1994; Jørgensen *et al.*, 2002; Boisen *et al.*, 2004; Boisen *et al.*, 2005), and Finnish newborn boys have larger testes and higher serum inhibin B levels (an indicator of testicular function)(Main *et al.*, 2006). Testicular cancer is rare in men of African descent (Jack *et al.*, 2007; Shah *et al.*, 2007), and in China and Japan (Parkin, 2005). Semen quality varies within as well as between countries (Fisch & Goluboff, 1996). The few high-quality studies of cryptorchidism show spatial variation that is compatible with that seen for semen quality (John Radcliffe Hospital Cryptorchidism Study Group, 1992; Berkowitz *et al.*, 1993; Boisen *et al.*, 2004).

Exposure to the factor(s) responsible for the deteriorating trends, and for the spatial/ethnic variation, must have the right characteristics to fit with the epidemiological observations. To find such a factor (or factors), it is also necessary to know what type of impairment would be expected; in other words, the pathogenic process(es) that would account for the epidemiological and clinical picture.

Can endocrine disruption explain the decline?

What pathogenic process could cause TDS? This will first be explored in relation to the hitherto dominant hypothesis, focusing on fetal or infantile exposure to environmental endocrine disruptors (EDs)(Sharpe & Skakkebaek,

1993; Skakkebaek *et al.*, 2001; Shen *et al.*, 2008). Two distinct questions need answering: (i) *can* EDs cause TDS? (ii) *did* EDs cause *the rise* in TDS? It is quite possible that something may increase the risk of a medical condition, but without playing a major role. For example, inorganic arsenic can cause lung cancer, e.g. via occupational exposure – the type of question posed in (i); but it was not the cause of the lung cancer epidemic of the twentieth century – equivalent to the question posed in (ii) – the answer is the rise of cigarette smoking.

The importance of (ii) is that whatever factor was responsible for this major increase is almost certainly still present and causing damage, and if identified it might be removable. In addition, (i) is important to prevent additional harm. Some epidemiological reports (Hardell *et al.*, 2003; Marsee *et al.*, 2006; Storgaard *et al.*, 2006) suggest that the answer to (i) may be "yes", but the effects are too weak to explain the marked contrasts seen in the descriptive epidemiology, even allowing for the possible effects of mixtures (Gray *et al.*, 2006).

In either case, the observed human outcomes would need to match those predicted by the ED hypothesis, and the proposed pathogenesis would need to correspond to that seen in human TDS patients. To answer question (ii), there are additional issues about the extent of exposure to known EDs, in relation to the dose required to produce effects, and in particular when and where such exposures could have occurred and whether this corresponds with the epidemiological observations; the conclusions will always be provisional, subject to the discovery of new EDs.

Some specific types of endocrine disruption are unconvincing as causes of TDS. The original hypothesis focusing on early-life estrogenic and/or anti-estrogenic exposure (Sharpe & Skakkebaek, 1993) is now discredited because:

(a) exposure to diethylstilbestrol (DES), a highly potent estrogen that was given to women in early pregnancy in pharmacological doses, is not consistently associated with any of the endpoints of TDS in epidemiological studies, with the possible exception of testicular cancer (Storgaard *et al.*, 2006), and effects only occurred at the highest doses (Sharpe, 2003);

(b) Chinese and Japanese men, who are exposed in utero to phytoestrogens derived from soy (Committee on Toxicity of Chemicals in Food, Consumer Products and the Environment, 2003), have a particularly low risk of testicular cancer (Parkin, 2005);

(c) other environmental estrogenic compounds are at least 100-fold less potent than these phytoestrogens (and present in smaller quantities), and at least 10^6-fold less potent than DES and estradiol (Safe, 1995).

Anti-androgen activity that could interfere with masculinisation is far more plausible biologically as a cause of TDS. Experimental evidence has shown that in utero exposure to certain phthalates such as dibutyl phthalate (DBP), which interferes with testosterone synthesis, can induce some features of TDS in rats (Fisher, 2004). In addition, reduced anogenital distance, a feature of the "phthalate syndrome" in rats, has been observed to be associated with phthalate exposure in human infants (Marsee *et al.*, 2006). It is therefore possible that these compounds can increase the risk of TDS. On the other hand, other features of phthalate syndrome are not seen in humans, notably agenesis or maldevelopment of the epididymis and other organs derived from the Wolffian duct, and delayed puberty (Foster, 2006). It is also unclear that these phthalates can cause testicular cancer, although high doses of DBP can cause transient delay in the entry of proliferating gonocytes into quiescence, which could possibly be related (Ferrara *et al.*, 2006).

If phthalates can cause TDS, or some features of it, could they be at least partly responsible for its rise? It is unlikely that they played an important role. First, it is doubtful whether human exposures are, or have ever been, high enough relative to doses used in rats. Secondly, these compounds were only introduced in the 1930s, too late for the initial increase in testicular cancer, although possibly relevant to the later rise in this and other TDS endpoints. This question of timing is more general, as the first three to four decades of rising testicular cancer incidence occurred before significant exposure to any known EDs was widespread.

Nevertheless, it is possible the phthalate model indicates that the unknown factor(s) is/are likely to be anti-androgenic. This is unlikely however, partly because it is not clear whether the mechanism by which DBP causes aspects of TDS in rats is anti-androgenic. But more important, high exposure levels to another anti-androgen, DDE, the major stable breakdown product of the insecticide DDT, which blocks the androgen receptor (Kelce *et al.*, 1995), has occurred throughout the tropics for several decades as a result of attempts at malaria control, without a testicular cancer epidemic following in its wake (Parkin, 2005). To conclude: no known anti-androgen or other ED can explain the epidemiological observations such as time trends that have given rise to concern.

Is there a genetic component to TDS?

There is considerable evidence on the heritability of testicular cancer and of male infertility, and a little on the other two conditions (recently reviewed by Lutke Holzik *et al.* (2004) and Joffe (2007)). Migrant studies show that the risk of testicular cancer that applied in the country of origin only changes in the second generation after migration, unlike most cancers (and other conditions such as ischemic heart disease), which adjust much more quickly (Parkin & Iskovich, 1997; Hemminki & Li, 2004; Montgomery *et al.*, 2005). Although only about 2% of testicular cancer cases are familial, the relative risk for someone who has an affected brother is approximately 9, and that for father–son pairs is about 4 (based on consistent findings in four separate studies), whereas for most cancers the relative risk among first-degree relatives is no more than 2 or 3 (Forman *et al.*, 1992; Li, 1993; Goldgar *et al.*, 1994). The strength of association means that a shared environmental factor is highly unlikely to be responsible, as the relative risk associated with such an exposure would have to be at least 50, considerably higher even than the increase in risk of lung cancer in heavy smokers (Khoury *et al.*, 1988). A twin study found that the relative risks of disease, given the presence of an affected monozygotic or dizygotic twin, were respectively 76.5 and 35.7, as would be expected with genetic causation, although this observation was not statistically significant as the study was small (Swerdlow *et al.*, 1997).

No specific genetic defect is known to be characteristic of testicular cancer patients. The International Testicular Cancer Linkage Consortium (ITCLC), which has the largest collection of testicular cancer pedigrees globally, has sought candidate genes, and the findings are suggestive of multiple alleles, each with a modest effect on risk (Rapley *et al.*, 2003; Rapley, 2007).

Testicular cancer is associated with subfertility in families as well as in individuals: men with testicular cancer have fewer siblings, suggesting parental subfertility (Richiardi *et al.*, 2004; Aschim *et al.*, 2008). Their brothers but not their sisters tend to have fewer children and a lower probability of dizygotic twinning, suggesting lower biological fertility among brothers as well (Richiardi & Akre, 2005).

Seven studies have found an aggregation of male infertility in families (Joffe, 2007), and there appear to be no such studies with negative findings. Relatives of men with fertility problems, especially brothers, tend to have impaired semen quality and/or to be involuntarily childless or subfertile (Czyglik *et al.*, 1986; Lilford *et al.*, 1994; Meschede *et al.*, 2000; Christensen *et al.*, 2003; Storgaard *et al.*, 2003; Gianotten *et al.*, 2004; Van Golde *et al.*, 2004). In families where more than one family member donated a semen sample, specific patterns of abnormality have been observed, e.g. a deficit primarily in sperm motility, or in density – again, a characteristic "fingerprint" – suggesting that different genetic abnormalities were involved in the different families (Lilford *et al.*, 1994). The emphasis here is on a characteristic ***pattern***; it is distinct from the type of specific defect, e.g. in sperm morphology, in which all spermatozoa are uniformly affected in the same way, a condition that is caused by a mutation at a specific locus (Andersson *et al.*, 2000).

In twin studies, intra-pair correlation of functional fertility – the risk of taking a long time to conceive – has been found between monozygotic but not dizygotic twins (Christensen *et al.*, 2003); and heritability has been observed of sperm density (>20%), morphology (41%) and sperm chromatin stability (68%)(Storgaard *et al.*, 2003). Suggestions of a recessive mode of inheritance (Czyglik *et al.*, 1986; Lilford *et al.*, 1994; Gianotten *et al.*, 2004) are contradicted by the finding that male fertility – as opposed to female fertility – is not associated with the degree of consanguinity among Hutterites (Ober *et al.*, 1999), and that semen quality is only weakly related to consanguinity (Baccetti *et al.*, 2001).

One study has estimated the odds ratio of cryptorchidism and hypospadias as 3.8 and 10.1 respectively, when a brother is affected (Weidner *et al.*, 1999). Although these two conditions not infrequently occur together in the same individual, they tend not to run in the same families (Weidner *et al.*, 1999). Another study has also found strong evidence of familial aggregation of hypospadias (Fredell *et al.*, 2002). Boys with hypospadias tend to be born after pregnancies that take a relatively long time to conceive (Czeizel & Tóth, 1990), suggesting couple infertility, and to have fathers with genital anomalies and impaired semen quality (Fritz & Czeizel, 1996; Asklund *et al.*, 2007).

Thus, it appears that all four endpoints have similar patterns of familial aggregation (Joffe, 2007); this provides a further indication that they share a common risk factor and pathogenetic mechanism. Some of the studies demonstrate a clearly genetic connection rather than a shared environmental factor (Joffe, 2007). The idea of a genetic aetiology for at least some TDS endpoints is supported by work on wildlife and domestic animals: for example in felids, inbreeding is related to teratospermia (which tends to be compensated by an increase in sperm number), and with cryptorchidism (Pukazhenthi *et al.*, 2006).

In general, both for testicular cancer and for infertility, the results suggest a highly heterogeneous picture, not implicating any specific locus, and with no single clear mode of inheritance. Infertility in the parents, and perhaps especially the father, appears to be a risk factor for testicular cancer and for hypospadias. Thus the process has a consistent unifying principle across families, but with family-specific features, and it does not originate in one or more specific loci.

How can infertility be heritable?

The evidence therefore clearly points to a heritable component, but the rapid time trends also imply a strong environmental element, with migrant and other studies pointing to origin in early life. How can these be reconciled? The traditional view, based on the equation of "genetic" with "inherited", is to apportion causation between genes and environment, and/or to think in terms of gene–environment interactions (Harland, 2000; Skakkebaek *et al.*, 2001). A third possibility may hold the key: that a genetic or epigenetic abnormality is caused, or made more frequent, by an environmental factor (Joffe, 2003; Joffe, 2007).

The idea of transmissible subfertility may seem absurd: evolution is driven by the probability of passing one's genes on to the next generation, so any gene that reduced fertility would rapidly be eliminated from a population (unless spermatogonial cells with mutations have a selective advantage (Goriely *et al.*, 2003; Zöllner *et al.*, 2004; Choi *et al.*, 2008)). A partial exception would be in the case of recessive genes, which would be eliminated more slowly, but as noted above recessivity does not appear to play a major role, at least in male infertility. The genetic predisposition towards TDS would therefore have to be constantly entering the population – which fits with the evidence of an environmental determinant, and also accords with the low proportion of familial cases.

This idea is complementary to Czeizel's "relaxed selection" hypothesis: that the decline of family size in the early twentieth century meant that the most biologically fertile couples ceased to dominate the proportion of births in the population, and also that more recently the availability of assisted reproductive techniques has increased the reproductive presence of couples with impaired fertility (Czeizel & Rothman, 2002). There are two reasons why Czeizel's hypothesis alone cannot account for the epidemiological observations. First, the calculations of Slama and Leridon show that the observed trend in low sperm density was too strong to be explained by this process, at least in France (Slama & Leridon, 2002); this would hold even more strongly for testicular cancer, which has had a stronger trend but for which the selection process would be indirect through associated subfertility. Second, the hypothesis depends on the existence of alleles for subfertility and the related conditions, but it does not explain how they came to be present in the population in the nineteenth century, before the relaxation in selection began. It thus requires the complementary hypothesis that postulates the generation of new damage.

The hypothesis

If TDS is primarily caused by environmentally induced genetic damage, the next question is, how does this occur? The remainder of this chapter puts forward a possible way of synthesising the available evidence. Other hypotheses are possible, and the different hypotheses are not necessarily mutually exclusive.

For example, gestational genetic damage to primordial germ cells has been experimentally induced using ethylnitrosourea (ENU), a highly potent mutagen (Brinkworth, 2000), and it is possible that pregnant women are exposed to similarly mutagenic agents e.g. in cigarette smoke. Maternal smoking during pregnancy is a possible risk factor for reduced semen quality (Jensen *et al.*, 2004b), although this finding is not universal (Richtoff *et al.*, 2007). However, seven studies agree that it does not predispose to testicular cancer (Pettersson *et al.*, 2007). Second, transient exposure of gestating female rats to certain endocrine disruptors has been reported to lead to transmissible infertility of the male offspring, and to other abnormalities, via an epigenetic mechanism (Anway *et al.*, 2005; Anway & Skinner, 2008). In both cases, it is unclear whether the mechanism involved would generate the range of abnormalities characteristic of TDS.

It is hypothesised that TDS has the following features:

- the target tissue is the germ cell;
- the defect is genetic and/or epigenetic;
- the focus is on disruption at cellular level;
- the key element is survival of cells that have undergone particular types of damage;
- the impairment is caused by or made more frequent by one or more environmental risk factors;
- it is already present in early life – either initiated then, or passed on through an intergenerational/inherited process – and therefore before meiosis starts.

The focus on the cell as the unit of analysis contrasts both with the search for a mutation at a specific locus (Rapley, 2007) and with the mode of toxicity of high-dose dibutyl phthalate, in which the tissue architecture is disrupted (Ferrara *et al.*, 2006). Because the target tissue is testicular germ cells, not Sertoli cells or other somatic cells, Sertoli cell-only tubules can readily be explained by a loss of germ cells.

A candidate causal factor would need to be absorbed (if a chemical agent), and to have exposure characteristics that correspond to the epidemiological observations in relation to time and space, etc. In principle, the defect could arise before or after conception. In the preconception or ***genetic*** case, exposure would damage the genome of a parent, or in familial cases of a grandparent or other ancestor. In the postconception or ***gestational*** case, maternal exposure to the fetus would be effective during the period of reproductive system differentiation. It would require an agent that crosses the placenta, reaches the fetus and affects the mitotically dividing gonocytes.

A possible mechanism

Many clues are available from clinical research and from emerging knowledge of basic cellular mechanisms of cell cycle control and carcinogenesis. However, this is a relatively new literature and many of the detailed findings require confirmation.

Several reports based on examination of sperm from male infertility patients and/or fertile controls, or from laboratory animals, have used techniques that detect DNA strand breaks or other lesions that lead to them: Comet assay (Morris *et al.*, 2002; Schmid *et al.*, 2003; Singh *et al.*, 2003), abnormal chromatin packaging (DFI (Schmid *et al.*, 2003; Wyrobek *et al.*, 2006)), and a heterologous mouse–human ICSI technique (HIGH (Derijck *et al.*, 2007)); other abnormalities that can be detected include immaturity (HDS (Wyrobek *et al.*, 2006)) and abnormal chromosome number or structure (FISH (Sloter *et al.*, 2000; Brinkworth & Schmid, 2003; Schmid *et al.*, 2003; Schmid *et al.*, 2004))(see box at the end of this chapter for explanation). It has also become possible to detect cell cycle delay, centrosome amplification and apoptosis in male germ cells more generally (Brinkworth & Schmid, 2003; Singh *et al.*, 2003; Sergerie *et al.*, 2005). In addition, a great deal of progress has been made in understanding the origin and consequences of chromosomal abnormalities in the male germline (Marchetti & Wyrobek, 2005).

The suggested hypothesis postulates that the DNA structure of the germline is damaged, starting with small duplications and deletions (D&D). The initial impairment is relatively minor, but the severity increases over time and across generations. In particular, the intricate process of meiosis is the stage at which the damage becomes more severe. This is because meiosis, unlike the simpler process of mitosis, starts with the pairing of homologous chromosomes. If the chromosomes do not have identical structure, irregularities such as unpaired loops occur (Figure 2.2 a and b), and synapsis is delayed by the process of synaptic adjustment (Moses & Poorman, 1981; Moses *et al.*, 1982). Evidence that meiosis is a more sensitive process than mitosis can be observed in strains of mice that have reciprocal or Robertsonian translocations: aberrations that cause slight or no interference with mitosis become more seriously disruptive in meiosis (Marchetti & Wyrobek, 2005).

The suggested model is thus primarily concerned with the structure and function of the DNA, and of the cell more generally, rather than the expression (or not) of specific genes. According to this view, the pattern of gene expression is secondary to structural changes in the early stages. However, as chromosomal instability increases, structural DNA damage leads to deletions or amplifications of specific genes or chromosome regions, introducing a positive feedback loop, and these can play a major part, e.g. in cancer spread. This model

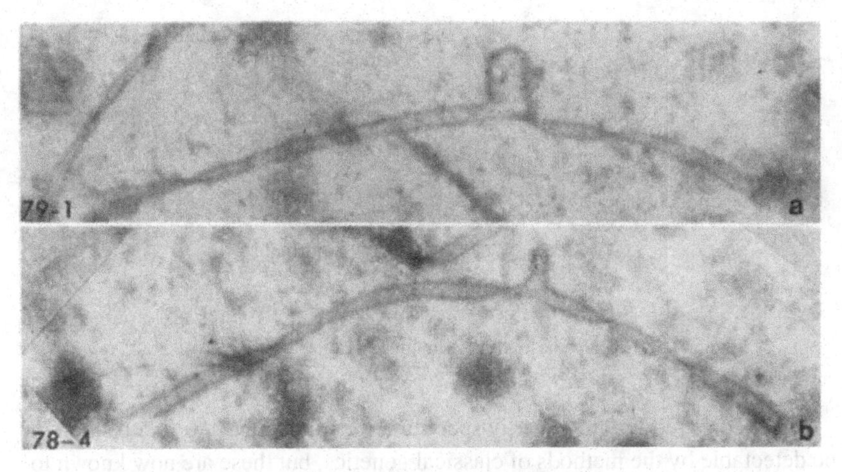

Figure 2.2. Impaired pairing in meiosis. Reproduced with kind permission from Springer Science+Business Media (Moses & Poorman, 1981).

corresponds with recent models of carcinogenesis in diverse tissues (Nigg, 2002; Gibbs, 2003; Pihan *et al.*, 2003; Storchova & Pellman, 2004; Kops *et al.*, 2005). The main difference from somatic carcinogenesis is the possibility in germ cells of a transgenerational process, which may explain the unusual age distribution of testicular cancer: the young adult who develops invasive carcinoma already had a DNA abnormality at the early embryonic stage, which was then exacerbated once meiosis started at puberty, and became frankly malignant one or two decades later. In this perspective, impaired semen quality in TDS cases is seen as a cancer-like but sub-malignant process, with zygotic abnormalities that are also cancer-like (Skotheim & Lothe, 2003; Chatzimeletiou *et al.*, 2005).

The process of ever-increasing DNA damage is not a totally deterministic one, so that different clones are differently affected. It is this 'scattergun' aspect, together with persistence and propagation of existing defects, that leads to the characteristic spectrum of abnormalities for each affected individual, or 'fingerprint', already referred to. According to this account the heterogeneity is between surviving clones, which accords with the well-recognised feature of TDS that different tubules have their own characteristic pattern and severity of abnormality.

At each cell division, especially later in the process, many abnormal cells are produced that are not viable, and the surviving cell lines are those able to pass through the cell checkpoints and other possible quality control mechanisms such as fertilisation. Thus, while the abnormality worsens over time, there is an upper limit to severity in surviving clones, as the more affected ones simply disappear (Figure 2.3).

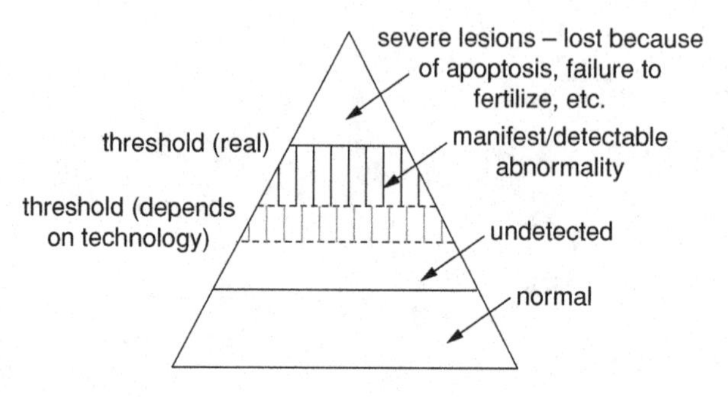

Figure 2.3. The hypothesised spectrum of severity.

At the lower end of the severity spectrum, the initial D&D were too small to be detectable by the methods of classical genetics, but these are now known to be widespread in the human genome, from studies of the HapMap collection (Redon *et al.*, 2006; De Ståhl *et al.*, 2007; Levy *et al.*, 2007) and from the sequencing of the diploid genome (Levy *et al.*, 2007), both of which show a hitherto unexpectedly high prevalence of copy number variations, especially those caused by duplications. The lower threshold of detectable abnormality thus depends on the current technology for its detection (Figure 2.3). Here the concern is with the ***newly arising*** small D&D that alter DNA structure and thereby interfere with meiosis.

The sensitivity of meiosis to aberrant DNA structure has an additional implication. When DNA damage is transmitted to the next generation, it is systemic, with all parts of the body having the same underlying anomalies. But the amplification that occurs with meiosis means that the ***apparent*** abnormalities are especially severe in the germ cells, as these are the only ones to undergo this process. This explains the apparent specificity of the target tissue, as men with the attributes of TDS generally tend to appear otherwise healthy. (The same factor(s) could however cause other health effects in different tissues, e.g. with different timing and route of exposure – many exposures are able to cause more than one disease.)

Figure 2.4 shows an attempt, confined to the male line, to synthesise the available evidence. The transgenerational process starts with DNA damage in spermatids of the F_0 generation, followed by faulty repair in the fertilised ovum, resulting in chromosomal aberrations. Although this account is built up from a large number of empirical studies, the authors of these studies might not necessarily agree with the role they have been allotted in Figure 2.4, which is a new synthesis.

Other versions of this transgenerational process may be possible, with events occurring in a different order, and/or with participation of the female line (see

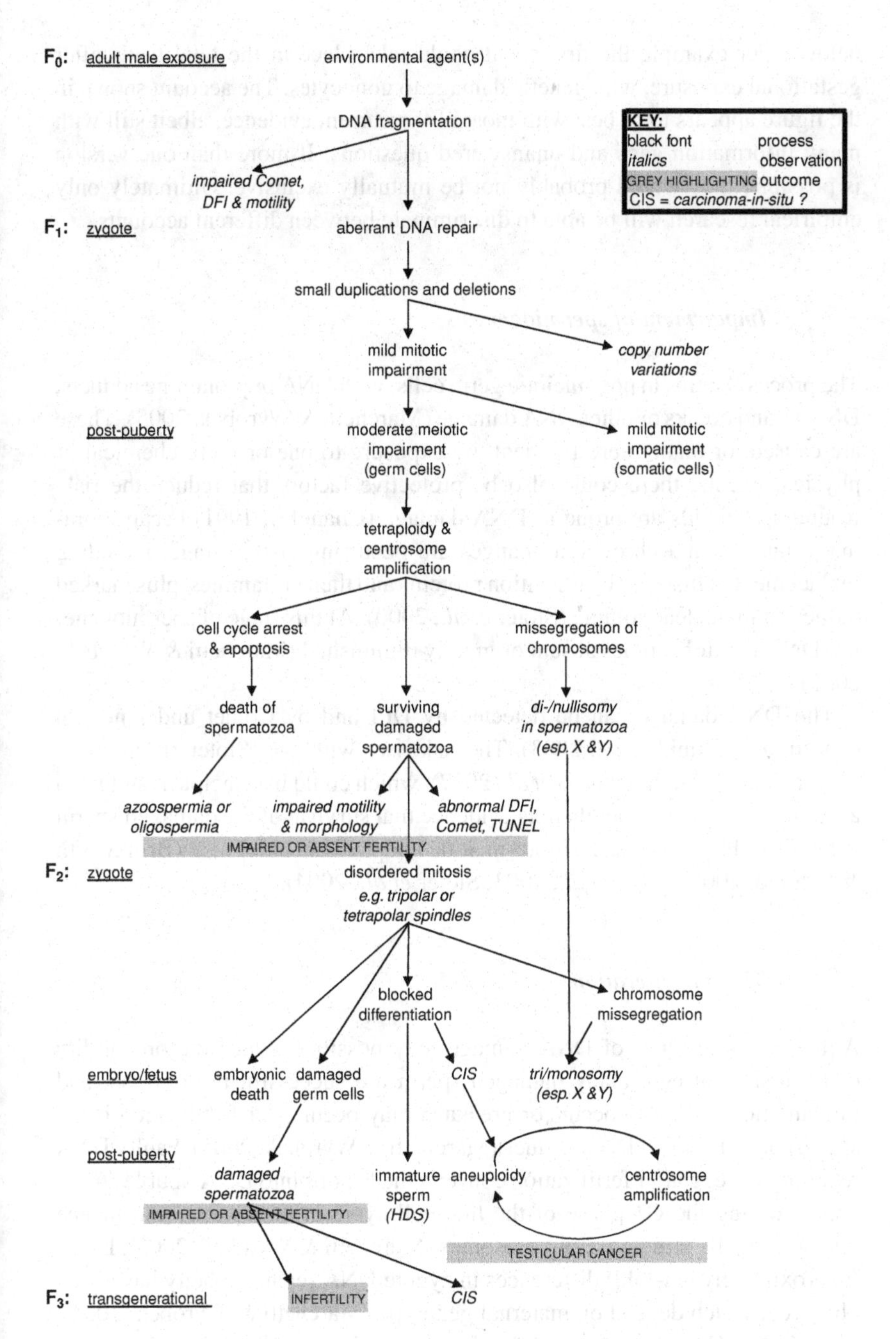

Figure 2.4. A new synthesis of the transgenerational process leading to TDS.

below). For example the first event could take place in the fetal testis after gestational exposure, with genetic damage to gonocytes. The account shown in the figure appears to fit best with most of the current evidence, albeit still with many information gaps and unanswered questions. If more than one version is possible, they would probably not be mutually exclusive. Ultimately only empirical research will be able to discriminate between different accounts.

Impairment of spermiogenesis

The process begins in post-meiotic germ cells, with DNA or protamine adducts, DNA strand breaks or other DNA damage (Marchetti & Wyrobek, 2005). These are caused, or made more frequent, by exposure to one or more chemical or physical agents; there could also be protective factors that reduce the risk. Round spermatids are prone to DNA damage (Chandley, 1991) because dramatic nuclear and chromatin changes are occurring at this stage, including replacement of histones by transition proteins and then protamines, plus marked reduction in nuclear volume (Sloter *et al.*, 2000). At this stage of spermiogenesis, DNA repair becomes absent or greatly diminished (Marchetti & Wyrobek, 2005).

The DNA damage can be detected by DFI and by Comet under neutral conditions (Schmid *et al.*, 2003). The risk rises with age (Sloter *et al.*, 2004; Wyrobek *et al.*, 2006; Schmid *et al.*, 2007), which could in principle result from accumulated damage, but there is evidence that survival of the damaged sperm is facilitated by lower rates of pre-meiotic apoptosis in older men (Brinkworth & Schmid, 2003; Singh *et al.*, 2003; Sloter *et al.*, 2004).

The F_1 generation

Although some types of DNA damage may possibly reduce the probability of fertilising an egg, many damaged sperm are successful in doing so, and implantation is able to occur (or breakage may occur after fertilization if the sperm carried a bulky DNA adduct)(Marchetti & Wyrobek, 2005). Faulty DNA repair by the egg after fertilization converts the lesions into DNA double strand breaks during the G1 phase of the first cell cycle in the zygote, and thence into abnormally structured chromosomes (Marchetti & Wyrobek, 2005). Large (approximately ten-fold) differences in zygote DNA repair capacity have been observed, which depend on maternal genotype (Marchetti & Wyrobek, 2005), and this could be relevant to explaining ethnic variation, e.g. of testicular cancer incidence.

Many such lesions (especially if unstable (Marchetti & Wyrobek, 2005)) are lethal once the embryo reaches the post-implantation stage when the embryonic genome is activated and stringent cell cycle checkpoints and apoptotic response occur, but many affected embryos survive. Some of these are individuals with clinically important DNA damage, but the F_1 generation also includes a larger number who have genetic aberrations that are smaller than the chromosomal defects that have hitherto been detectable but larger than base-pair mutations (non-SNP(single nucleotide polymorphism) events). For example, even using FISH staining at three points on chromosome 1 (the ACM method (Sloter *et al.*, 2000; Schmid *et al.*, 2004)), only relatively large D&D can be seen, far bigger than the ones now known to be commonplace through newer techniques (Redon *et al.*, 2006; Korbel *et al.*, 2007; Levy *et al.*, 2007).

As the individual develops, the small D&D hardly affect mitosis, and typically have no clinical consequences (Sloter *et al.*, 2000). The resulting copy number variations have sufficiently subtle effects that their existence has only recently been recognised. However, after puberty when meiosis starts (in the male case), it is disrupted by the D&D as described above. This may explain the observation of extended asynapsis in the testes of severely subfertile men (Guichaoua *et al.*, 2005), and the high frequency of synaptic anomalies and delay observed in spermatocytes, especially in infertile men (Vendrell *et al.*, 1999; Codina-Pascual *et al.*, 2006). The meiotic checkpoints may trigger apoptosis (Sciurano *et al.*, 2007), possibly because of insufficient tension at chiasmata (Petronczki *et al.*, 2003), with consequent reduced number/density of spermatozoa, i.e. oligospermia. In the surviving cells (Guichaoua *et al.*, 2005; Derijck *et al.*, 2007), the checkpoint(s) may delay the completion of meiosis, as in azoospermic men who have been found to have decreased mRNA transcripts of MPF (M-phase promoting factor) and its regulators (Lin *et al.*, 2006), i.e. a change in gene expression may occur secondary to the structural impairment.

The cell cycle delay may lead to centrosome amplification under the influence of STK15 (AuroraA) over-expression (Mayer *et al.*, 2003)(another secondary gene expression event), which correlates strongly with the degree of impairment of spermatogenesis (Mayer *et al.*, 2003). Tetraploidy and loss of normal cellular architecture result, leading to altered morphology and motility of the resulting spermatozoa, i.e. teratospermia and asthenospermia. It has been observed that immotile sperm tend to have a large number of double DNA strand breaks (Derijck *et al.*, 2007). A further possible outcome is the missegregation of chromosomes, mainly the sex chromosomes as they share only a small pseudoautosomal region (Schmid *et al.*, 2003), resulting in sperm disomy or nullisomy (Petronczki *et al.*, 2003), with a lower rate of aneuploidy in ejaculate compared with testicular sperm, reflecting selective loss (Gianaroli *et al.*, 2005). As this description suggests, different sperm cells are differently affected, producing a scattergun effect with consequences for the spermatozoa that are diverse in type, as well as producing a range of severity (Morris *et al.*, 2002; Schmid *et al.*, 2003).

The F_2 generation

In the F_2 generation, the spermatozoon's cellular disorganisation continues within the zygote but potentially causes more severe impairment, with even mitosis becoming delayed and disrupted (Sakkas, 1999; Chatzimeletiou *et al.*, 2005; Borini *et al.*, 2006). Specifically, in humans (and other mammals but not rodents) the zygotic centrosome is paternally inherited, and the multiple centrosomes result in tri- and tetrapolar spindles that lead to "genetic instability similar to that observed with human tumour cells" (Chatzimeletiou *et al.*, 2005; Chatzimeletiou *et al.*, 2007). Cellular architecture is distorted, and cytokinesis may be absent or asymmetric, so that binucleate cells are common, and chromosomes may missegregate. Affected cells are likely to survive until the third post-fertilization day (4–8 cell stage) as there appears to be no checkpoint before then (Chatzimeletiou *et al.*, 2005). Embryos fathered by certain men with poor semen quality repeatedly fail to develop beyond the early stages, owing to a paternal effect (Tesaryk *et al.*, 2002).

Thus, aneuploidy may arise within embryonic development as well as from fertilization with an aneuploid sperm (see below). Depending on the severity, instability/aneuploidy may lead sooner or later to embryonic death, manifest to the woman as a menstrual cycle without conception, or as recognised fetal loss (miscarriage) respectively. Sperm chromosome aneuploidy has been observed in the husbands of women with unexplained recurrent pregnancy loss (Carrell *et al.*, 2003).

However, the embryos with the milder anomalies survive. The scattergun effect now operates in mitosis. Each cell has progeny that replicate its defects, but imperfectly, with new defects that are diverse in type as well as in severity. This leads to mosaicism in the embryo/fetus, placenta and membranes (Chatzimeletiou *et al.*, 2005). Subsequently the history of that particular spectrum of anomalies, subject to differential clonal survival, becomes embodied in the adult as their characteristic fingerprint.

As already stated, abnormal chromosome numbers may arise from di-/nullisomy in the fertilising sperm or from missegregation in the F_2 generation, leading to trisomy or monosomy, mainly involving a sex chromosome. In the case of an aneuploid sperm the whole embryo is likely to be affected, leading to 47 XXY (Klinefelter syndrome), 47 XYY or 45 X (Turner syndrome), which are commonly of male origin (Sloter *et al.*, 2004). Probably more often, the anomaly arises in this F_2 generation, affecting only a few cell lines. This can be seen in somatic cells such as lymphocytes as well as in germ cells (Gazvani *et al.*, 2000; Rubes *et al.*, 2002; De Palma *et al.*, 2005).

In the fetal testis, some severely affected yet still surviving tetraploid clones with supernumerary centrosomes undergo aberrant mitosis, followed by chromosomal instability and chromosomal loss, leading to aneuploidy together with a block in differentiation (Mayer *et al.*, 2003; Frigyesi *et al.*, 2004; Rajpert-De Meyts, 2006). These are the characteristics of CIS (carcinoma-in-situ) cells, which are aneuploid germ cells that have failed to differentiate into normal gonocytes (Rajpert-De Meyts, 2006), retaining some features of stem cells including pluripotency (Høi-Hansen *et al.*, 2004; Høi-Hansen *et al.*, 2005; Almstrup *et al.*, 2006; Rajpert-De Meyts, 2006). Thus the cause of the failure to differentiate is here seen as structural, with the expression of genes such as OCT3/4 (POU5F1) and AP2γ (markers of pluripotency)(Oosterhuis & Looijenga, 2005; Høi-Hansen *et al.*, 2007) as a secondary manifestation. At this stage it is unclear how this lack of differentiation also affects the methylation status and micro-RNA expression that also characterise these cells (Looijenga *et al.*, 2007).

Centrosome amplification and aneuploidy are now present as part of the established cellular structure, and STK15 is no longer over-expressed as it was in the early stages (Mayer *et al.*, 2003). The mosaicism-generating process explains why CIS tubules as well as other abnormalities are so heterogeneous (Rajpert-De Meyts, 2006), with a focal pattern within the testis, and the frequent occurrence of cancers that are unilateral but with dysgenetic changes in the contralateral testis (Rajpert-De Meyts, 2006). This is reminiscent of the scattergun pattern seen in spermatogenesis abnormalities.

The tendency to mitotic instability continues after puberty, and is again accentuated during meiosis leading to a life-long tendency to poor semen quality as in the previous generation, with a person-specific range of sperm abnormalities including some combination of impaired density, motility and/or morphology (Lilford *et al.*, 1994), which is paralleled by stable person-specific patterns of DNA fragmentation (Sergerie *et al.*, 2005), distributions of break-point locations (Sloter *et al.*, 2000), and frequency of chromosomally abnormal sperm (Sloter *et al.*, 2000). The spermatozoa may be particularly susceptible to oxidative damage (Aitken *et al.*, 2004; Sergerie *et al.*, 2005), introducing a positive feedback loop into the process. The block in differentiation also leads to immature spermatozoa, as detected by HDS (Wyrobek *et al.*, 2006).

In addition, the onset of meiosis accentuates the cellular abnormalities in CIS tubules. In some cells this is followed by growth arrest and cell death (Bartkova *et al.*, 2005). In others, testicular germ cell cancer ensues: a positive feedback loop of further centrosome amplification and ever-greater degrees of aneuploidy develop, as well as an associated progressive loss of cell cycle-related genes (Bartkova *et al.*, 2003) that further accentuates chromosomal instability (Nigg,

2002). Carcinomas tend to have CIS tissue adjacent, probably because they are clones that arise from CIS-affected cells (Mayer *et al.*, 2003). The age of onset and the type of disease reflect the severity of the underlying cellular disturbance, as well as possible epigenetic status, with earlier onset for less-differentiated tumours (various types of non-seminoma) compared with seminomas (Oosterhuis & Looijenga, 2005). The cellular process is essentially random, not involving particular chromosomes apart from the already mentioned specific properties of sex chromosomes; however, once a clone develops a particular characteristic its properties might well change, e.g. chromosome region 12p appears to enable germ cells to survive outside their Sertoli cell-associated niche, facilitating tumour spread (Oosterhuis & Looijenga, 2005).

Comments on the mechanism

This model of testicular cancer accords with several observations in testicular cancer patients. These include a deregulated cell cycle (Bartkova *et al.*, 2003), structural chromosomal abnormalities and hyperploidy (Ottesen *et al.*, 2004), and the absence of known specific-gene mutations (Rajpert-De Meyts *et al.*, 2002; Crockford *et al.*, 2006), Y-chromosome haplotypes (Rajpert-De Meyts, 2006) or chromosome-regional defects (Frydelund-Larsen *et al.*, 2003). Its similarity to carcinogenesis in other tissues means that it is unnecessary to postulate an endocrine-dependent mechanism of carcinogenesis specifically for the testis (Skakkebaek *et al.*, 2001).

Thus, TDS is seen as a transgenerational condition with relatively mild sub-malignant and more severe malignant forms. Subfertility features in each generation, but with varying pathogenesis; only in the F_2 generation is it part of TDS, before that it could be considered as a risk factor for TDS. In addition, much male subfertility may be unrelated to TDS. These two considerations may explain the heterogeneity of findings that are based on samples of infertile men (Morris *et al.*, 2002; Wyrobek *et al.*, 2006). The testicular dysgenesis syndrome can also be transmitted to the F_3 and future generations, in which case it retains some of the individual's fingerprint as a familial characteristic (Lilford *et al.*, 1994). Within this family transmission, the manifestations of TDS tend to worsen in each generation, which could explain the earlier onset in familial cases, and why the association in risk is greater between brothers than between fathers and sons. However, this is limited by a ceiling of severity as already explained (Figure 2.3), because cellular disruption beyond a certain level would not survive the embryonic stage.

This account, starting with DNA damage and leading on to structural aberrations which finally predispose to chromosome loss or gain, would explain why

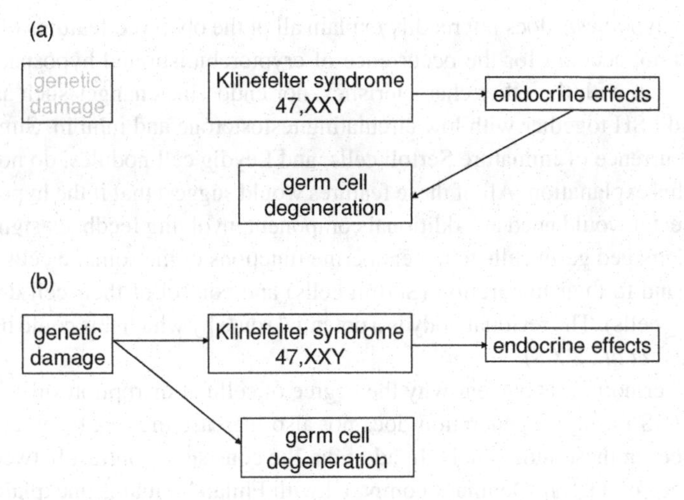

Figure 2.5. Alternative conceptions of the relationship between aneuploidy and TDS: (a) primary lesion is aneuploidy, TDS is generated via an endocrine mechanism; (b) genetic damage is the primary lesion, with aneuploidy and TDS both being consequences.

in healthy men aneuploidies (1–3%) are observed less frequently than D&D (5–13%), which in turn are less frequent than breaks and fragments (Sloter *et al.*, 2000; Schmid *et al.*, 2004), which make up over 75% of all chromosomal aberrations (Sloter *et al.*, 2000). The focus on structural abnormality is compatible with the observation that in the context of ICSI (intracytoplasmic sperm injection), selection of spermatozoa for a smooth, symmetric and oval configuration, with no extrusion or invagination of the nuclear chromatin mass, produces higher pregnancy rates and lower miscarriage rates compared with the standard procedure (Berkovitz *et al.*, 2006). It also implies that the association of Klinefelter syndrome with germ cell degeneration (Lanfranco *et al.*, 2004) should be interpreted in terms of the process that gives rise to both, not as direct causation of one by the other (see Figure 2.5).

The focus here on genetic damage and on transmissibililty to the next generation is distinct from male germline mutation at a specific locus as with achondroplasia and Apert's syndrome (Crow, 2003). It should also not be confused with "genetic" in the usual sense of "inherited". The primary mechanism is genetic damage that is usually of environmental origin but is inherited in a minority of cases; in addition this pathogenic process can be modified by inherited factors that might be manifest as, for example, ethnic variation. One candidate is the zygote's repair capacity, which is maternally mediated, and exhibits variability of sufficient magnitude. In addition, epigenetic processes such as hypomethylation and/or micro-RNA could play an important role.

This hypothesis does not readily explain all of the observed features of TDS. It does not account for the occurrence of cryptorchidism and hypospadias in association with the other characteristics, nor endocrine changes such as high LH and FSH together with low circulating testosterone and inhibin. Similarly, the occurrence of immature Sertoli cells, and Leydig cell nodules, do not flow from this explanation. All of these features would suggest that if the hypothesis is correct it would need an additional component involving feedback signalling from damaged germ cells to the endocrine functions of the somatic cells of the testis, and to their maturation (Sertoli cells) and control of their cell division (Leydig cells). The residual body is a potential route by which this could happen (Rolland *et al.*, 2008).

Furthermore, it is unclear why the degree of cellular disruption sufficient to cause CIS in the F2 generation does not also increase the risk of other types of cancer in these individuals. In addition, the consistent contrast between the incidence of TDS in Denmark compared with Finland remains unexplained.

Environmental factors

What factor(s) might be responsible? A candidate would need to correspond with the type of pathogenic process, i.e. if this model is correct it would be able to cause DNA damage in the germline, as well as having exposure characteristics that correspond with the descriptive epidemiology; the process could also be influenced by protective and exacerbating factors, for example paternal occupation (Knight & Marrett, 1997).

As the risk of DNA damage in the F_0 generation increases with age, one hypothesis is that increasing age at fatherhood has played a part in the rise of TDS, but this idea conflicts with the epidemiological findings (Henderson *et al.*, 1997). Damage to DNA also increases with genital tract infections, but again the time-course of changes in such infections in the twentieth century does not correspond with the epidemiology of testicular cancer (Aitken, 2004).

Evidence is accumulating that post-meiotic male germ cells can be affected by a wide variety of chemical agents. Possible agents include:

- glycidamide, a metabolic product of acrylamide, which is present in many foods, especially starchy food cooked at high temperature (Adler *et al.*, 2000; Tyl *et al.*, 2000) as well as in tobacco smoke (Sloter *et al.*, 2000), metabolism being carried out by CYP2E1, which may be ethnically highly variable (Ghanayem *et al.*, 2005);
- *cis*-unsaturated fatty acids (Aitken *et al.*, 2006);
- cigarettes (possibly), e.g. owing to benzo[*a*]pyrene adducts, which are transmissible to the zygote (Zenzes, 2000).

One possible mechanism is oxidative stress, which damages the DNA (Aitken *et al.*, 2006), and also oxidises the sperm plasma membrane making fertilization less likely (Aitken, 2004). Protective factors could therefore include antioxidants, for which there is some epidemiological support (Eskenazi *et al.*, 2005).

Alternatively, the initiating feature could be a physical rather than a chemical insult: heat. Mild scrotal heat stress of 42°C for 30 minutes leads to abnormal DFI and Comet, likely owing to oxidative stress (Banks *et al.*, 2005), and to decreased survival of embryos (Zhu *et al.*, 2004). This is interesting as sedentary occupations have increased over time and are associated with raised intra-scrotal temperature (Bujan *et al.*, 2000; Mieusset *et al.*, 2007).

What about females?

The existence of environmentally caused transgenerational damage invites the question whether a germline, meiosis-amplified process has been occurring in both sexes. What impact might such a pathogenic process have in females, who lack the advantage of easily accessible gametes and are more difficult to study? Clearly there is no simple parallel as the epidemiology of analogous female conditions is quite different from that seen in males; for example, there has been no general rising trend in ovarian cancer (Gonzalez-Diego *et al.*, 2000).

A similar process, starting with male exposure and sperm DNA fragmentation but with female F_1 and/or F_2 generations, might be expected to have consequences that are partly similar and partly sex-specific. Reproduction in the human female line displays certain rather similar defects to that in the male, e.g. mosaicism, cell cycle delay, and chromosome gain and loss (Delhanty, 2001; Hassold & Hunt, 2001; Coonen *et al.*, 2004; Daphnis *et al.*, 2005). On the other hand, female reproduction differs importantly from male reproduction, e.g. early-life onset of meiosis and subsequent gamete quiescence, and maternal centrosomes not being passed on to the zygote (although it does contain maternal transcripts). In female mice, age-related embryonal aneuploidy appears to be related to progressive shortening of the meiotic prophase, leaving less time for accurate attachment of chromosomes (Chandley, 1991), which fits well with the early part of the model described here.

Conclusion

There is undeniable evidence that testicular cancer incidence greatly increased during the twentieth century in certain types of population, and it is highly likely that male fertility also deteriorated during this period, at least in northwestern Europe. Cryptorchidism and hypospadias may also have increased.

It is unlikely that environmental endocrine disruptors can explain much of these epidemiological observations. In addition, clinical research and other evidence show that genetic abnormalities occur in such patients, and this fits with consistent observations that these conditions are heritable, albeit weakly. Evolutionary and other arguments strongly suggest that the genetic damage has been newly introduced into these populations as a result of an environmental factor.

A transgenerational process is suggested as the pathogenesis of testicular cancer and impaired spermatogenesis, which would lead to progressively worsening cellular abnormalities in the male germline. The resulting germ cell defects are hypothesised to lead to the observed features of male infertility in the sub-malignant form of this condition, and in addition to carcinoma-in-situ and then invasive carcinoma in the malignant form.

Acknowledgements

I would like to thank Fernanda Almeida, Sue Barlow, Alan Boobis, Martin Brinkworth, Leendert Looijenga, Ian Morris, Frank Sullivan and Allen Wilcox for discussion of the issues dealt with in this chapter. They do not necessarily agree with the views expressed in it.

Box

- **Comet** under **neutral** conditions: identifies double DNA strand breaks (Morris *et al.*, 2002; Schmid *et al.*, 2003; Singh *et al.*, 2003; Guichaoua *et al.*, 2005)
- **Comet** under **alkaline** conditions: identifies single DNA strand breaks (Guichaoua *et al.*, 2005)
- Sperm Chromatin Structure Assay (**SCSA**) measures the susceptibility of DNA to acid denaturation
  - DNA Fragmentation Index (**DFI**), previously called COMP αT: identifies DNA fragmentation (Schmid *et al.*, 2003; Wyrobek *et al.*, 2006)
  - High DNA Stainability (**HDS**): identifies DNA immaturity (Wyrobek *et al.*, 2006)
- Human/mouse heterologous ICSI and γ H2AX staining (**HIGH**) assay measures double DNA strand breaks, and can be used for selected individual spermatozoa (Derijck *et al.*, 2007)
- Terminal deoxynucleotidyl transferase-mediated dUDP nick-end labelling (**TUNEL**): identifies apoptosis, and may also indicate DNA strand breaks (Brinkworth & Schmid, 2003; Sergerie *et al.*, 2005)

- Fluorescence *in-situ* hybridisation (**FISH**): identifies individual chromosomes, and can therefore detect abnormal chromosome number (Brinkworth & Schmid, 2003); also, if multiple probes per chromosome are used, can identify large deletions and duplications (Sloter *et al.*, 2000; Schmid *et al.*, 2004)

References

Adami, H. O., Bergström, R., Möhner, M. *et al.* (1994). Testicular cancer in nine northern European countries. *International Journal of Cancer*, 59, 33–8.

Adler, I. D., Baumgartner, A., Gonda, H., Friedman, M. A. & Skerhut, M. (2000). 1-aminobenzotriazole inhibits acrylamide-induced dominant lethal effects in spermatids of male mice. *Mutagenesis*, 15, 133–6.

Aitken, R. J. (2004). Founders' Lecture. Human spermatozoa: fruits of creation, seeds of doubt. *Reproduction, Fertility, and Development*, 16, 655–64.

Aitken, R. J., Koopman, P. & Lewis, S. E. M. (2004). Seeds of concern. *Nature*, 432, 48–52.

Aitken, R. J., Wingate, J. K., De Iuliis, G. N., Koppers, A. J. & McLaughlin, E. A. (2006). Cis-unsaturated fatty acids stimulate reactive oxygen species generation and lipid peroxidation in human spermatozoa. *Journal of Clinical Endocrinology and Metabolism*, 91, 4154–63.

Almstrup, K., Brask Sonne, S., Høi-Hansen, C. *et al.* (2006). From embryonic stem cells to testicular germ cell cancer – should we be concerned? *International Journal of Andrology*, 29, 211–18.

Andersson, M., Oat Peltoniemi, Makinen, A., Sukura, A. & Rodriguez-Martinez, H. (2000). The hereditary "short tail" sperm defect – a new reproductive problem in Yorkshire boars. *Reproduction in Domestic Animals*, 35, 59–63.

Anway, M. D., Cupp, A. S., Uzumcu, M. & Skinner, M. K. (2005). Epigenetic transgenerational actions of endocrine disruptors and male fertility. *Science*, 308, 1466–9.

Anway, M. D. & Skinner, M. K. (2008). Epigenetic programming of the germ line: effects of endocrine disruptors on the development of transgenerational disease. *Reproductive Biomedicine Online*, 16, 23–5.

Aschim, E. L., Haugen, T. B., Tretli, S. & Grotmol, T. (2008). Subfertility among parents of men diagnosed with testicular cancer. *International Journal of Andrology*, 31, 588–94.

Asklund, C., Jørgensen, N., Skakkebaek, N. E. & Jensen, T. K. (2007). Increased frequency of reproductive health problems among fathers of boys with hypospadias. *Human Reproduction*, 22, 2639–46.

Auger, J., Kunstmann, J. M., Czyglik, F. *et al.* (1995). Decline in semen quality among fertile men in Paris during the past 20 years. *New England Journal of Medicine*, 332, 281–5.

Baccetti, B., Capitani, S., Collodel, G. *et al.* (2001). Genetic sperm defects and consanguinity. *Human Reproduction*, 16, 1365–71.

Banks, S., King, S. A., Irvine, D. S. & Saunders P. T. K. (2005). Impact of a mild scrotal heat stress on DNA integrity in murine spermatozoa. *Reproduction*, 129, 505–14.

Bartkova, J., Horejsí, Z., Koed, K. *et al.* (2005). DNA damage response as a candidate anti-cancer barrier in early human tumorigenesis. *Nature*, 434, 864–70.

Bartkova, J., Rajpert-De Meyts, E., Skakkebaek, N. E., Lukas, J. & Bartek, J. (2003). Deregulation of the G1/S-phase control in human testicular germ cell tumours. *Acta Pathologica, Microbiologica, et Immunologica Scandinavica*, 111, 252–66.

Bergström, R., Adami, H. O., Möhner, M. *et al.* (1996) Increase in testicular cancer incidence in six European countries: a birth cohort phenomenon. *Journal of the National Cancer Institute*, 88, 727–33.

Berkovitz, A., Eltes, F., Lederman, H. *et al.* (2006). How to improve IVF-ICSI outcome by sperm selection. *Reproductive Biomedicine Online*, 12, 634–8.

Berkowitz, G. S., Lapinski, R. H., Dolgin, S. E. *et al.* (1993). Prevalence and natural history of cryptorchidism. *Pediatrics*, 92, 44–9.

Boisen, K. A., Chellakooty, M., Schmidt, I. *et al.* (2005). Hypospadias in a cohort of 1072 Danish newborn boys: prevalence and relationship to placental weight, anthropometrical measurements at birth, and reproductive hormone levels at 3 months of age. *Journal of Clinical Endocrinology and Metabolism*, 90, 4041–6.

Boisen, K. A., Kaleva, M., Main, K. M. *et al.* (2004). Difference in prevalence of congenital cryptorchidism in infants between two Nordic countries. *Lancet*, 363, 1264–9.

Borini, A., Tarozzi, N., Bizzaro, D. *et al.* (2006). Sperm DNA fragmentation: paternal effect on early post-implantation embryo development to ART. *Human Reproduction*, 21, 2876–81.

Brinkworth, M. H. (2000). Paternal transmission of genetic damage: findings in animals and humans. *International Journal of Andrology*, 23, 123–35.

Brinkworth, M. H. & Schmid, T. E. (2003). Effect of age on testicular germ cell apoptosis and sperm aneuploidy in MF-1 mice. *Teratogenesis, Carcinogenesis, and Mutagenesis Supplement*, 2, 103–9.

Bujan, L., Daudin, M., Charlet, J. P., Thonneau, P. & Mieusset, R. (2000). Increase in scrotal temperature in car drivers. *Human Reproduction*, 15, 1355–7.

Carlsen, E., Giwercman, A., Keiding, N., & Skakkebaek, N. E. (1992). Evidence for decreasing quality of semen during past 50 years. *BMJ (Clinical research ed.)*, 305, 609–13.

Carrell, D. T., Wilcox, A. L., Lowy, L. *et al.* (2003). Elevated sperm chromosome aneuploidy and apoptosis in patients with unexplained recurrent pregnancy loss. *Obstetrics and Gynaecology*, 101, 1229–35.

Chandley, A. C. (1991).On the parental origin of de novo mutations in man. *Journal of Medical Genetics*, 28, 217–23.

Chatzimeletiou, K., Morrison, E. E., Prapas, N., Prapas,Y. & Handyside, A. H. (2005). Spindle abnormalities in normally developing and arrested human implantation embryos in vitro identified by confocal laser scanning microscopy. *Human Reproduction*, 20, 672–82.

Chatzimeletiou, K., Rutherford, A. J., Griffin, D. K. & Handyside, A. H. (2007). Is the sperm centrosome to blame for the complex polyploidy chromosome patterns observed in cleavage-stage embryos from an oligoasthenoteratozoospermia (OAT) patient? *Zygote*, 15, 81–90.

Choi, S.-K., Yoon, S.-R., Calabrese, P. & Arnheim, N. (2008). A germ-line-selective advantage rather than an increased mutation rate can explain some unexpectedly common human disease mutations. *Proceedings of the National Academy of Sciences of the United States of America*, 105, 10143–8.

Christensen, K., Kohler, H.-P., Basso, O. *et al.* (2003). The correlation of fecundability among twins: evidence of a genetic effect on fertility? *Epidemiology*, 14, 60–4.

Codina-Pascual, M., Navarro, J., Oliver-Bonet, M. *et al.* (2006). Behaviour of human heterochromatic regions during the synapsis of homologous chromosomes. *Human Reproduction*, 21, 1490–7.

Committee on Toxicity of Chemicals in Food, Consumer Products and the Environment (2003). *Phytoestrogens and Health*. London: Food Standards Agency.

Coonen, E., Derhaag, J. G., Dumoulin, J. C. M. *et al.* (2004). Anaphase lagging mainly explains chromosomal mosaicism in human preimplantation embryos. *Human Reproduction*, 19, 316–24.

Crockford, G. P., Linker, R., Hockley, S. *et al.* (2006). Genome-wide linkage screen for testicular germ cell tumour susceptibility loci. *Human Molecular Genetics*, 15, 443–51.

Crow, J. F. (2003). There's something curious about paternal-age effects. *Science*, 301, 606–67.

Czeizel, A. & Tóth, J. (1990). Correlation between the birth prevalence of isolated hypospadias and parental subfertility. *Teratology*, 41, 167–72.

Czeizel, A. E. & Rothman, K. J. (2002). Does relaxed reproductive selection explain the decline in male reproductive health? A new hypothesis. *Epidemiology*, 13, 113–14.

Czyglik, F., Mayaux, M.-J., Guihard-Moscato, M.-L., David, G. & Schwartz, D. (1986). Lower sperm characteristics in 36 brothers of infertile men, compared with 545 controls. *Fertility and Sterility*, 45, 255–8.

Daphnis, D. D., Delhanty, J. D. A., Jerkovic, S. *et al.* (2005). Detailed FISH analysis of day 5 human embryos reveals the mechanisms leading to mosaic aneuploidy. *Human Reproduction*, 20, 129–37.

Davies, J. M. (1981). Testicular cancer in England and Wales: some epidemiological aspects. *Lancet*, 1, 928–32.

Davies, T. W., Palmer, C. R., Ruja, E. & Lipscombe, J. M. (1996). Adolescent milk, dairy product and fruit consumption and testicular cancer. *British Journal of Cancer*, 74, 657–60.

De Palma, A., Burrello, N., Barone, N. *et al.* (2005). Patients with abnormal sperm parameters have an increased sex chromosome aneuploidy rate in peripheral leukocytes. *Human Reproduction*, 20, 2153–6.

De Ståhl, T. D., Sandgren, T., Piotrowski, A. *et al.* (2008). Profiling of copy number variations (CNVs) in healthy individuals from three ethnic groups using a human genome 32K BAC-clone-based array. *Human Mutation*, 29, 398–408.

Delhanty, J. D. A. (2001). Preimplantation genetics: an explanation for poor human fertility? *Annals of Human Genetics*, 65, 331–8.

Derijck, A. A. H. A., Van Der Heijden, G. W., Ramos, L. *et al.* (2007). Motile human normozoospermic and oligospermic semen samples show a difference in double-stranded DNA break incidence. *Human Reproduction*, 22, 2368–76.

Erenpreiss, J., Bungum, M., Spanó, M. *et al.* (2006). Intra-individual variation in sperm chromatin structure assay parameters in men from infertile couples: clinical implications. *Human Reproduction*, 21, 2061–4.

Eskenazi, B., Kidd, S. A., Marks, A. R. *et al.* (2005). Antioxidant intake is associated with semen quality in healthy men. *Human Reproduction*, 20, 1006–12.

Ferrara, D., Hallmark, N., Scott, H. *et al.* (2006). Acute and long-term effects of in utero exposure of rats to di(n-butyl) phthalate on testicular germ cell development and proliferation. *Endocrinology*, 147, 5352–62.

Fisch, H. & Goluboff, E. T. (1996). Geographic variations in sperm counts: a potential cause of bias in studies of semen quality. *Fertility and Sterility*, 65, 1044–6.

Fisch, H., Goluboff, E. T., Olson, J. H. *et al.* (1996). Semen analyses in 1,283 men from the United States over a 25-year period: no decline in quality. *Fertility and Sterility*, 65, 1009–14.

Fisher, J. S. (2004). Environmental anti-androgens and male reproductive health: focus on phthalates and testicular dysgenesis syndrome. *Reproduction*, 127, 305–15.

Forman, D., Oliver, R. T., Brett, A. R. *et al.* (1992). Familial testicular cancer: a report of the UK family register, estimation of risk and an HLA class 1 sib-pair analysis. *British Journal of Cancer*, 65, 255–62.

Foster, P. M. D. (2006). Disruption of reproductive development in male rat offspring following in utero exposure to phthalate esters. *International Journal of Andrology*, 29, 140–7.

França, L. R., Avelar, G. F. & Almeida, F. F. L. (2005). Spermatogenesis and sperm transit through the epididymis in mammals with emphasis on pigs. *Theriogenology*, 63, 300–18.

Fredell, L., Iselius, L., Collins, A. *et al.* (2002). Complex segregation analysis of hypospadias. *Human Genetics*, 111, 231–4.

Frigyesi, A., Gisselsson, D., Hansen, G. B. *et al.* (2004). A model for karyotypic evolution in testicular germ cell tumors. *Genes Chromosomes Cancer*, 40, 172–8.

Fritz, G. & Czeizel, A. E. (1996). Abnormal sperm morphology and function in the fathers of hypospadiacs. *Journal of Reproduction and Fertility*, 106, 63–6.

Frydelund-Larsen, L., Vogt, P. H., Leffers, H. *et al.* (2003). No AZF deletion in 160 patients with testicular germ cell neoplasia. *Molecular Human Reproduction*, 9, 517–21.

Gazvani, M. R., Wilson, E. D. A., Richmond, D. H. *et al.* (2000). Role of mitotic control in spermatogenesis. *Fertility and Sterility*, 74, 251–6.

Ghanayem, B. I., Witt, K. L., El-Hadri, L. *et al.* (2005). Comparison of germ cell mutagenicity in male CYP2E1-null and wild-type mice treated with acrylamide: evidence supporting a glycidamide-mediated effect. *Biology of Reproduction*, 72, 157–63.

Gianaroli, L., Magli, M. C., Cavallini, G. *et al.* (2005). Frequency of aneuploidy in sperm from patients with extremely severe male factor infertility. *Human Reproduction*, 20, 2140–52.

Gianotten, J., Westerveld, G. H., Leschot, N. J. *et al.* (2004). Familial clustering of impaired spermatogenesis: no evidence for a common genetic inheritance pattern. *Human Reproduction*, 19, 71–6.

Gibbs, W. W. (2003). Roots of cancer. *Scientific American*, July, 56–65.

Goldgar, D. E., Easton, D. F., Cannon-Albright, L. A. & Skolnick, M. H. (1994). Systematic population-based assessment of cancer risk in first-degree relatives of cancer patients. *Journal of the National Cancer Institute*, 86, 1600–8.

Gonzalez-Diego, P., Lopez-Abente, G., Pollan, M. & Ruiz, M. (2000). Time trends in ovarian cancer mortality in Europe (1955–1993) – effect of age, birth cohort and period of death. *European Journal of Cancer*, 36, 1816–24.

Goriely, A., McVean, G. A. T., Röjmyr, M., Ingemarsson, B. & Wilkie, A. O. M. (2003). Evidence for selective advantage of pathogenic FGFR2 mutations in the male germ line. *Science*, 301, 643–6.

Gray, L. E., Wilson, V. S., Stoker, T. *et al.* (2006). Adverse effects of environmental antiandrogens and androgens on reproductive development in mammals. *International Journal of Andrology*, 29, 96–104.

Guichaoua, M. R., Perrin, J., Metzler-Guillemain, C. *et al.* (2005). Meiotic anomalies in infertile men with severe spermatogenic defects. *Human Reproduction*, 20, 1897–902.

Hardell, L., van Bavel, B., Lindstrom, G. *et al.* (2003). Increased concentrations of polychlorinated biphenyls, hexachlorobenzene, and chlordanes in mothers of men with testicular cancer. *Environmental Health Perspectives*, 111, 930–4.

Harland, S. J. (2000). Conundrum of the hereditary component of testicular cancer. *Lancet*, 356, 1455–6.

Hassold, T. & Hunt, P. (2001). To err (meiotically) is human: the genesis of human aneuploidy. *Genetics (Nature reviews)*, 2, 280–91.

Hemminki, K. & Li, X. (2004). Familial risk in testicular cancer as a clue to a heritable and environmental aetiology. *British Journal of Cancer*, 90, 1765–70.

Henderson, B. E., Ross, R. K., Yu, M. C. & Bernstein, L. (1997). An explanation for the increasing incidence of testis cancer: decreasing age at first full-term pregnancy. *Journal of the National Cancer Institute*, 89, 818–19.

Høi-Hansen, C. E., Almstrup, K., Nielsen, J. E. *et al.* (2005). Stem cell pluripotency factor NANOG is expressed in human fetal gonocytes, testicular carcinoma in situ and germ cell tumours. *Histopathology*, 47, 48–56.

Høi-Hansen, C. E., Nielsen, J. E., Almstrup, K. *et al.* (2004). Identification of genes differentially expressed in testes containing carcinoma in situ. *Molecular Human Reproduction*, 10, 423–31.

Høi-Hansen, C. E., Olesen, I. A., Jørgensen, N. *et al.* (2007). Current approaches for detection of carcinoma in situ testis. *International Journal of Andrology*, 30, 398–405.

Irvine, S., Cawood, E., Richardson, D., MacDonald, E. & Aitken, J. (1996). Evidence of deteriorating semen quality in the United Kingdom: birth cohort study in 577 men in Scotland over 11 years. *BMJ (Clinical research ed.)*, 312, 467–71.

Jack, R. H., Davies, E. A. & Møller, H. (2007). Testis and prostate cancer incidence in ethnic groups in South East England. *International Journal of Andrology*, 30, 215–22.

Jensen, T. K., Andersson, A.-M., Jørgensen, N. *et al.* (2004a). Body mass index in relation to semen quality and reproductive hormones among 1,558 Danish men. *Fertility and Sterility*, 82, 863–70.

Jensen, T. K., Jørgensen, N., Punab, M. *et al.* (2004b). Association of in utero exposure to maternal smoking with reduced semen quality and testis size in adulthood: a cross-sectional study of 1,770 young men from the general population in five European countries. *American Journal of Epidemiology*, 159, 49–58.

Joffe, M. (2001). Are problems with male reproductive health caused by endocrine disruption? *Occupational and Environmental Medicine*, 58, 281–8.

Joffe, M. (2003). Infertility and environmental pollutants. *British Medical Bulletin*, 68, 47–70.

Joffe, M. (2007). What harms the developing male reproductive system? In *Male-mediated Developmental Toxicology*, ed. D. Anderson and M. H. Brinkworth. RSC.

John Radcliffe Hospital Cryptorchidism Study Group. (1992). Cryptorchidism: a prospective study of 7500 consecutive male births, 1984–8. *Archives of Disease in Childhood*, 67, 892–9.

Jørgensen, N., Carlsen, E., Nermoen, I. *et al.* (2002). East–West gradient in semen quality in the Nordic–Baltic area: a study of men from the general population in Denmark, Norway, Estonia and Finland. *Human Reproduction*, 17, 2199–208.

Kelce, W. R., Stone, C. R., Laws, S. C. *et al.* (1995). Persistent DDT metabolite p,p'-DDE is a potent androgen receptor antagonist. *Nature*, 15, 581–5.

Khoury, M. J., Beaty, T. H. & Liang, K. Y. (1988). Can familial aggregation of disease be explained by familial aggregation of environmental risk factors? *American Journal of Epidemiology*, 127, 674–83.

Knight, J. A. & Marrett, L. D. (1997). Parental occupational exposure and the risk of testicular cancer in Ontario. *Journal of Occupational and Environmental Medicine*, 39, 333–8.

Kops, G. J. P. L., Weaver, B. A. A. & Cleveland, D. W. (2005). On the road to cancer: aneuploidy and the mitotic checkpoint. *Nature reviews. Cancer*, 5, 773–85.

Korbel, J. O., Urban, A. E., Affourtit, J. P. *et al.* (2007). Paired-end mapping reveals extensive structural variation in the human genome. *Science*, 318, 420–6.

Lanfranco, F., Kamischke, A., Zitzmann, M. & Nieschlag, E. (2004). Klinefelter's syndrome. *Lancet*, 364, 273–83.

Levy, S., Sutton, G., Ng, P. C. *et al.* (2007). The diploid genome sequence of an individual human. *PLoS Biology*, 5, 0001–32.

Li, F. P. (1993). Molecular epidemiology studies of families. *British Journal of Cancer*, 68, 217–19.

Lilford, R., Jones, A. M., Bishop, D. T., Thornton, J. & Mueller, R. (1994). Case-control study of whether subfertility in men is familial. *BMJ (Clinical research ed.)*, 309, 570–3.

Lin, Y. M., Teng, Y. N., Chung, C. L. *et al.* (2006). Decreased mRNA transcripts of M-phase promoting factor and its regulators in the testes of infertile men. *Human Reproduction*, 21, 138–44.

Looijenga, L. H. J., Gillis, A. J. M., Stoop, H. J., Hersmus, R. & Oosterhuis, J. W. (2007). Chromosomes and expression in human testicular germ-cell tumors. Insight into their cell of origin and pathogenesis. *Annals of the New York Academy of Sciences*, 1120, 187–214.

Lutke Holzik, M. F., Rapley, E. A., Hoekstra, H. J. *et al.* (2004). *The Lancet Oncology*, 5, 363–71.

Main, K. M., Toppari, J., Suomi, A. M. *et al.* (2006). Larger testes and higher inhibin B levels in Finnish than in Danish newborn boys. *Journal of Clinical Endocrinology and Metabolism*, 91, 2732–7.

Marchetti, F. & Wyrobek, A. J. (2005). Mechanisms and consequences of paternally-transmitted chromosomal abnormalities. *Birth Defects Research (Part C)*, 75, 112–29.

Marsee, K., Woodruff, T. J., Axelrad, D. A., Calafat, A. M. & Swan, S. H. (2006). Estimated daily phthalate exposures in a population of mothers of male infants exhibiting reduced anogenital distance. *Environmental Health Perspectives*, 114, 805–9.

Mayer, F., Stoop, H., Sen, S. *et al.* (2003). Aneuploidy of human testicular germ cell tumors is associated with amplification of centrosomes. *Oncogene*, 22, 3859–66.

Meschede, D., Lemke, B., Behre, H. M. *et al.* (2000). Clustering of male infertility in the families of couples treated with intracytoplasmic sperm injection. *Human Reproduction*, 15, 1604–08.

Mieusset, R., Bengoudifa, B. & Bujan, L. (2007). Effect of posture and clothing on scrotal temperature in fertile men. *Journal of Andrology*, 28, 170–5.

Møller Jensen, O., Carstensen, B., Glattre, E. *et al.* (1988). *Atlas of Cancer Incidence in the Nordic Countries*. Helsinki: Nordic Cancer Union.

Montgomery, S. M., Granath, F., Ehlin, A., Sparen, P. & Ekbom, A. (2005). Germ-cell testicular cancer in offspring of Finnish immigrants to Sweden. *Cancer Epidemiology, Biomarkers & Prevention*, 14, 280–2.

Morris, I. D., Ilott, S., Dixon, L. & Brison, D. R. (2002). The spectrum of DNA damage in human sperm assessed by single cell gel electrophoresis (Comet assay) and its relationship to fertilization and embryo development. *Human Reproduction*, 17, 990–8.

Moses, M. J. & Poorman, P. A. (1981). Synaptosomal complex analysis of mouse chromosomal rearrangements. II. Synaptic adjustment in a tandem duplication. *Chromosoma*, 81, 519–35.

Moses, M. J., Poorman, P. A., Roderick, T. H. & Davisson, M. T. (1982). Synaptonemal complex analysis of mouse chromosomal rearrangements. IV. Synapsis and synaptic adjustment in two paracentric inversions. *Chromosoma*, 84, 457–74.

Nigg, E. A. (2002). Centrosome aberrations: cause or consequence of cancer progression? *Nature reviews. Cancer*, 2, 1–11.

Ober, C., Hyslop, T. & Hauck, W. W. (1999). Inbreeding effects on fertility in humans: evidence for reproductive compensation. *American Journal of Human Genetics*, 64, 225–31.

Oosterhuis, J. W. & Looijenga, L. H. J. (2005). Testicular germ-cell tumours in a broader perspective. *Nature reviews. Cancer*, 5, 210–22.

Ottesen, A. M., Larsen, J., Gerdes, T. *et al.* (2004). Cytogenetic investigation of testicular carcinoma in situ and early seminoma by high-resolution comparative genomic hybridization analysis of subpopulations flow sorted according to DNA content. *Cancer Genetics and Cytogenetics*, 149, 89–97.

Parkin, D. M. (2005). *Cancer Incidence in Five Continents*. Lyon, France: IARC Press.

Parkin, D. M. & Iscovich, J. (1997). Risk of cancer in migrants and their descendants in Israel. II. Carcinomas and germ-cell tumours. *International Journal of Cancer*, 70, 654–60.

Paulsen, C. A., Berman, N. G. & Wang, C. (1996). Data from men in greater Seattle area reveals no downward trend in semen quality: further evidence that deterioration of semen quality is not geographically uniform. *Fertility and Sterility*, 65, 1015–20.

Petronczki, M., Siomos, M. F. & Nasmyth, K. (2003). Un ménage à quatre: the molecular biology of chromosome segregation in meiosis. *Science*, 112, 423–40.

Pettersson, A., Akre, O., Richiardi, L., Ekbom, A. & Kaijser, M. (2007). Maternal smoking and the epidemic of testicular cancer – a nested case-control study. *International Journal of Cancer*, 120, 2044–6.

Pihan, G. A., Wallace, J., Zhou, Y. & Doxsey, S. J. (2003). Centrosome abnormalities and chromosome instability occur together in pre-invasive carcinomas. *Cancer Research*, 63, 1398–404.

Pukazhenthi, B. S., Neubauer, K., Jewgenow, K., Howard, J. & Wildt, D. E. (2006). The impact and potential etiology of teratospermia in the domestic cat and its wild relatives. *Theriogenology*, 66, 112–21.

Rajpert-De Meyts, E. (2006). Developmental model for the pathogenesis of testicular carcinoma in situ: genetic and environmental aspects. *Human Reproduction Update*, 12, 303–23.

Rajpert-De Meyts, E., Leffers, H., Petersen, J. H. *et al.* (2002). CAG repeat length in androgen-receptor gene and reproductive variables in fertile and infertile men. *Lancet*, 359, 44–6.

Rapley, E. A. (2007). Susceptibility alleles for testicular germ cell tumour: a review. *International Journal of Andrology*, 30, 242–50.

Rapley, E. A., Crockford, G. P., Easton, D. F., Stratton, M. R. & Bishop, D. T. (2003). Localisation of susceptibility genes for familial testicular germ cell tumour. *Acta Pathologica, Microbiologica, et Immunologica Scandinavica*, 111, 128–35.

Redon, R., Ishikawa, S., Fitch, K. R. *et al.* (2006). Global variation in copy number in the human genome. *Nature*, 444, 444–54.

Richiardi, L. & Akre, O. (2005). Fertility among brothers of patients with testicular cancer. *Cancer Epidemiology Biomarkers & Prevention*, 14, 2557–62.

Richiardi, L., Akre, O., Lambe, M. *et al.* (2004). Birth order, sibship size, and risk for germ-cell testicular cancer. *Epidemiology*, 15, 323–9.

Richtoff, J., Elzanaty, S., Rylander, L., Hagmar, L. & Giwercman, A. (2007). Association between tobacco exposure and reproductive parameters in adolescent males. *International Journal of Andrology*, 31, 31–9.

Rolland, A. D., Chalmel, F., Cavel, P., Coiffec-Dorval, I. & Jégou, B. (2008). Residual bodies are regulators of Sertoli cell function. Presented at: 15th European Testis Workshop, Naantali, Finland, 2–6 May 2008. Miniposter IV.98.

Rubes, J., Vozdova, M., Robbins, W. A. *et al.* (2002). *American Journal of Human Genetics*, 70, 1507–19.

Safe, S. H. (1995). Environmental and dietary estrogens and human health: is there a problem? *Environmental Health Perspectives*, 103, 346–51.

Sakkas, D. (1999). The need to detect DNA damage in human spermatozoa: possible consequences on embryo development. In *The Male Gamete*, ed. C. Gagnon. Vienna, Illinois: Cache River Press.

Schmid, T. E., Brinkworth, M. H., Hill, F. *et al.* (2004). Detection of structural and numerical chromosomal abnormalities by ACM-FISH analysis in sperm of oligozoospermic infertility patients. *Human Reproduction*, 19, 1395–400.

Schmid, T. E., Eskenazi, B., Baumgartner, A. *et al.* (2007). The effects of male age on sperm DNA damage in healthy non-smokers. *Human Reproduction*, 22, 180–7.

Schmid, T. E., Kamischke, A., Bollwein, H., Nieschlag, E. & Brinkworth, M. H. (2003). Genetic damage in oligospermic patients detected by fluorescence in-situ hybridization, inverse restriction site mutation assay, sperm chromatin structure assay and the Comet assay. *Human Reproduction*, 18, 1474–80.

Sciurano, R., Rahn, M., Rey-Valzacchi, G. & Solari, A. J. (2007). The asynaptic chromatin in spermatocytes of translocation carriers contains the histone variant gamma-H2AX and associates with the XY body. *Human Reproduction*, 22, 142–50.

Sergerie, M., Laforest, G., Boulanger, K., Bissonnette, F. & Bleau, G. (2005). Longitudinal study of sperm DNA fragmentation as measured by terminal uridine nick end-labelling assay. *Human Reproduction*, 20, 1921–7.

Setchell, B. P. (1998). The Parkes Lecture. Heat and the testis. *Journal of Reproduction and Fertility*, 114, 179–94.

Shah, M. N., Devesa, S. S., Zhu, K. & McGlynn, K. A. (2007). Trends in testicular germ cell tumours by ethnic group in the United States. *International Journal of Andrology*, 30, 206–14.

Sharpe, R. M. (1994). Regulation of spermatogenesis. In *The Physiology of Reproduction*, ed. E. Knobil and J. D. Neale, 2nd edition. New York NY: Raven Press, Ltd.

Sharpe, R. M. (2003). The 'oestrogen hypothesis' – where do we stand now? *International Journal of Andrology*, 26, 2–15.

Sharpe, R. M. & Skakkebaek, N. E. (1993). Are oestrogens involved in falling sperm counts and disorders of the male reproductive tract? *Lancet*, 341, 1392–5.

Shen, H., Main, K. M., Andersson, A.-M. *et al.* (2008). Concentrations of persistent organochlorine compounds in human milk and placenta are higher in Denmark than in Finland. *Human Reproduction*, 23, 201–10.

Singh, N. P., Muller, C. H. & Berger, R. E. (2003). Effects of age on DNA double-strand breaks and apoptosis in human sperm. *Fertility and Sterility*, 80, 1420–30.

Skakkebaek, N. E. (2007). Testicular cancer trends as 'whistle blower' of testicular development problems in populations. *International Journal of Andrology*, 30, 198–205.

Skakkebaek, N. E., Berthelsen, J. G., Giwercman, A. & Muller, J. (1987). Carcinoma-in-situ of the testis: possible origin from gonocytes and precursors of all types of germ cell tumours except spermatocytoma. *International Journal of Andrology*, 10, 19–28.

Skakkebaek, N. E., Rajperts-de Meyts, E. & Main, K. M. (2001). Testicular dysgenesis syndrome: an increasingly common developmental disorder with environmental aspects. *Human Reproduction*, 16, 972–8.

Skotheim, R. I. & Lothe, R. (2003). The testicular germ cell tumour genome. *Acta Pathologica, Microbiologica, et Immunologica Scandinavica*, 111, 136–51.

Slama, R. & Leridon, H. (2002). How much of the decline in sperm counts can be explained by relaxed reproductive selection? *Epidemiology*, 13, 613–15.

Sloter, E. D., Low, X., Moore, D. H., Nath, J. & Wyrobek, A. J. (2000). Multicolor FISH analysis of chromosomal breaks, duplications, deletions, and numerical abnormalities in the sperm of healthy men. *American Journal of Human Genetics*, 67, 862–72.

Sloter, E., Nath, J., Eskenazi, B. & Wyrobek, A. J. (2004). Effects of male age on the frequencies of germinal and heritable chromosomal abnormalities in humans and rodents. *Fertility and Sterility*, 81, 925–43.

Storchova, Z. & Pellman, D. (2004). From polyploidy to aneuploidy, genome instability and cancer. *Nature Review of Molecular Cell Biology*, 5, 45–54.

Storgaard, L., Bonde, J. P., Ernst, E. *et al.* (2003). The impact of genes and environment on semen quality: an epidemiological twin study. In Storgaard, L., *Genetical and Prenatal Determinants for Semen Quality: An Epidemiological Twin Study (PhD thesis)*. Aarhus: University of Aarhus.

Storgaard, L., Bonde, J. P. & Olsen, J. (2006). Male reproductive disorders in humans and prenatal indicators of estrogen exposure: a review of published epidemiological studies. *Reproductive Toxicology*, 21, 4–15.

Swerdlow, A. J., De Stavola, B. L., Swanwick, M. A. & Maconochie, N. E. S. (1997). Risks of breast and testicular cancers in young adult twins in England and Wales: evidence on prenatal and genetic aetiology. *Lancet*, 350, 1723–8.

Tesaryk, J., Mendoza, C. & Greco, E. (2002). Paternal effects acting during the first cell cycle of human preimplantation development after ICSI. *Human Reproduction*, 17, 184–9.

Toppari, J., Kaleva, M. & Virtanen, H. E. (2001). Trends in the incidence of cryptorchidism and hypospadias, and methodological limitations of registry-based data. *Human Reproduction Update*, 7, 282–6.

Tyl, R. W., Marr, M. C., Myers, C. B., Ross, W. P. & Friedman, M. A. (2000). Relationship between acrylamide reproductive and neurotoxicity in male rats. *Reproductive Toxicology*, 14, 147–57.

Van Golde, R. J., van der Avoort, I. A., Tuerlings, J. H. *et al.* (2004). Phenotyic characteristics of male subfertility and its familial occurrence. *Journal of Andrology*, 25, 819–23.

Van Waeleghem, K., De Clercq, N., Vermeulen, L., Schoonjans, F. & Comhaire, F. (1996). Deterioration of sperm quality in young healthy Belgian men. *Human Reproduction*, 11, 325–9.

Vendrell, J. M., Garcia, F., Veiga, A. *et al.* (1999). Meiotic abnormalities and spermatogenic parameters in severe oligozoospermia. *Human Reproduction*, 14, 375–8.

Vierula, M., Niemi, M., Keiski, A. *et al.* (1996). High and unchanged sperm counts of Finnish men. *International Journal of Andrology*, 19, 11–17.

Weidner, I. S., Møller, H., Jensen, T. K. & Skakkebaek, N. E. (1999). Risk factors for cryptorchidism and hypospadias. *Journal of Urology*. 161, 1606–9.

Wittmaack, F. M. & Shapiro, S. S. (1992). Longitudinal study of semen quality in Wisconsin men over a decade. *Wisconsin Medical Journal*, 91, 477–9.

Wyrobek, A. J., Eskenazi, B., Young, S. *et al.* (2006). Advancing age has differential effects on DNA damage, chromatin integrity, gene mutations, and aneuploidies in sperm. *Proceedings of the National Academy of Sciences*, 103, 9601–6.

Zenzes, M. T. (2000). Smoking and reproduction: gene damage to human gametes and embryos. *Human Reproduction Update*, 6, 122–31.

Zhu, B., Walker, S. K., Oakey, H., Setchell, B. P. & Maddocks, S. (2004). Effect of paternal health stress on the development in vitro of preimplantation embryos in the mouse. *Andrologia*, 36, 384–94.

Zöllner, S., Wen, X., Hanchard, N. A. *et al.* (2004). Evidence for extensive transmission distortion in the human genome. *American Journal of Human Genetics*, 74, 62–72.

3 *The microenvironment in health and cancer of the mammary gland*

JOHN P. WIEBE

Introduction

Mammary glands do not normally feature prominently in discussions of reproduction. Instead, discussions of reproduction deservedly emphasize the primary sex organs, the reproductive tract and the cardinal hormones (gonadotrophins and gonadal steroids) involved. However, newborn mammals cannot process foraged food and only the mammary gland is capable of synthesizing renewable food, in the form of milk, which is digestible by the neonates. Milk is a complex nutrient, consisting mainly of milk sugar (lactose), lipids, the milk proteins (casein and whey) as well as monovalent and divalent cations and immune antibodies. The female mammary glands are markedly altered functionally and morphologically during pregnancy in preparation for the production of this complex liquid food mixture for delivery to the neonates. The mammary gland thus forms an important adjunct to the mammalian reproductive system and is essential for the survival of mammalian species.

The conversion from inactive to active milk-producing mammary tissues requires extensive surges in cell proliferation as well as metabolic modifications. These alterations occur regularly during each ovulatory cycle in preparation for a potential pregnancy and the post-partum lactational activity.

In the absence of pregnancy, the proliferative rate returns to basal levels in the normal breast. The regulation of these proliferative oscillations depends on the steroid hormone milieu within microregions of the breast. It is argued that dysregulation of this microenvironment can lead to conditions within regions of the breast that result in foci of neoplasia. In particular, progesterone metabolites have been identified in the breast microenvironment, which act as autocrines/paracrines with the ability to either regulate or deregulate cell proliferation and cell death, and thus determine normalcy or progression to neoplasia.

Reproduction and Adaptation, eds. C. G. Nicholas Mascie-Taylor and Lyliane Rosetta.
Published by Cambridge University Press.

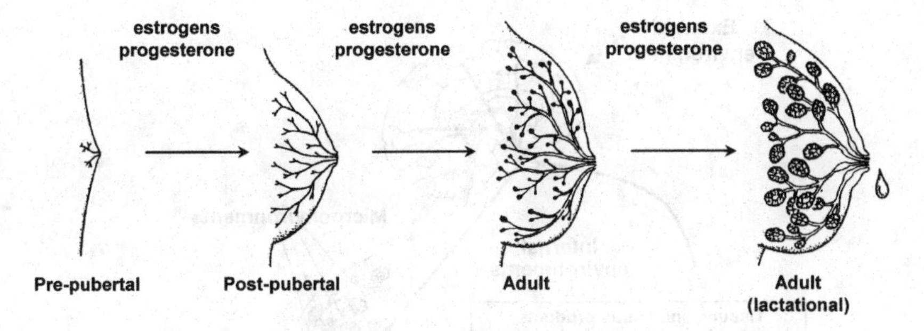

Figure 3.1. Development of the mammaries and involvement of estrogen and progesterone.

Steroid hormones in mammary gland development, differentiation and cyclical changes

Early in fetal life, two ectodermal ridges appear on either side of the ventral midline. In females, localized thickenings become the mammary buds which grow into the underlying dermis to form the primary mammary cords, the precursors to the duct systems. Growth of the undeveloped mammae is initiated at the onset of puberty as a result of increased cyclical ovarian production of estrogen and progesterone. Growth and branching of the ducts occur and a lobular system is developed. Figure 3.1 depicts the development of the mammary glands and emphasizes that both estrogen and progesterone are involved in the changes from immature pre-pubertal structures to the mature adult and functional lactational breast. Most of the increase in size of the mammary glands is owing to interlobular fat deposition, largely as a result of estrogen action. While both estrogen and progesterone play a role in the development of lobules and alveoli, it is primarily progesterone that causes marked proliferation in these structures (Going *et al.*, 1988; Pike *et al.*, 1993). Progesterone is also largely responsible for the proliferative changes in the breasts that occur during each menstrual cycle of a woman (Potten *et al.*, 1988). At pregnancy, duct growth is further stimulated by estrogens, whereas proliferation of the lobular–alveolar system is greatly enhanced by the actions of progesterone. During the reproductive years, the primary site of estradiol and progesterone production is the cycling ovary. In addition, the placenta produces large amounts of progesterone during pregnancy, while the adrenals provide progesterone as a precursor for the corticosteroids, progesterone metabolites, androgens and estrogens throughout life.

It should be noted that besides the requirement for estrogens and progesterone, other hormones such as growth hormone (somatotrophin), thyroid

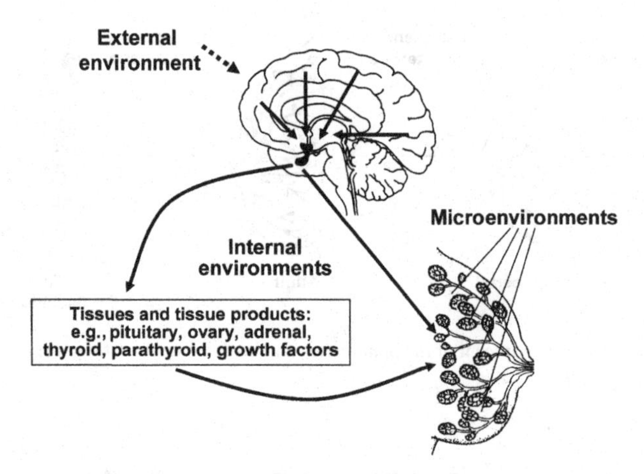

Figure 3.2. Environmental (external and internal) influences on the microenvironments in the breast.

hormones, corticosteroids (glucocorticosteroids), insulin and prolactin are also involved in attaining the full development and functioning of the lactating mammary glands (Lyons *et al.*, 1958; Topper & Freeman, 1980). The production of gonadal steroids (ovarian estradiol and progesterone) as well as the hormones of the pituitary, adrenals, and thyroid are under the control of the hypothalamus–pituitary–target tissue axes, which in turn are modulated by internal and external environmental factors. All of these contribute to creating a unique and variable microenvironment within the breast, as illustrated in Figure 3.2.

During the reproductive years, the female mammary gland shows regular cyclical changes. Surges in cell proliferation in the breast are followed by apoptotic activity during every normal cycle. These oscillations are controlled by the gonadal steroids and closely parallel the cyclical changes in the ovary. The cyclical increases in cell numbers are highly predictable and regulated and, what is more important, they are reversible in the normal mammary gland, at the end of every cycle. Thus, a careful balance between proliferation and apoptosis maintains general normalcy in cell numbers in the tissues. When that balance becomes deregulated, such that either cell proliferation is increased or apoptosis is decreased, or both, without a return to the normal condition, cells may become neoplastic and tumors can develop and grow. The primary cause of this deregulation has generally been attributed to prolonged heightened exposure to the sex steroids, particularly estradiol. Many epidemiological studies (see below) have suggested a correlation between sampled blood estrogen levels and breast cancer risk. However, the biological relevance of such a correlation may be questionable.

Serum steroid hormone levels and breast cancer risk: a critique

Numerous studies have attempted to correlate serum estrogen levels with altered risk of breast cancer. Although there are several studies that failed to identify any significant associations between breast cancer and sex steroids (Bulbrook *et al.*, 1986; Wysowski *et al.*, 1987; Garland *et al.*, 1992; Helzlsouer *et al.*, 1994; Kabuto *et al.*, 2000), the majority of reports conclude that such an association exists. Much of the evidence on which these correlations have been built, comes from measures of blood samples from selected populations of "control" (no breast cancer) and cancer cases. Table 3.1 shows calculated serum steroid hormone concentrations from several selected epidemiological studies. The examples are chosen from numerous other similar reports in which correlations are made between reported statistically significantly higher serum estrogen levels and increased risk of breast cancer. The basis for attempting to establish such a link comes from a range of in vitro and in vivo studies which have suggested that estrogens may have mitogenic actions in mammary cells (for example, Foster & Wimalasena, 1996; Prall *et al.*, 1997; Castro-Rivera *et al.*, 2001) and stimulate development and growth of mammary tumors in animal models (Leung *et al.*, 1975; Nandi *et al.*, 1995). However, closer examination of the methods and the data suggest that the biological relevance of such determinations is tenuous.

It can be argued that the measurements of blood hormone levels in epidemiological studies do not offer a biologically sound explanation of localized breast cell deregulation and subsequent development of steroid-induced breast cancer. Invariably, such measurements of a population of women are based on single, or at best a few, samples from each woman (see references in Table 3.1) It is well established that there can be marked variability between blood samplings, not only between individuals but also from the same person. For example, Muti *et al.* (1996) collected two blood samples after 12 hours of fasting, one year apart in the same month, day, hour and minute of the day, and for premenopausal women, on the presumptive identical day of the luteal phase of the menstrual cycle. In spite of these strict inclusion criteria and highly standardized blood draws – which would be impossible in epidemiological studies – the authors concluded that in premenopausal women total estradiol showed "very poor reliability between the two determinations."

Table 3.1 *Examples of blood measurements of estradiol, estrone, estrone sulfate, progesterone, dehydroepiandrosterone (DHEA) and DHEA sulfate (DHEAS), in control and matched breast cancer cases in premenopausal and postmenopausal women (in pg/ml, except for DHEAS, which is in ng/ml).*

Hormones	Control	Breast cancer	p	References
Premenopausal				Baulieu, 1990; Milgrom, 1990; Yen, 1990
Estradiol: proliferative phase	60			
5 days before ovulation	100			
2–3 days before ovulation	300			
24 hours before ovulation	600			
Premenopausal	*mean (95% CI)*	*mean (95% CI)*		Sturgeon *et al.*, 2004
Estradiol: early follicular	44 (37–51)	50 (41–61)		
Estradiol: late follicular	144 (121–172)	119 (100–142)		
Estradiol: lluteal	111 (98–125)	101 (86–118)		
Progesterone	3680 (2600–5217)	3633 (2442–5404)		
DHEA	3383 (3095–3694)	3830 (3490–4204)		
Postmenopausal (Malmö, Sweden)	*mean (range)*	*mean (range)*		Manjer *et al.*, 2003
Estradiol	3.7 (0.7–461)	1.9 (0.7–79)		
Eestrone	22 (4–168)	27 (10–335)		
DHEAS (ng/ml)	1056 (37–4379)	1119 (88–3595)		
(Umeå, Sweden)				
Estradiol	1.5 (0.7–24)	2.9 (0.7–163)		
Estrone	21 (7–54)	23 (12–163)		
DHEAS (ng/ml)	1229 (37–3481)	1167 (37–3334)		
Premenopausal	*median (range)*	*median (range)*		Yu *et al.*, 2003
Estradiol	45 (3–339)	41 (3–1024)	0.969	
Estrone	29 (0.4–158)	33 (0.1–706)	0.016	
Postmenopausal				
Estradiol	5.7 (3–55)	5.7 (1–84)	0.95	
Estrone	18 (1–326)	20 (1–215)	0.003	
Progesterone	1200 (300–28,500)	1200 (300–87,400)	0.489	
Premenopausal	*mean (95% CI)*	*mean (95% CI)*		Kaaks *et al.*, 2005
Estradiol	104 (97–111)	109 (99–119)	0.32	
Progesterone	3755 (3325–4189)	2970 (2371–3570)	0.01	
DHEAS (ng/ml)	1340 (1288–1395)	1446 (1372–1523)	0.03	
Premenopausal	*median (range)*	*median (range)*		Eliasson *et al.*, 2006
Estradiol (follicular)	44 (22–88)	48 (28–101)	0.01	
Estradiol (luteal)	120 (69–192)	125 (76–182)	0.35	
Progesterone (luteal)	14,470 (4720–25,140)	15,720 (4800–24,910)	0.8	

Table 3.1 (*cont.*)

Hormones	Control	Breast cancer	p	References
Postmenopausal	*median (range)*	*median (range)*		Cauley *et al.*, 1999
Estradiol	6 (2–56)	8 (3–22)	0.001	
Estrone sulfate	161 (0–1088)	221 (42–1035)	0.004	
DHEAS (ng/ml)	363 (0–3326)	750 (88–3565)	0.04	
Postmenopausal	*median (range)*	*median (range)*		Missmer *et al.*, 2004
Estradiol	6 (4–13)	7 (4–15)	0.001	
Estrone	23 (14–38)	26 (15–43)	0.001	
Estrone sulfate	280 (136–600)	339 (154–823)	0.001	
Progesterone	40 (15–100)	40 (15–100)	0.64	
DHEA	2480 (1160–4730)	2830 (1270–5570)	0.01	
DHEAS (ng/ml)	850 (350–1690)	900 (440–2050)	0.001	
Postmenopausal	*median (range)*	*median (range)*		Hankinson *et al.*, 1998
Estradiol	7 (4–14)	8 (4–16)	0.04	
Estrone	28 (17–45)	31 (20–51)	0.02	
Eestrone sulfate	192 (97–420)	232 (102–593)	0.02	
DHEA	2050 (990–3660)	2100 (970–4340)	0.48	
DHEAS (ng/ml)	790 (340–1630)	870 (420–2000)	0.01	

Measurements of steroid hormone levels in the peripheral blood may be highly variable. Blood levels of a steroid hormone such as estradiol can vary three to ten fold within days or even one to two hours prior to ovulation owing to the cyclical, circadian and short-term rhythms and oscillations in gonadotropin releasing (GnRH) and luteinizing (LH) hormones which regulate rhythmic changes in steroid hormone levels (Table 3.1, Baulieu, 1990; Milgrom, 1990; Yen, 1990). The rhythmic changes are also affected by internal (psychic, alcohol, drugs, health) and external (photoperiod, diet, stressors) environmental factors. The differences between individuals are even greater because of variability in month-by-month cycle length within and between individuals (Lenton *et al.*, 1984a, 1984b), and because of the improbability of drawing all samples at precisely the same point in the cycle. While the cyclical changes may not be of concern in postmenopausal samples, large variations still remain between assays (Table 3.1, Hankinson *et al.*, 1998; Cauley *et al.*, 1999; Manjer *et al.*, 2003, Yu *et al.*, 2003; Missmer *et al.*, 2004). Because of the large variations within and between individuals, along with the shortcomings of the sampling and assay methods, the range in sample hormone concentration values may be very large. It is doubtful that means (calculated from one or two samples per individual) which differ by only a few pg/ml but show several-fold to several thousand-fold differences in range of values, have biological relevance, statistical significance notwithstanding (Table 3.1, Hankinson *et al.*, 1998; Cauley *et al.*, 1999; Yu *et al.*, 2003; Missmer *et al.*, 2004; Eliasson *et al.*, 2006).

Another reason to question the relevance of serum hormone measurements relates to the nature of the assays. Most steroid hormone levels are determined with assays based on antibodies generated against a bovine serum albumin (BSA)-steroid conjugate. Different antibodies vary in level of specificity; none are completely specific for only one steroid and therefore varying levels of cross-reaction occur with different steroids. Even if the level of "cross-talk" for another steroid such as dehydroepiandrosterone (DHEA) or DHEA-sulfate (DHEAS) is less than 1% compared to the steroid of interest (e.g. estradiol), the results can be greatly skewed if the former is present at many-fold higher concentrations than the latter. For example, whereas estradiol and estrone are generally present in the low picomolar range, DHEAS is present in the blood at micromolar levels (See Table 3.1, Hankinson *et al.*, 1998; Cauley *et al.*, 1999; Manjer *et al.*, 2003) and progesterone is often found at nanomolar levels (Sturgeon *et al.*, 2004; Kaaks *et al.*, 2005; Pinheiro *et al.*, 2005; Eliasson *et al.*, 2006). Therefore, if whole serum is analyzed without extracting and separating the compound of interest from all others – which is usually the method of choice – wrong measures may be obtained. On the other hand, because of differences in solubility of steroids as well as of organic solvents, the employment of steroid extraction procedures can result in marked differences in recovery between different steroids and therefore can also lead to errors in concentration estimates.

Finally, steroids such as estrone sulfate, DHEA and DHEAS (which may be present in the nano- to micro-molar range) can be readily converted to estrone and estradiol in many tissues including breast, while androgens, estrogens and progesterone can be metabolized to other active steroid hormones. Therefore, the concentrations in blood samplings may bear no relation to concentrations in local regions within the breast where the biological actions occur. For example, levels of estradiol, progesterone, estrone sulfate, and DHEAS may be, respectively, 7-fold, 6-fold, 130->500-fold, and >1000-fold higher in nipple aspirate fluid than in serum (Chatterton *et al.*, 2003, 2004, 2005; Gann *et al.*, 2006). Thus, regardless of statistically significant associations between isolated blood steroid determinations and breast cancer, changes in blood levels do not reflect the breast microenvironment, where the potential hormone-sensitive, neoplastic dysregulations actually take place. In addition, recent evidence shows that normalcy and neoplasia in the mammary gland may depend, not on the blood levels, and not even on the local (within breast) levels of hormones such as estradiol or progesterone, but on levels of certain active steroid metabolites produced within the mammary tissues.

Role of the microenvironment in normal and neoplastic breast

The normal development of the breast as well as the cyclical changes that take place are regulated by the milieu within regions of the tissues and between the cells (the microenvironment). Microenvironments and gradients exist within normal tissues because different cell types within an organ/tissue may differ in the expression of metabolizing enzymes, co-factors, receptors, etc., and because of differing levels of vascularity and various locally produced regulatory factors (Sutherland, 1988). Likewise, when determining the role of any factor(s) in maintaining breast cell/tissue normalcy or in changes that lead to neoplasia and ultimately to metastasis, it is the composition/concentration in the microenvironment that needs to be considered. Measures of circulating factors such as hormones determined in blood obtained from a distant peripheral sampling site may bear little or no relevance to the concentrations within the tissue microenvironment, as evidenced by the high levels of steroids in breast nipple aspirate fluid. The contribution of the microenvironment to tumorigenic processes is underscored by findings that breast tumors implanted into different tissue sites (e.g. primary breast tumors, axillary lymph nodes, mammary fat pads, cranium) show marked differences in the extent and nature of the angiogenic response (Edel *et al.*, 2000; Monsky *et al.*, 2002; Boudreau & Myers, 2003; Schneider & Miller, 2005). Also, the invasion of metastatic cells and growth of tumors in distant tissues depends, in part, on the interaction between particular migrating tumor cells and specific target organ environments (Radinsky & Ellis, 1996; Chambers *et al.*, 2002). Thus, breast cancers metastasize to bone, brain, liver and kidney more frequently than to other organs/tissues.

Variations also exist within tumors. Tumors are organ-like structures that are heterogeneous and complex. They comprise stromal cells and cancer cells that are embedded in an extracellular matrix and nourished by a vascular network; each of these components may vary from one location to another in the same tumor (Trédan *et al.*, 2007). Tumor microregions can be considered as niches in which there may be major gradients of critical metabolites (e.g. oxygen, glucose, lactate, H^+ ions), hormonal products and growth and necrosis factors (Sutherland, 1988). The role of the microenvironment both in maintaining breast tissue normalcy as well as in permitting progression to neoplasia and metastasis is fundamental. Therefore, the identification of factors and changes in the breast microenvironment that control the transformation and progression from normalcy to cancer is essential for a better understanding of breast cancer and for the development of effective therapies. What follows is an account of the changes in the breast microenvironment of the levels of two progesterone metabolites, one cancer-promoting and the other cancer-inhibiting, and how these in turn control (stimulate or suppress) transformation and progression from normalcy to cancer. The evidence suggests that the promotion of breast cancers is related to relative changes in *in situ* concentrations of the cancer-inhibiting and cancer-promoting hormones, not to estradiol levels.

The role of progesterone metabolites in regulating breast tissue normalcy and cancer

Direct or indirect roles for progesterone (or synthetic progestin) involvement in mammary gland tumorigenesis have been suggested for some time, but results of treatments have been conflicting, some reports showing no effects, others showing either stimulation or inhibition of growth of human tumors (Horwitz *et al.*, 1985; Anderson *et al.*, 1989; Santen *et al.*, 1990) or human breast cancer cell lines (Braunsberg *et al.*, 1987; Clark & Sutherland, 1990; Cappelletti *et al.*, 1991; King, 1991; Clark *et al.*, 1994; Musgrove & Sutherland, 1994; Groshong *et al.*, 1997). Similarly, in dogs (Segaloff, 1975; Mol *et al.*, 1996) and in rodents, progestins may stimulate or inhibit tumor growth (Welsh, 1982), including chemically induced tumors (Jabara, 1967; Luo *et al.*, 1997; Lanari & Molinolo, 2002). These conflicting results, in addition to the lack of evidence that progressive changes in tumor development are related to circulating levels of steroids, led us to speculate about the potential importance of further metabolism of steroids occurring locally within the tumor and its adjacent host tissue. Such local metabolism might markedly change the biological potency of steroid hormones and create a microenvironment around and within tumor cells that promotes cancer. We hypothesized that progesterone may be converted within breast tissue to two types of metabolites: those that stimulate and those that inhibit cell proliferation and tumorigenesis. According to this hypothesis, progesterone would serve as precursor (or pro-hormone) and the metabolites as the active hormones. The state or progression of mammary tumors then could depend on the ratio of cancer-promoting to cancer-inhibiting progesterone metabolites. If such progesterone metabolites could be shown to exist, they might also provide the basis of an endocrine explanation for all those human breast cancer cases which are unresponsive to the anti-estrogen therapy.

Progesterone metabolism in breast tissues and breast cell lines

Many metabolism studies from a large number of tissues and various species have indicated that in addition to the gonads and adrenals, perhaps most, if not all, tissues have some capacity to convert progesterone to other products (Wiebe,

2006). These studies demonstrated the presence in tissues and cells of one or more enzymes capable of acting on various sites in the progesterone molecule, and leading to the formation of various classes of 21-carbon steroids, in addition to the known corticosteroids, androgens and estrogens (Wiebe, 2006). These progesterone metabolizing enzymes (PMEs) include 5α-reductase, 5β-reductase, 3α-hydroxysteroid oxidoreductase (3α-HSO), 3β-HSO, 20α-HSO, 20β-HSO, 6α(β)-, 11β-, 17-, and 21-hydroxylase, and C_{17-20}-lyase. In spite of this large number of specific site-directed enzymes capable of local transformation of progesterone, the 21-carbon progesterone metabolites, for the most part, have been considered to be waste products and the PMEs a means of controlling the local (in tissue) concentrations of progesterone.

In terms of mammary cancer, the presence of progesterone metabolizing enzymes had been demonstrated in dimethylbenz(a)anthracene-induced rat mammary tumors (Mori *et al.*, 1978; Mori & Tamaoki, 1980; Eechaute *et al.*, 1983), human breast tissues (Lloyd, 1979; Miller, 1990) and modified breast cancer cell lines (T47Dco) (Fennessey *et al.*, 1986; Horwitz *et al.*, 1986). Although selective differences in progesterone metabolizing enzyme activities between normal and tumor tissues were noted in some of these studies, they were not linked to any potential effects of the metabolites themselves on cancer induction or promotion, prior to our studies (Wiebe *et al.*, 2000).

Formation of the progesterone metabolites, 3α-dihydroprogesterone (3αHP) and 5α-dihydroprogesterone (5αP), in human breast tissues and cell lines

The progesterone metabolites involved in breast cancer modulation were first identified using ^{14}C-labeled progesterone and human breast tissues. Studies of paired (normal and tumor) breast tissues (Wiebe *et al.*, 2000), each pair from the same breast, showed that progesterone is readily converted into two classes of metabolites: those with a double bond in the carbon-4 position of ring A (delta-4-pregnenes; 4-pregnenes) and those that are 5α-reduced (5α-pregnanes), illustrated in Figure 3.3. Reduction of progesterone to 5α-pregnanes is catalyzed by 5α-reductase (5α-R) and the first 5α-reduced metabolite is 5α-pregnane-3,20-dione (also known as 5α-dihydroprogesterone; 5αP). The conversion of progesterone to the 4-pregnene, 3α-hydroxy-4-pregnene-20-one (also known as 3α-dihydroprogesterone; 3αHP), results from the action of 3α-hydroxysteroid oxidoreductase (3α-HSO). The conversions have been demonstrated in all breast tissues examined, regardless of the age of the women, subtypes and grades of carcinomas, and whether the tissues are estrogen receptor (ER) and progesterone receptor (PR) positive and/or negative (Wiebe *et al.*, 2000). The

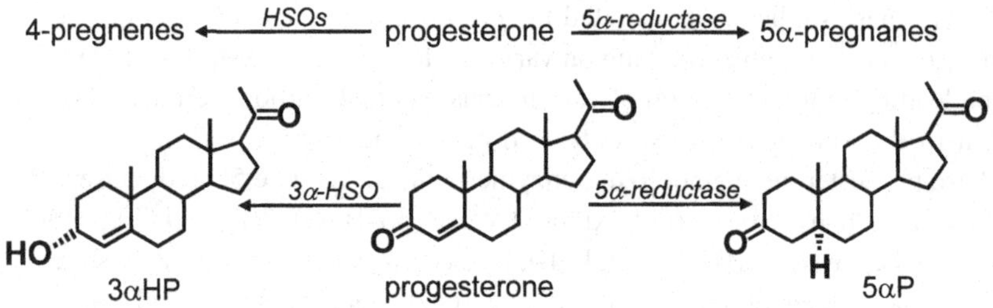

Figure 3.3. Progesterone conversion to the 4-pregnene, 3αHP, and the 5α-pregnene, 5αP, as a result of the actions of 3α-HSO and 5α-reductase in human breast tissue and cell lines.

same metabolic pathways also have been demonstrated in a variety of human breast cell lines (Wiebe & Lewis, 2003).

Production of 5α-pregnanes is higher and production of 4-pregnenes is lower in tumor issues and in tumorigenic cell lines than in normal tissue and non-tumorigenic cell lines

Although both normal and tumorous breast tissues convert progesterone to the two classes of metabolites, there are significant quantitative differences. In normal breast tissue, the levels of 4-pregnenes (especially 3αHP) greatly exceed the levels of 5α-pregnanes, whereas in tumorous tissue, 5α-pregnanes (especially 5αP) greatly exceed 4-pregnenes (Wiebe *et al.*, 2000, 2005). The differences in amounts of 4-pregnenes and 5α-pregnanes are mainly owing to changes in the amounts of 5αP and 3αHP, and the ratio of 5αP:3αHP is nearly 30-fold higher in tumorous than in normal breast tissues (Figure 3.4a). Measurements of actual amounts of progesterone metabolites present in breast tissues, using radioimmunoassays and mass spectrometry, also showed that the ratio of 5αP:3αHP is markedly higher (15- to 25-fold) in tumor than in normal tissues, validating the large shift shown in the [^{14}C]progesterone metabolism studies (Wiebe *et al.*, 2000). Evidence of changes in concentration ratios of 5αP:3αHP in the breast microenvironment was also provided by breast nipple aspirate fluid samples (Wiebe, 2006); samples from tumorous breasts had, on average, five-fold more 5αP than 3αHP. Moreover, the nipple aspirate fluid samples provided evidence that these progesterone metabolites (and progesterone itself) can be present at micromolar (about 1–10 μM) concentrations. The differences in levels suggest active metabolism of the locally available progesterone and the ability of the cells to alter the microenvironment in terms of the progesterone metabolites.

Additional evidence that changes in progesterone metabolism may be related to tumorigenesis was obtained from studies on human breast cell lines with

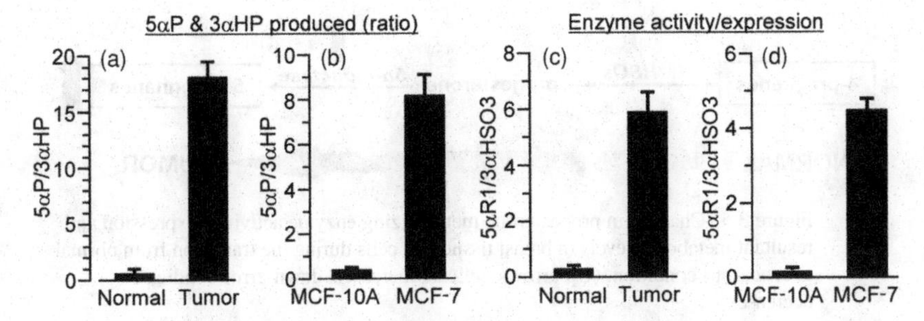

Figure 3.4. Comparison of progesterone metabolism ((a) and (b)) and progesterone metabolizing enzyme activities/expression ((c) and (d)) between normal and tumorous breast tissues and between non-tumorigenic (MCF10A) and tumorigeic (MCF-7) human breast cell lines. Progesterone metabolism is depicted as the ratio of the amounts of 5αP to 3αHP produced (5αP/3αHP); enzyme activity/expression levels are shown as the ratio of 5α-reductase type 1 to 3α-HSO type 3 (5αR1/3αHSO3).

varying characteristics (Wiebe & Lewis, 2003). Three of the cell lines (MCF-7, MDA-MB-231, T47D) are known to be tumorigenic in immunosuppressed mice (Anderson *et al.*, 1984; Soto *et al.*, 1986); among these, MCF-7 and T47D cells are ER/PR-positive (Horwitz *et al.*, 1975) and estrogen-dependent for tumorigenicity, whereas MDA-MB-231 cells are ER/PR-negative and develop tumors spontaneously without estrogen. The fourth cell line, MCF-10A, is ER/PR-negative and considered to be non-tumorigenic (Soule *et al.*, 1990). The results showed that the production of 5α-pregnanes (such as 5αP) is higher and that of 4-pregnenes (such as 3αHP) is lower in tumorigenic than in non-tumorigenic human breast cell lines (Figure 3.4b)(Wiebe & Lewis, 2003, and unpublished results); the ratios of 5αP:3αHP were 14- to 24-fold higher in the tumorigenic cell lines, providing essentially the same pattern of results as the tissues. The progesterone metabolism studies on human breast tissues and cell lines showed that a marked increase in 5αP and decrease in 3αHP production accompany the shift toward breast cell tumorigenicity and neoplasia. Such a shift would occur in the immediate environment bathing the cells – allowing the hormones to act as autocrines and/or paracrines – and need have no bearing on circulating levels of either progesterone or its metabolites.

The activities and mRNA expression of 5α-reductase are increased, whereas those of 3α-HSO are decreased in breast carcinoma and tumorigenic breast cell lines

The metabolic studies, as well as in vitro enzyme kinetic studies, showed that the activity of 5α-reductase is higher, whereas that of the hydroxysteroid

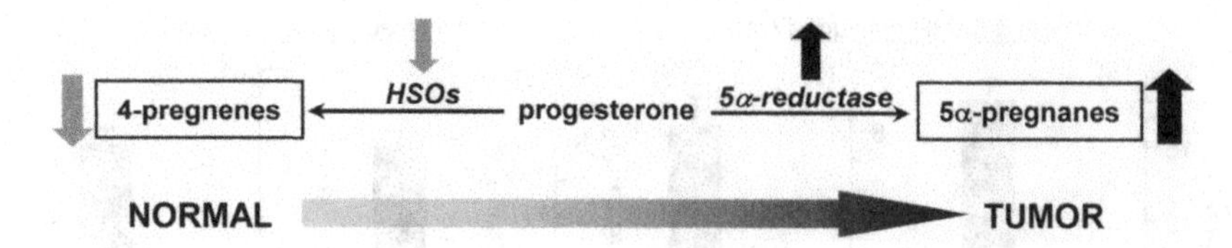

Figure 3.5. Changes in progesterone metabolizing enzyme activities/expression and resultant metabolite levels in breast tissue and cells during the transition from normal to cancerous condition. (Up-arrows indicate increases; down-arrows indicate decreases.)

oxidoreductases (such as 3α-HSO) is lower in tumor than in normal breast tissue (Figure 3.4c) (Wiebe *et al.*, 2000; Wiebe, 2005). Reverse transcription polymerase chain reaction (RT-PCR) studies on tissues from 38 patients also showed significantly higher ($p < 0.0001$) levels of expression of 5α-reductase (*SRD5A1/2*) mRNA and significantly lower ($p < 0.0002$) levels of expression of the HSO (*AKR1C1, AKR1C2, AKR1C3*) mRNAs in the tumor tissues than in the normal tissues (Lewis *et al.*, 2004), essentially paralleling the enzyme activity results. Similarly, studies on breast cell lines showed that 5α-reductase activity and gene expression are significantly higher, whereas the HSO activities and gene expression are lower ($p < 0.001$) in tumorigenic than in non-tumorigenic cell lines (Figure 3.4d) (Wiebe & Lewis, 2003). Overall, the enzyme activity and expression studies strongly suggest that in the transition from normal to cancerous conditions, 5α-reductase has been stimulated, whereas the HSOs have been suppressed (illustrated in Figure 3.5).

Role of the progesterone metabolites, 5αP and 3αHP, in regulation of breast cancer

Opposing effects of 5αP and 3αHP on cell proliferation. Uncontrolled replication is one of the hallmarks of cancer (Cohen & Ellwein, 1990; Pike *et al.*, 1993). The effects of 3αHP and 5αP on the proliferation of five different breast cell lines (MCF-7, MDA-MB-231, T47D, ZR75–1, MCF-10A) were studied using cell count and [^{3}H]thymidine incorporation assays (Wiebe *et al.*, 2000, 2005). The results showed that 3αHP and 5αP have opposing actions: 3αHP suppresses, whereas 5αP stimulates proliferation in a dose-dependent manner (Figure 3.6) at physiological levels. The effects occur in all five cell lines tested, regardless of presence or absence of ER, PR or tumorigenicity.

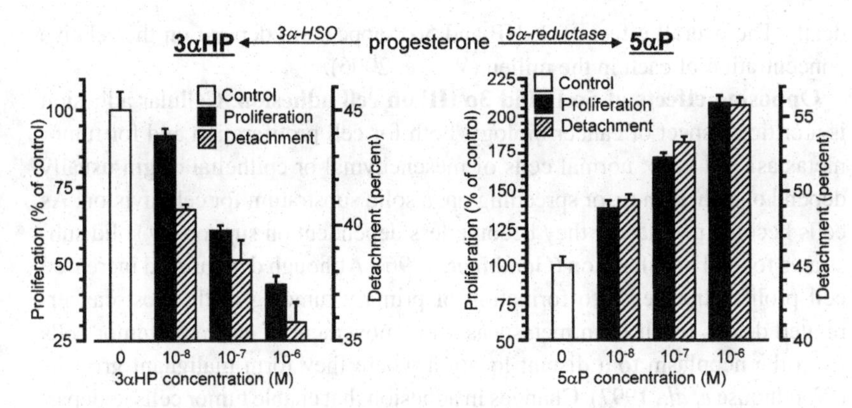

Figure 3.6. Effects of 3αHP and 5αP on proliferation (solid bars) and detachment (hatched bars) of MCF-7 human breast cell lines. Essentially the same results were obtained with all other breast cell lines that were studied.

As proof-of-principle that it is not progesterone but the 5α-reduced metabolite, 5αP, that leads to increased cell proliferation and detachment, MCF-7 cells were treated with progesterone in the presence/absence of dutasteride, a known 5α-reductase inhibitor (Wiebe *et al.*, 2006). Treatment of cells with progesterone without medium change for 72 hours resulted in significant conversion to 5α-pregnanes and increases in cell proliferation and detachment. Dutasteride inhibited >95% of the 5α-reduction and blocked the increases in proliferation and detachment. Concomitant treatment with 5αP abrogated the effects of dutasteride and restored the increases in proliferation and detachment, providing proof-of-principle that the effects were owing, not to progesterone but to the 5α-reduced metabolite.

Opposing actions of 5αP and 3αHP on DNA synthesis, mitosis, apoptosis. Increases in cell numbers result not only from increased rates of cell division, but also from decreases in rate of cell attrition via programmed cell death (apoptosis) (Thompson, 1995). A balance between proliferation and apoptosis provides the homeostasis in normal tissues, whereas alteration in this balance can set off a series of changes that ultimately leads to malignancy. Recent studies on human breast cell lines (Wiebe *et al.*, 2010), using several methods of evaluating apoptosis and mitosis, showed that 3αHP results in significant increases in apoptosis and decreases in mitosis, leading to decreases in total cell numbers. In contrast, treatment with 5αP resulted in decreases in apoptosis and increases in mitosis. When cells were treated simultaneously with both 3αHP and 5αP, the independent effects of the individual hormones were abrogated. Thus, with respect to cell proliferation, the results indicate that the actions of 3αHP and 5αP are diametrically opposed and involve both cell division and cell death. The overall effect of 3αHP and 5αP appears to depend on the relative concentration of each in the milieu (Wiebe, 2006).

Opposing effects of 5αP and 3αHP on cell adhesion. Cellular adhesion is a critical aspect of cancer biology, both for cell proliferation and for tumor metastasis. In vitro, normal cells of mesenchymal or epithelial origin usually depend on adhesion to, or spreading on, a solid substratum for cell division. As cells become neoplastic, they become less dependent on support by solid substrates for cell proliferation (Gumbiner, 1996). Although deregulated increased cell proliferation leads to formation of primary tumor growth, most cancer-related deaths result from metastasis – the movement of activated tumor cells from the neoplasm to a distant location where they form malignant growths (Woodhouse *et al.*, 1997). Changes in adhesion that enable tumor cells to depart from the primary site of growth constitute the first step toward invasion and cancer metastasis (Raz, 1988; Steeg, 2006). Once cells have escaped from the stable location, by altering their anchorage, they may become mobile and invasive.

To determine whether 5αP and 3αHP play a role in the acquisition of metastatic potential, their effects on cell adhesion were examined by attachment and detachment assays (Wiebe *et al.*, 2000; Wiebe & Muzia, 2001; Pawlak *et al.*, 2005; Wiebe *et al.*, 2005). The results showed that 5αP causes significant dose-dependent decreases in attachment to, and increases in detachment from, the substratum; the opposite effect is observed with 3αHP, which promotes cell attachment and decreases cell detachment (Figure 3.6). Data from studies in which cells were treated simultaneously with both 3αHP and 5αP showed that the independent effects of the individual hormones on adhesion are cancelled when present in equal concentrations (Pawlak *et al.*, 2005) and support the view that the overall effects depend on the relative concentration of each in the milieu (Wiebe, 2006). Again, the studies were conducted on both tumorigenic and non-tumorigenic human breast cells with essentially the same results.

Effect of 5αP on adhesion plaques, vinculin expression and polymerized F-actin in breast cancer cells. The cytoskeletal organization differs between normal and cancerous cells (Ben-Ze'ev, 1985) and between highly metastatic and low metastatic cells (Suzuki *et al.*, 1998). In normal cells, actin may be present in a more highly polymerized form (Bershadsky *et al.*, 1995), and then, during transformation to the metastatic condition, may be accompanied by disruption and/or visible disappearance of actin filaments (Suzuki *et al.*, 1998). Similarly, vinculin, a protein known to be associated with cell–cell and cell–substrate adhesion sites (Humphries & Newham, 1998), may be readily detected in normal cells, while in highly malignant cell lines vinculin expression may be substantially suppressed (Sadano *et al.*, 1992). Treatment of human breast cancer cells with 5αP resulted in dose-dependent decreases in polymerized actin stress fibers and ratio of insoluble/soluble actin, as well as decreases in vinculin expression and vinculin-containing adhesion plaques (Wiebe & Muzia, 2001). The results suggest that the changes in adhesion and proliferation following 5αP treatment may be related to depolymerization of actin and decreased expression of vinculin. On the other hand, 3αHP maintained high levels of vinculin expression and actin polymerization.

Other differences between the progesterone metabolites, 5αP and 3αHP

Separate and specific receptors for 5αP and 3αHP are associated with the plasma membrane. Binding studies using [^{3}H]-labeled 5αP and 3αHP have demonstrated separate, specific, high affinity, saturable binding sites (receptors) in the plasma membrane fractions of MCF-7 and MCF-10A cells (Weiler & Wiebe, 2000; Pawlak *et al.*, 2005). The binding of 5αP and 3αHP is not to progesterone-, androgen-, estrogen- or corticosteroid-receptors. 5αP does not displace bound 3αHP, and 3αHP does not displace bound 5αP; neither 5αP nor 3αHP are displaced by 200-fold concentrations of progesterone, estradiol, testosterone, dihydrotestosterone, cortisol and other similar steroids. The location of the receptors in the plasma membrane suggests that cell signaling pathways may be involved in the mechanisms of action of 5αP and 3αHP. Indeed, studies with MCF-7 cells (Wiebe *et al.*, 2005) have shown that 5αP results in rapid activation of Erk1/2, which is suppressed by the MEK inhibitor, PD98059, and that 5αP-induced increases in cell proliferation and detachment are linked to the 5αP-induced activation of the MAPK pathway. The 5αP receptor numbers per cell can be up-regulated by estradiol or 5αP, and down-regulated by 3αHP in MCF-7 and MCF-10A cells (Weiler & Wiebe, 2000; Pawlak *et al.*, 2005). The receptor findings further illustrate that marked differences between a parent compound (progesterone) and its metabolites, as well as between metabolites, can exist.

Opposing actions of 5αP and 3αHP on estrogen receptor expression. In addition to the actions described above, 5αP and 3αHP have opposing regulatory effects on estrogen receptor (ER) levels in ER-positive MCF-7 cells (Pawlak & Wiebe, 2007). Estrogen receptor numbers are selectively up-regulated by 5αP and down-regulated by 3αHP. The suggested implications for breast cancer are that progesterone metabolites may also significantly affect estrogen responses in estrogen-targeted cells, in addition to their own specific effects on both ER-positive and ER-negative cells. Thus changes in concentrations of progesterone metabolites in the breast microenvironment may not only affect direct responses to the metabolites but also regulate responses to

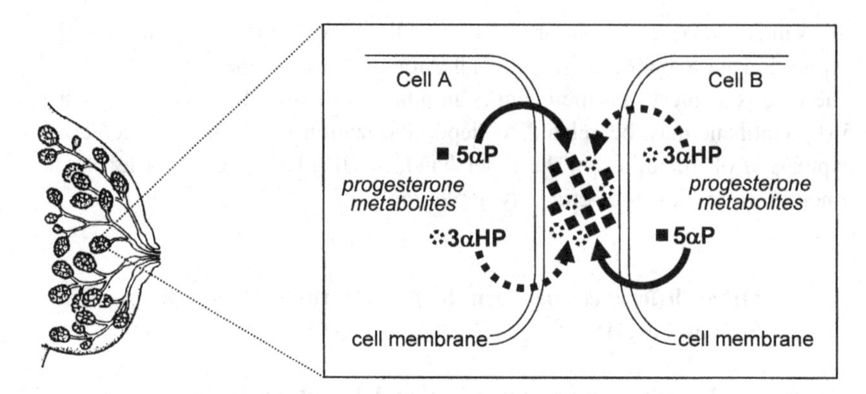

Figure 3.7. Diagram indicating changes in concentrations of 5αP and 3αHP in the microenvironment between (and within) two neighbouring breast cells.

estrogens by modulatory effects on estrogen receptor levels. And in each case, the actions of 5αP are opposite to those of 3αHP.

Model of regulation of breast cancer by 3αHP and 5αP in the breast microenvironment

Current hormone-based therapies for breast cancer involve suppression of estrogen levels and actions. However, only a portion of all breast cancers respond to current estrogen-based endocrine therapy, and in time nearly all become unresponsive to this treatment. Ultimately, therefore, for the majority of breast cancers there is currently no adequate hormonal explanation or therapy. Nevertheless, estrogens figure prominently in breast cancer research, and many epidemiological studies have tried to link increased risk with blood estrogen levels. Observations of the opposing actions of 3αHP and 5αP indicate that changing levels of these progesterone metabolites within the breast tissue microenvironment can provide a better explanation (as well as predictor) of breast cancer than levels of circulating hormones such as estradiol. These studies suggest that local changes in concentrations of 5αP and 3αHP, resulting from selective up- or down-regulation of progesterone metabolizing enzymes, could determine the promotion and inhibition of human breast cancers (Wiebe, 2006).

Only small changes in enzyme activity/expression are needed to result in significant local concentration changes. For example, either a slight elevation of 5α-reductase activity, or a reduction of 3α-HSO activity, in one or more cells could lead to an increase in the ratio of 5αP: 3αHP in the immediate intra- and extracellular environment (Figure 3.7). The effects on the cells in the local area

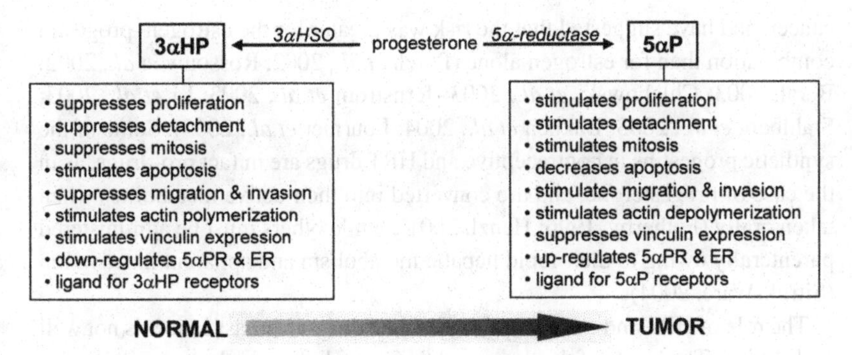

Figure 3.8. Diagram showing the opposing actions of 3αHP and 5αP.

could be substantial, since the two hormones exhibit strong opposing regulatory actions on cancer-related processes. The opposing actions of 3αHP and 5αP (summarized in Figure 3.8) have been demonstrated with respect to DNA synthesis, mitosis, apoptosis, adhesion and adhesion/cytoskeletal molecules, signaling mechanisms, separate specific receptors, regulation of 5αP receptor levels and estradiol receptor levels, and induction and growth of tumors. Therefore a small shift toward elevation of the cancer-promoting 5αP could trigger and promote progression toward breast cancer. On the other hand, higher levels of cancer-suppressing 3αHP, could ensure maintenance of normalcy. Numerous in vitro studies have shown that the opposing actions of 5αP and 3αHP are exhibited in all breast cell lines examined, whether normal or tumorigenic, estrogen-responsive or estrogen-unresponsive, and estrogen receptor and progesterone receptor positive or negative. The progesterone metabolites thus can provide a unifying theory that may explain not only the changes that lead to various forms of breast cancers, but also how the majority of breast tissues remain normal (non-tumorous).

Potential effects of exogenous hormones on the mammary gland microenvironment and cancer

Hormone replacement therapy (HRT) drugs and progestin-based contraceptives. For decades women in their reproductive years have added to their body loads of steroid hormones by using hormonal contraceptives containing synthetic progestins, and in the case of many menopausal and post-menopausal women by using hormone replacement therapy (HRT). A number of epidemiological studies, in several countries, have linked prolonged estrogen–progestin oral contraceptive and estrogen–progestin HRT with increased risk of breast cancer, and have suggested that the risk was greater for the estrogen–progestin combination than for estrogen alone (Porch *et al.*, 2002; Rossouw *et al.*, 2002; Beral, 2003; Chlebowski *et al.*, 2003; Jernstrom *et al.*, 2003; Li *et al.*, 2003; Stahlberg *et al.*, 2003; Bakken *et al.*, 2004; Fournier *et al.*, 2005). Some of the synthetic progestins in contraceptive and HRT drugs are in fact pro-drugs, as in the case of progesterone, and are converted into their active metabolites when taken orally (Fotherby, 1996; Henzl, 2001), while others must be administered parenterally owing to their rapid hepatic metabolism and apparent inactivation (Sitruk-Ware, 2004).

The role of HRT and contraceptive steroidal drugs in breast cancer is not well understood. The levels of the drugs and their metabolites in the breast microenvironment as well as their effects on the levels and actions of the progesterone metabolites have not been determined. In light of the findings that the progesterone metabolites have potent actions with respect to breast cancer, it is tempting to speculate that the drugs might have effects on progesterone metabolism and/or on the actions of the progesterone metabolites. For example, actions of the drugs (or their metabolites) on progesterone metabolizing enzyme activities/expression could result in a shift from predominantly anti-cancer 3αHP to predominantly pro-cancer 5αP hormone levels in the microenvironment, while up- or down-regulation of the receptors for either 3αHP or 5αP could significantly alter sensitivity levels and biological response. Additionally, effects of these drugs on the microenvironment within other tissues such as bone, brain, liver, and lung could enhance breast cancer metastasis to the target tissues. To ascertain the possible role of the drugs in breast cancer regulation via the progesterone metabolites, it will be necessary to measure their levels and composition, and to determine the effects of the drugs on progesterone metabolism and on the actions of the progesterone metabolites.

Growth promoting hormones. Various non-European Union (EU) countries (including Argentina, Australia, Brazil, Canada, USA, and South Africa) permit the use of natural steroid hormones and synthetic hormone-like substances in various combinations with the aim of improving weight gain and feed efficiency in beef cattle and other farm animals. In the EU Member States, the practice is not allowed, nor is the import of meats or meat products from countries where it is allowed. The most commonly used commercial products contain different combinations of the natural hormones, estradiol, testosterone and progesterone, and the synthetic compounds, zeranol (a phytoestrogen with high affinity for estrogen receptors), trenbolone acetate (known to act as a potent androgen), and melengestrol acetate (MGA) (which resembles natural progestins and binds to progesterone receptors). The application occurs in the form of small implants containing the active hormones into the subcutaneous tissue of the ears, or, in the case of MGA, as addition to the daily feed-concentrate supplies (Galbraith,

2002). Pharmaceutically, these implants represent slow-release devices, containing relatively large quantities of hormones intended for very low release per day over a period of several months.

Measurements of the growth promoting hormones and their metabolites (residues), as well as natural steroid hormones, in edible tissues from treated and non-treated animals have shown significant amounts of the synthetic compounds in the meats of treated animals and also significantly higher levels of the natural hormones (EFSA, 2007). For example, residue levels of estradiol experimentally implanted into animals were significantly higher in muscle tissue (about 4-fold) and liver (15-fold) in treated animals than in non-treated animals, and zeranol residues in muscle and adipose tissue were about 12 times the concentration of estradiol (Paris *et al.*, 2006). Similarly, increases in trenbolone actetate (Yoshioka *et al.*, 2000; Henricks *et al.*, 2001; Taguchi *et al.*, 2001; MacNeil *et al.*, 2003; Tsai *et al.*, 2004) and melengestrol acetate (Daxenberger *et al.*, 1999; Hageleit *et al.*, 2001), or their residues have been detected in consumable tissues from treated animals.

In terms of breast cancer, it is noteworthy that the synthetic growth promoting hormones and/or their metabolites may exert cancer-promoting actions. Thus zeranol or its metabolites have been shown to suppress apoptosis in MCF-7 breast cancer cells (Ahamed *et al.*, 2001) and in endothelial cells (Duan *et al.*, 2006), to suppress expression of PTPγ, a tumor suppressor gene (Liu *et al.*, 2004), to interact with estrogen receptors (Martin *et al.*, 1978), to stimulate cell proliferation and ER-β mRNA expression to the same extent as estradiol (Liu & Lin, 2004; Wollenhaupt *et al.*, 2004), to cause significant reduction in adhesion of cultured porcine ovarian cells (Tiemann *et al.*, 2003), and to affect not only ER-positive cancerous (MCF-7), but also ER-negative "normal" (MCF-10A) human breast cells (Irshaid *et al.*, 1999; Lin *et al.*, 2000). Of particular interest in the context of the progesterone metabolites, is the evidence that these synthetic compounds can interfere with progesterone synthesizing/metabolizing enzyme expression/activity (Tiemann *et al.*, 2003) and thus could be exerting effects on breast cancer by altering the 3αHP: 5αP ratios in the breast microenvironment.

The synthetic growth promoting hormones can exhibit high affinities for known steroid receptors. For example, the affinity of the trenbolone acetate metabolite, 17β-trenbolone, for the androgen receptor is similar to that of dihydrotestosterone, the most potent endogenous androgen, and its affinity for the progesterone receptor is higher than progesterone itself (Bauer *et al.*, 2000). Likewise, melengestrol acetate has a high affinity (as much as 5-fold greater than progesterone itself) for the progesterone receptor (Bauer *et al.*, 2000; Le Guevel & Pakdel, 2001), interacts with estrogen receptors and stimulates proliferation of MCF-7 human breast cell lines (Perry *et al.*, 2005). These actions are similar to those of the cancer-promoting progesterone metabolite

5αP (Pawlak & Wiebe, 2007) and it is interesting to speculate that the synthetic growth promoting adjuncts in meat might be interacting with the progesterone metabolite pathways at the level of the breast microenvironment to influence breast cancer. However, no studies have been conducted to determine either the levels of these substances in the breast, or their potential effects on progesterone metabolism and metabolite actions in the breast microenvironment.

Acknowledgements

I thank Dr. Kathleen Hill and Wendy Cladman for reading and commenting on a draft of the manuscript. The research was supported by grants from the Canadian Institutes of Health Research, Canadian Breast Cancer Research Alliance and The Susan G. Komen Breast Cancer Foundation (Susan G. Komen for the Cure).

References

Ahamed, S., Foster, J.S., Bukovsky, A. & Wimalasena, J. (2001). Signal transduction through the ras/erk pathway is essential for the mycoestrogen zearalenone-induced cell-cycle progression in MCF-7 cells. *Molecular Carcinogenesis*, 30, 88–98.

Anderson, T.J., Battersby, S., King, R.J.B., McPherson, K. & Going, J.J. (1989). Oral contraceptive use influences resting breast proliferation. *Human Pathology*, 20, 1139–44.

Anderson, W.A., Perotti, M.E., McManaway, M., Lindsey, S. & Eckberg, W.R. (1984). Similarities and differences in the ultrastructure of two hormone- dependent and one independent human breast carcinoma grown in athymic nude mice: comparison with the rat DMBA-induced tumor and normal secretory mammocytes. *Journal of Submicroscopic Cytology*, 16, 673–90.

Bakken, K., Alsaker, E., Eggen, A.E. & Lund, E. (2004). Hormone replacement therapy and incidence of hormone-dependent cancers in the Norwegian Women and Cancer study. *International Journal of Cancer*, 112, 130–4.

Bauer, E.R., Daxenberger, A., Petri, T., Sauerwein, H. & Meyer, H.H. (2000). Characterization of the affinity of different anabolic and synthetic hormones to the human androgen receptor, human sex hormone binding globulin and to the bovine progestin receptor. *Acta Pathologica, Microbiologica, et Immunologica Scandinavica*, 108, 838–46.

Baulieu, E.-E. (1990). Hormones: a complex communication network. In *Hormones: From Molecules to Diseases*, ed. E-E. Baulieu & P.A. Kelly. Paris: Hermann, pp. 38.

Ben-Ze'ev, A. (1985). The cytoskeleton in cancer cells. *Biochimica et Biophysica Acta*, 780, 197–212.

Beral, V. & Million Women Study Collaborators. (2003). Breast cancer and hormone replacement therapy in the Million Women Study. *Lancet*, 362, 419–27.

Bershadsky, A.D., Gluck, U., Denisenko, O.N., Sklyarova, T.V. *et al.* (1995). The state of actin assembly regulates actin and vinculin expression by a feedback loop. *Journal of Cell Science*, 108, 183–93.

Boudreau, N. & Myers, C. (2003). Breast cancer-induced angiogenesis: multiple mechanisms and the role of the microenvironment. *Breast Cancer Research*, 5, 140–6.

Braunsberg, H., Coldham, N.G., Leake, R.E., Cowan, S.K. & Wong, W.R. (1987). Action of a progestogen on human breast cancer cells: mechanism of growth stimulation and inhibition. *European Journal of Cancer and Clinical Oncology*, 23, 563–72.

Bulbrook, R.D., Moore, J.W., Clark, G.M. *et al.* (1986). Relation between risk of breast cancer and biological availability of estradiol in the blood: prospective study in Guernsey. *Annals of the New York Academy of Sciences*, 464, 378–88.

Cappelletti, V., Ruedl, C., Granata, G. *et al.* (1991). Interaction between hormone-dependent and hormone-independent human breast cancer cells. *European Journal of Cancer*, 27, 1154–7.

Castro-Rivera, E., Samudio, I. & Safe, S. (2001). Estrogen regulation of cyclin D1 gene expression in ZR-75 breast cancer cells involves multiple enhancer elements. *The Journal of Biological Chemistry*, 276, 30853–61.

Cauley, J.A., Lucas, F.L., Kuller, L.H. *et al.* (1999). Elevated serum estradiol and testosterone concentrations are associated with a high risk for breast cancer. *Annals of Internal Medicine*, 130, 270–7.

Chambers, A.F., Groom A.C. & MacDonald, I.C. (2002). Dissemination and growth of cancer cells in metastatic sites. *Nature Reviews. Cancer*, 2, 563–73.

Chatterton, R.T. Jr., Geiger, A.S., Gann, P.H. & Khan, S.A. (2003). Formation of estrone and estradiol from estrone sulfate by normal breast parenchymal tissue. *Journal of Steroid Biochemistry and Molecular Biology*, 86, 159–66.

Chatterton, R.T. Jr., Geiger, A.S., Khan, S.A. *et al.* (2004). Variation in estradiol, estradiol precursors, and estrogen-related products in nipple aspirate fluid from normal premenopausal women. *Cancer Epidemiology, Biomarkers and Prevention*, 13, 928–35.

Chatterton, R.T. Jr., Geiger, A.S., Mateo, E.T., Helenowski, I.B. & Gann, P.H. (2005). Comparison of hormone levels in nipple aspirate fluid of pre- and postmenopausal women: effect of oral contraceptives and hormone replacement. *Journal of Clinical Endocrinology and Metabolism*, 90, 1686–91.

Chlebowski, R.T., Hendrix, S.L., Langer, R.D. *et al.* (2003). Influence of oestrogen plus progestin on breast cancer and mammography in healthy postmenopausal women: the Women's Health Randomized Trial. *Journal of the American Medical Association*, 289, 3243–53.

Clark, C.L. & Sutherland, R.L. (1990). Progestin regulation of cellular proliferation. *Endocrine Reviews*, 11, 266–302.

Clark, R., Skaar, T., Baumann, K. *et al.* (1994). Hormonal carcinogenesis in breast cancer: cellular and molecular studies of malignant progression. *Breast Cancer Research and Treatment*, 31, 237–48.

Cohen, S.M. & Ellwein, L.B. (1990). Cell proliferation in carcinogenesis. *Science*, 249, 1007–11.

Daxenberger, A., Meyer, K., Hageleit, M. & Meyer, H.H.D. (1999). Detection of melengestrol acetate residues in plasma and edible tissues of heifers. *Veterinary Quarterly*, 21, 154–8.

Duan, J., Dai, S., Fang, C.X. *et al.* (2006). Phytoestrogen α-zearalanol antagonizes homocysteine-induced imbalance of nitric oxide/endothelin-1 and apoptosis in human umbilical vein endothelial cells. *Cell Biochemistry and Biophysics*, 45, 137–45.

Edel, M.J., Harvey, J.M. & Papadimitriou, J.M. (2000). Comparison of vascularity and angiogenesis in primary invasive mammary carcinomas and in their respective axillary lymph node metastases. *Clinical and Experimental Metastasis*, 18, 695–702.

Eechaute, W., de Thibault de Boesinghe, L. & Lacroix, E. (1983). Steroid metabolism and steroid receptors in methylbenz(a)anthracene-induced rat mammary tumors. *Cancer Research* 43, 4260–5.

EFSA (2007). Opinion of the Scientific Panel on contaminants in the food chain on a request from the Commission related to hormone residues in bovine meat and meat products. Adopted on 12 June, 2007. *The EFSA Journal*, 510, 1–62. (www.efsa.europa.eu)

Eliasson, A.H., Missmer, S.A., Tworoger, S.S. *et al.* (2006). Endogenous steroid hormone concentrations and risk of breast cancer among postmenopausal women. *Journal of the National Cancer Institute*, 98, 1406–15.

Fennessey, P.V., Pike, A.W., Gonzalez-Aller, C. & Horwitz, K.B. (1986). Progesterone metabolism in T47Dco human breast cancer cells–I. 5α-pregnan-3β,6α-diol-20-one is the secreted product. *Journal of Steroid Biochemistry*, 25, 641–8.

Foster, J.S. & Wimalasena, J. (1996). Estrogen regulates activity of cyclin-dependent kinases and retinoblastoma protein phosphorylation in breast cancer cells. *Molecular Endocrinology*, 10, 488–98.

Fotherby, K. (1996). Bioavailability of orally administered sex steroids used in oral contraception and hormone replacement therapy. *Contraception*, 54, 59–69.

Fournier, A., Berrino, F., Riboli, E., Avenel, V. & Clavel-Chapelon, F. (2005). Breast cancer risk in relation to different types of hormone replacement therapy in the E3N-EPIC cohort. *International Journal of Cancer*, 114, 448–54.

Galbraith, H. (2002). Hormones in international meat production: biological, sociological and consumer issues. *Nutrition Research Reviews*, 15, 243–314.

Gann, P.H., Geiger, A.S., Helenowski, I.B., Vonesh, E.F. & Chatterton, R.T. (2006). Estrogen and progesterone levels in nipple aspirate fluid of healthy premenopausal women: relationship to steroid precursors and response proteins. *Cancer Epidemiology, Biomarkers and Prevention*, 15, 39–44.

Garland, C.F., Friedlander, N.J., Barrett-Connor, E. & Khaw, K.T. (1992). Sex hormones and postmenopausal breast cancer: a prospective study in an adult community. *American Journal of Epidemiology*, 135, 1220–30.

Going, J.J., Anderson, T.J., Battersby, S. & MacIntyre, C.C.A. (1988). Proliferative and secretory activity in human breast during natural and artificial menstrual cycles. *American Journal of Pathology*, 130, 193–204.

Groshong, S.D., Owen, G. I., Grimison, B. *et al.* (1997). Biphasic regulation of breast cancer cell growth by progesterone: role of the cyclin-dependent kinase inhibitors, p21 and p27Kip1. *Molecular Endocrinology*, 11, 1593–607.

Gumbiner, B.M. (1996). Cell adhesion: the molecular basis of tissue architecture and morphogenesis. *Cell*, 84, 345–57.

Hageleit, M., Daxenberger, A. & Meyer, H.H.D. (2001). A sensitive enzyme immunoassay (EIA) for the determination of melengestrol acetate (MGA) in adipose and muscle tissues. *Food Additives and Contaminants*, 18, 285–91.

Hankinson, S.E., Willett, W.C., Manson, J.E. *et al.* (1998). Plasma sex steroid hormone levels and risk of breast cancer in postmenopausal women. *Journal of the National Cancer Institute*, 90, 1292–9.

Helzlsouer, K.J., Alberg, A.J., Bush, T.L. *et al.* (1994). A prospective study of endogenous hormones and breast cancer. *Cancer Detection and Prevention*, 18, 79–85.

Henricks, D.M., Gray, S.L., Owenby, J.J. & Lackey, B.R. (2001). Residues from anabolic preparations after good veterinary practice. *Acta Pathologica, Microbiologica et Immunologica Scandinavica*, 109, S345–55.

Henzl, M.R. (2001). Norgestimate. From the laboratory to three clinical indications. *The Journal of Reproductive Medicine*, 46, 647–61.

Horwitz, K.B., Costlow, M.E. & McGuire, W.L. (1975). MCF-7: a human breast cancer cell line with estrogen, androgen, progesterone, and glucocorticoid receptors. *Steroids*, 26, 785–95.

Horwitz, K.B., Pike, A.W., Gonzalez-Aller, C. & Fennessey, P.V. (1986). Progesterone metabolism in T47Dco human breast cancer cells–II. Intracellular metabolic path of progesterone and synthetic progestins. *Journal of Steroid Biochemistry*, 25, 911–6.

Horwitz, K.B., Wei, L.L., Sedlacek, S.M. & D'Arville, C.N. (1985). Progestin action and progesterone receptor structure in human breast cancer: a review. *Recent Progress in Hormone Research*, 41, 249–316.

Humphries, M.J. & Newham, P. (1998). The structure of cell-adhesion molecules. *Trends in Cell Biology*, 8, 78–83.

Irshaid, F., Kulp, S.K., Sugimoto, Y., Lee, K. & Lin, Y.C. (1999). Zeranol stimulates estrogen-regulated gene expression on MCF-7 human breast cancer cells and normal human breast epithelial cells. *Biology of Reproduction*, 60 (Suppl. 1), 234–5.

Jabara, A.G. (1967). Effects of progesterone on 9,10-dimethyl-1,2-benzanthracene-induced mammary tumours in Sprague-Dawley rats. *British Journal of Cancer*, 21, 418–29.

Jernstrom, H., Bendhal, P.O., Lidfeldt, J. *et al.* (2003). A prospective study of different types of hormone replacement therapy use and the risk of subsequent breast cancer: the women's health in the Lund area (WHILA) study (Sweden). *Cancer Causes and Control*, 14, 673–80.

Kaaks, R., Berrino, F., Key, T. *et al.* (2005). Serum sex steroids in premenopausal women and breast cancer risk within the European Prospective Investigation into Cancer and Nutrition (EPIC). *Journal of the National Cancer Institute*, 97, 755–65.

Kabuto, M., Akiba, S., Stevens, R.G., Neriishi, K. & Land, C.E. (2000). A prospective study of estradiol and breast cancer in Japanese women. *Cancer Epidemiology, Biomarkers and Prevention*, 9, 575–9.

King, R.J.B. (1991). A discussion of the roles of estrogen and progestin in human mammary carcinogenesis. *Journal of Steroid Biochemistry and Molecular Biology*, 39, 811–18.

Lanari, C. & Molinolo, A.A. (2002). Progesterone receptors – animal models and cell signaling in breast cancer. Diverse activation pathways for the progesterone receptor: possible implications for breast biology and cancer. *Breast Cancer Research*, 4, 240–3.

Le Guevel, R. & Pakdel, F. (2001). Assessment of oestrogenic potency of chemicals used as growth promoter by in-vitro methods. *Human Reproduction*, 16, 1030–6.

Lenton, E.A., Landgren, B.M. & Sexton, L. (1984a). Normal variation in the length of the luteal phase of menstrual cycle: identification of the short luteal phase. *British Journal of Obstetrics and Gynaecology*, 91, 685–9.

Lenton, E.A., Landgren, B.M., Sexton, L. & Harper, R. (1984b). Normal variation in the length of the follicular phase of the menstrual cycle: effect of chronological age. *British Journal of Obstetrics and Gynaecology*, 91, 681–4.

Leung, B.S., Sasaki, G.H. & Leung, J.S. (1975). Estrogen-prolactin dependency in 7, 12-dimethylbenz(a) anthracene-induced tumors. *Cancer Research*, 35, 621–7.

Lewis, M.J., Wiebe, J.P. & Heathcote, J.G. (2004). Expression of progesterone metabolizing enzyme genes (AKR1C1, AKR1C2, AKR1C3, SRD5A1, SRD5A2) is altered in human breast carcinoma. *BMC Cancer*, 4,(27), 1–12. (http://www.biomedcentral.com/1471–2407/4/27.)

Li, C.I., Malone, K.E., Porter, P.L. *et al.* (2003). Relationship between long durations and different regimens of hormone therapy and risk of breast cancer. *Journal of the American Medical Association*, 289, 3254–63.

Lin, Y.C., Kulp, S.K., Sugimoto, Y. & Brueggemeier, R.W. (2000). Potential risk of growth promoter in beef for breast cancer growth. Era of Hope, *Department of Defence Breast Cancer Research Program Meeting Proceedings* 2000 II, 480.

Liu, S. & Lin, Y.C. (2004). Transformation of MCF-10A human breast epithelial cells by zeranol and estradiol-17beta. *Breast Journal*, 10, 514–21.

Liu, S., Sugimoto, Y., Sorio, C., Tecchio, C. & Lin, Y.C. (2004). Function analysis of estrogenically regulated protein tyrosine phosphatase γ (PTPγ) in human breast cancer cell line MCF-7. *Oncogene*, 23, 1256–62.

Lloyd, R.V. (1979). Studies on the progesterone receptor content and steroid metabolism in normal and pathological human breast tissues. *Journal of Clinical Endocrinology and Metabolism*, 48, 585–93.

Luo, S., Stojanovic, M. & Labrie, F. (1997). Inhibitory effect of the novel anti-estrogen EM-800 and medroxyprogesterone acetate on estrone-stimulated growth of dimethylbenz[*a*]anthracene-induced mammary carcinoma in rats. *International Journal of Cancer*, 73, 580–6.

Lyons, W.R., Li, C.H. & Johnson, R.E. (1958). The hormonal control of mammary growth and lactation. *Recent Progress in Hormone Research*, 14, 219–54.

MacNeil, J.D., Reid, J., Neiser, C.D. & Fesser, A.C.E. (2003). Single-laboratory validation of a modified liquid chromatographic method with UV detection for determination of trenbolone residues in bovine liver and muscle. *Journal of the AOAC International*, 86, 916–24.

Manjer, J., Johansson, R., Berglund, G. *et al.* (2003). Postmenopausal breast cancer risk in relation to sex steroid hormones, prolactin and SHBG (Sweden). *Cancer Causes and Control*, 14, 599–607.

Martin, P.M., Horwitz, K.B., Ryan, D.S. & McGuire, W.L. (1978). Phytoestrogen interaction with estrogen receptors in human breast cancer cells. *Endocrinology*, 103, 1860–7.

Milgrom, E. (1990). Steroid hormones. In *Hormones: From Molecules to Diseases*, ed. E-E. Baulieu and P.A. Kelly. Paris: Hermann, pp. 425.

Miller, W.R. (1990). Pathways of hormone metabolism in normal and non-neoplastic breast tissue. *Annals of the New York Academy of Sciences*, 586, 53–59.

Missmer, S.A., Eliassen, A.H., Barbieri, R.L. & Hankinson, S.E. (2004). Endogenous estrogen, androgen, and progesterone concentrations and breast cancer risk among postmenopausal women. *Journal of the National Cancer Institute*, 96, 1856–65.

Mol, J.A., van Garderen, E., Rutteman, G.R. & Rijnberk, A. (1996). New insights into the molecular mechanism of progestin-induced proliferation of mammary epithelium: induction of the local biosynthesis of growth hormone (GH) in the mammary gland of dogs, cats and humans. *Journal of Steroid Biochemistry and Molecular Biology*, 57, 67–71.

Monsky, W.L., Carreira, C.M., Tsuzuki, Y. *et al.* (2002). Role of host microenvironment in angiogenesis and microvascular functions in human breast cancer xenografts: mammary fat pad *versus* cranial tumors. *Clinical Cancer Research*, 8, 1008–13.

Mori, M., Tominaga, T. & Tamaoki, B. (1978). Steroid metabolism in the normal mammary gland and in the dimethylbenzthracene-induced mammary tumour of rats. *Endocrinolog*, 102, 1387–97.

Mori, M. & Tamaoki, B. (1980). In vitro metabolism of progesterone in the mammary tumour and the normal mammary gland of GRS/A strain of mice and dependency of some steroid-metabolizing enzyme activities upon ovarian function. *European Journal of Cancer*, 16, 185–93.

Musgrove, E.A. & Sutherland, R.L. (1994). Cell cycle control by steroid hormones. *Seminars in Cancer Biology*, 5, 381–9.

Muti, P., Trevisan, M., Micheli, A. *et al.* (1996). Reliability of serum hormones in premenopausal and postmenopausal women over a one-year period. *Cancer Epidemiology, Biomarkers and Prevention*, 5, 917–22.

Nandi, S., Guzman, R.C. & Yang, J. (1995). Hormones and mammary carcinogenesis in mice, rats, and humans: a unifying hypothesis. *Proceedings of the National Academy of Sciences of the United States of America*, 92, 3650–7.

Paris, A., Andre, F., Antignac, J.-P. *et al.* (2006). Hormones et promoteurs de croissance en productions animales: de la physiologie à l'évaluation du risque. *INRA Productions Animales*, 19, 149–240.

Pawlak, K.J. & Wiebe, J.P. (2007). Regulation of estrogen receptor (ER) levels in MCF-7 cells by progesterone metabolites. *Journal of Steroid Biochemistry Molecular Biology*, 107, 172–9.

Pawlak, K.J., Zhang, G. & Wiebe, J.P. (2005). Membrane 5α-pregnane-3,20-dione (5αP) receptors in MCF-7 and MCF-10A breast cancer cells are up-regulated by estradiol and 5αP and down-regulated by the progesterone metabolites, 3α-dihydroprogesterone and 20α-dihydroprogesterone, with associated changes in cell proliferation and detachment. *Journal of Steroid Biochemistry and Molecular Biology*, 97, 278–88.

Perry, G.A., Welshons, W.V., Bott, R.C. & Smith, M.F. (2005). Basis of melengestrol acetate action as a progestin. *Domestic Animal Endocrinology*, 28, 147–61.

Pike, M.C., Spicer, D.V., Dahmoush, L. & Press, M.F. (1993). Estrogens, progestogens, normal breast cell proliferation, and breast cancer risk. *Epidemiologic Reviews*, 15, 17–35.

Pinheiro, S.P, Holmes, M.D., Pollak, M.N., Barbieri, R.L. & Hankinson, S.E. (2005). Racial differences in premenopausal endogenous hormones. *Cancer Epidemiology, Biomarkers and Prevention*, 14, 2147–53.

Porch, J.V., Lee, I.M., Cook, N.R., Rexrode, K.M. & Buring, J.E. 2002. Oestrogen–progestin replacement therapy and breast cancer risk: the Women's Health Study (United States). *Cancer Causes and Control*, 13, 847–54.

Potten, C.S., Watson, R.J., Williams, G.T. *et al.* (1988). The effect of age and menstrual cycle upon proliferative activity of the normal human breast. *British Journal of Cancer*, 58, 163–70.

Prall, O.W., Sarcevic, B., Musgrove, E.A., Watts, C.K. & Sutherland, R.L. (1997). Estrogen-induced activation of Cdk4 and Cdk2 during G1-S phase progression is accompanied by increased cyclin D1 expression and decreased cyclin-dependent kinase inhibitor association with cyclin E-Cdk2. *The Journal of Biological Chemistry*, 272, 10882–94.

Radinsky, R. & Ellis, L.M. (1996). Molecular determinants in the biology of liver metastasis. *Surgical Oncology Clinics of North America*, 5, 215–29.

Raz, A. (1988). Adhesive properties of metastasizing tumor cells. *Ciba Foundation Symposia*, 141, 109–22.

Rossouw, J.E., Anderson, G.L., Prentice, R.L. *et al.* (2002). Risks and benefits of oestrogen plus progestin in healthy postmenopausal women: principal results from the Women's Health Initiative randomized controlled trial. *Journal of the American Medical Association*, 288, 321–33.

Sadano, H., Inoue, M. & Taniguchi, S. (1992). Differential expression of vinculin between weakly and highly metastatic B16-melanoma cell lines. *Japanese Journal of Cancer Research*, 83, 625–30.

Santen, R.J., Manni, H., Harvey, H. & Redmond, C. (1990). Endocrine treatment of breast cancer in women. *Endocrine Reviews*, 11, 221–65.

Schneider, B.P. & Miller, K.D. (2005). Angiogenesis of breast cancer. *Journal of Clinical Oncology*, 23, 1782–90.

Segaloff, A. (1975). Steroids and carcinogenesis. *Journal of Steroid Biochemistry*, 6, 171–5.

Sitruk-Ware, R. (2004). Pharmacological profile of progestins. *Maturitas*, 47, 277–83.

Soto, A.M., Murai, J.T., Siiteri, P.K. & Sonnenschein, C. (1986). Control of cell proliferation: evidence for negative control on estrogen-sensitive T47D human breast cancer cells. *Cancer Research*, 46, 2271–5.

Soule, H.D., Maloney, T.M. *et al.* (1990). Isolation and characterization of a spontaneously immortalized human breast epithelial cell line, MCF-10. *Cancer Research*, 50, 6075–86.

Stahlberg, C., Pederson, A.T., Lynge, E. & Ottesen, B. (2003). Hormone replacement therapy and risk of breast cancer: the role of progestins. *Acta Obstetrica et Gynecologica Scandinavica*, 82, 335–44.

Steeg, P.S. (2006). Tumor metastasis: mechanistic insights and clinical challenges. *Nature Medicine*, 12, 895–904.

Sturgeon, S.R., Potischman, N., Malone, K.E. *et al.* (2004). Serum levels of sex hormones and breast cancer risk in premenopausal women: a case-control study (USA). *Cancer Causes and Control*, 15, 45–53.

Sutherland, R.M. (1988). Cell and environment interactions in tumor microregions: the multicell spheroid model. *Science*, 240, 177–84.

Suzuki, H., Nagata, H., Shimada, Y. & Konno, A. (1998). Decrease in gamma-actin expression, disruption of actin microfilaments and alterations in cell adhesion systems associated with acquisition of metastatic capacity in human salivary gland adenocarcinoma cell clones. *International Journal of Oncology*, 12, 1079–84.

Taguchi, S., Yoshida, S., Tanaka, Y. & Hori, S. (2001). Simple and rapid analysis of trenbolone and zeranol residues in cattle muscle and liver by stack-cartridge solid-phase extraction and HPLC using on-line clean-up with EC and UV detection. *Journal of the Food Hygienic Society of Japan*, 42, 226–30.

Thompson, C.B. (1995). Apoptosis in the pathogenesis and treatment of disease. *Science*, 267, 1456–62.

Tiemann, U., Tomek, W., Schneider, F. & Vanselow, J. (2003). Effects of the mycotoxins α- and β-zearalenol on regulation of progesterone synthesis in cultured cells from porcine ovaries. *Reproductive Toxicology*, 17, 673–81.

Topper, Y.J. & Freeman, C.S. (1980). Multiple hormone interactions in the developmental biology of the mammary gland. *Physiological Reviews*, 60, 1049–106.

Trédan, O., Galmarini, C.M., Patel, K. & Tannock, I.F. (2007). Drug resistance and the solid tumor microenvironment. *Journal of the National Cancer Institute*, 99, 1441–54.

Tsai, C.F., Chang, M.H., Pan, J.Q. & Chou S.S. (2004). A method for the detection of trenbolone in bovine muscle and liver. *Journal of Food and Drug Analysis*, 12, 353–7.

Welsh, C. W. (1982). Hormones and murine mammary tumorigenesis: an historical view. In *Hormonal Regulation of Mammary Tumors*, ed. B.S. Leung. Montreal: Eden Press, pp. 1–29.

Wiebe, J.P. (2005). Role of progesterone metabolites in mammary cancer. In Mammary Development, Function and Cancer, Special Issue of *The Journal of Dairy Research*, 72, 51–7.

Wiebe, J.P. (2006). Progesterone metabolites in breast cancer. *Endocrine-Related Cancer*, 13, 717–38.

Wiebe, J.P., Beausoleil, M., Zhang, G. & Cialacu, V. (2010). Opposing actions of the progesterone metabolites, 5α-dihydroprogesterone (5αP) and 3α-dihydroprogesterone (3αHP) on mitosis, apoptosis, and expression of Bcl-2, Bax and p21 in human breast cell lines. *Journal of Steroid Biochemistry and Molecular Biology*, 118, 125–32.

Wiebe, J.P., Souter, L. & Zhang, G. (2006). Dutasteride affects progesterone metabolizing enzyme activity/expression in human breast cell lines resulting in suppression of cell proliferation and detachment. *Journal of Steroid Biochemistry and Molecular Biology*, 100, 129–40.

Wiebe, J.P., Muzia, D., Hu, J., Szwajcer, D. *et al.* (2000). The 4-pregnene and 5α-pregnane progesterone metabolites formed in nontumorous and tumorous breast tissue have opposite effects on breast cell proliferation and adhesion. *Cancer Research*, 50, 936–43.

Weiler, P.J. & Wiebe, J.P. (2000). Plasma membrane receptors for the cancer-regulating progesterone metabolites, 5α-pregnane-3, 20-dione (5αP) and 3α-hydroxy-4-pregnen-20-one (3αHP) in MCF-7 breast cancer cells. *Biochemical and Biophysical Research Communications*, 272, 731–7.

Wiebe, J.P. & Muzia, D. (2001). The endogenous progesterone metabolite, 5α-pregnene-3,20-dione, decreases cell-substrate attachment, adhesion plaques, vinculin expression, and polymerized F-actin in MCF-7 breast cancer cells. *Endocrine*, 16, 7–14.

Wiebe, J.P. & Lewis, M.J. (2003). Activity and expression of progesterone metabolizing 5α-reductase, 20α-hydroxysteroid oxidoreductase and 3α(β)-hydroxysteroid oxidoreductases in tumorigenic (MCF-7, MDA-MB-231, T-47D) and nontumorigenic (MCF-10A) breast cell lines. *BMC Cancer*, 3, 9.

Wiebe, J.P., Lewis, M.J., Cialacu, V., Pawlak, K.J. *et al.* (2005). The role of progesterone metabolites in breast cancer: Potential for new diagnostics and therapeutics. *Journal of Steroid Biochemistry and Molecular Biology*, 93, 201–8.

Wollenhaupt, K., Jonas, L., Tiemann, U. & Tomek, W. (2004). Influence of the mycotoxins α- and β-zearalenol (ZOL) on regulators of cap-dependent translation control in pig endometrial cells. *Reproductive Toxicology*, 19, 189–99.

Woodhouse, E.C., Chuaqui, R.F. & Liotta, L.A. (1997). General mechanisms of metastasis. *Cancer*, 80, 1529–37.

Wysowski, D.K., Comstock, G.W., Helsing, K.J. & Lau, H.L. (1987). Sex hormone levels in serum in relation to the development of breast cancer. *American Journal of Epidemiology*, 125, 791–9.

Yen, S. (1990). Clinical endocrinology of reproduction. In *Hormones: From Molecules to Diseases*, ed. E-E. Baulieu & P.A. Kelly. Paris: 1Hermann, pp. 454–62.

Yoshioka, N., Akiyama, Y. & Takeda, N. (2000). Determination of alpha- and beta-trenbolone in bovine muscle and liver by liquid chromatography with fluorescence detection. *Journal of Chromatography*, B 739, 363–7.

Yu, H., Shu, X.O., Shi, R., Dai, Q. *et al.* (2003). Plasma sex steroid hormones and breast cancer risk in Chinese women. *International Journal of Cancer*, 105, 92–7.

4 *The energetic cost of physical activity and the regulation of reproduction*

DARNA L. DUFOUR

Introduction

The association of physical activity with the suppression of reproductive function in women has been recognized for a number of years. The relationship seems clearest in the association of strenuous exercise and amenorrhea (the absence of menstrual cycling owing to complete ovarian suppression). However, we now recognize that exercise in recreational, as well as competitive athletes, is associated with reproductive dysfunction along a continuum, ranging from normal ovulatory cycles to subtle luteal phase defects (LPD), anovulatory cycles, and finally to amenorrhea (De Souza, 2003; Prior & Vigna, 1985). The most common menstrual cycle anomaly associated with exercise is LPD (De Souza *et al.*, 1998), although amenorrhea is the most dramatic and best known.

There is a growing consensus that the primary cause of these menstrual cycle abnormalities is a deficit of dietary energy intake (EI) (Loucks, 2005), which can lead to a temporary state of negative energy balance, the most obvious consequence of which is the loss of body weight, as tissues (both protein and fat) are catabolised for fuel. Less obvious consequences are the decreases in metabolic rate per kg FFM (fat-free mass, i.e. muscle and organs), the differential loss of tissues with different metabolic rates, and the accompanying reduction in core temperature (Elia, 1997). Behavioural changes also occur, such as reductions in the duration and intensity of physical activities and the adoption of more resting postures (Keys, 1950). The decreases in metabolic rate and the changes in behaviour are considered adaptive responses to energy deficit as they serve to reduce total daily energy expenditure (EE) and hence restore energy balance (EB) albeit at a lower level of body weight and energy flux.

In addition, reproductive function is suppressed under conditions of negative EB in both animals and humans (Wade & Jones, 2004). This is the endocrine-level response, and is also considered an adaptive response, but the evolutionary

Reproduction and Adaptation, eds. C. G. Nicholas Mascie-Taylor and Lyliane Rosetta.
Published by Cambridge University Press.

explanation is different. The argument is that under conditions of energy deficiency animals adjust energetic priorities, i.e. how they use available energy, in order to promote individual survival, and since reproduction is not necessary for individual survival it receives very low priority (Wade & Jones, 2004). Although this argument is compelling from a theoretical point of view, actually measuring day-to-day EB in order to assess its effect on reproductive function in humans is challenging methodologically.

The goals of this chapter are to review the evidence that suppression of reproductive function in humans is associated with energy deficit, not exercise or physical activity itself, to discuss briefly some of the mechanisms involved, and to consider some of the issues that await resolution.

Evidence for the role of low energy availability in reproductive suppression

The idea that suppression of reproductive function is the result of an energy deficit and not physical activity per se is based on evidence from a number of different kinds of studies: (1) cross-sectional studies of reproductive dysfunction in athletes and exercising women; (2) observations of the reversal of amenorrhea in athletes; (3) prospective studies of normally cycling women; (4) animal models; (5) observational studies of women in subsistence-based economies.

Cross-sectional studies. Warren (1980) was probably the first to propose that the suppression of reproductive function in females who exercised was an endocrine response to energy deficit or, as she stated, an "energy drain". She found that the incidence of amenorrhea in adolescent ballet dancers was positively correlated with the hours of exercise per week, and that the onset of menarche coincided with a decrease in exercise (or a rest period owing to injury) in 67% of the dancers, rather than any change in body weight. This suggests that the physiological priority is to re-engage reproductive function before adding body tissue as evidenced by weight gain.

A number of cross-sectional studies have found that amenorrhoea in women runners of stable body weight was associated with lower levels of energy intake (EI) than expected, and lower, although not always significantly so, than in eumenorrhoeic runners with similar body sizes, compositions and training regimes (Deuster *et al.*, 1986; Drinkwater *et al.*, 1984; Kaiserauer *et al.*, 1989; Nelson *et al.*, 1986; Wilmore *et al.*, 1992). Indeed, Nelson *et al.* (1986) suggested that amenorrhoea was an adaptive response to negative EB. Although these studies are suggestive, the accuracy of the dietary intake data is questionable.

More compelling is the work of researchers like Marcus *et al.* (1985), Lebenstedt *et al.* (1999), and Myerson *et al.* (1991) in demonstrating that amenorrhoeic runners show physiological evidence of a hypometabolic state in comparison to eumenorrhoeic runners and sedentary controls. Myerson *et al.* (1991) found that resting metabolic rate (RMR) per kg and circulating levels of T_3 (tri-iodothyronine) were lower in amenorrhoeic runners than eumenorrhoeic runners and controls. The amenorrhoeic and eumenorrhoeic runners were similar in age, body size, percentage body fat and training regime. The controls were similar in age, but sedentary and had a higher percentage of body fat. The differences in RMR cannot be attributed to differences in body composition because RMR per kg FFM was also lower. De Souza *et al.* (2007) also found lower RMR per kg FFM and lower T_3 in amenorrhoeic women who exercised more than 2 hours per week in comparison to ovulatory women who exercised a similar amount. Further, they found that exercising women with evidence of subtle menstrual disturbances (inconsistent and anovulatory cycles) also had lower RMR per kg FFM. In contrast Wilmore *et al.* (1992) failed to find a lower RMR per kg FFM in amenorrhoeic versus eumenorrhoeic elite runners. Reasons for this contradictory finding are unclear, but the sample sizes were very small ($n = 8$ and $n = 5$), and data from controls were not available.

The lower RMR in amenorrhoeic athletes and exercising women with menstrual disturbances indicates that they were currently, or had been, in negative EB. If they were in negative EB at the time the studies were done, they would have been losing weight. However, since none of the studies reported weight loss, the assumption is that the women were weight stable, and hence in EB. In this case the low RMR can be taken as evidence of adaptation to a period of negative EB in the past during which they had catabolized body tissue for energy, reduced the metabolic rate of FFM, and perhaps made behavioural changes to re-establish EB at a lower level of energy flux. This process was clearly demonstrated by the Minnesota semi-starvation experiment in which men were able to re-establish EB at 50% of their baseline EI (Keys, 1950). A lower than expected RMR has also been reported in some, but not all, cases of life-long chronic undernutrition in which individuals are weight stable for long periods of time (Shetty, 1984; Ferro-Luzzi *et al.*, 1990). When EB is at a relatively low level of energy flux, it is referred to as a hypometabolic state.

Low levels of the thyroid hormone T_3 are used as an endocrine marker of a hypometabolic state, as T_3 levels are decreased by negative EB (Lopresti *et al.*, 1991; Loucks & Callister 1993; Myerson *et al.*, 1991; Zanker & Swaine, 1998). Loucks & Callister (1993) demonstrated that an increase in exercise EE without a compensatory increase in EI resulted in a significant reduction in T_3 levels in young sedentary women, while an increase in exercise EE with adequate EI did not affect T_3 levels. A number of studies have reported lower T_3 in athletes with reproductive irregularities (see, for example, Marcus *et al.*, 1985 and Tomten & Høstmark, 2006).

Observations of reversals of reproductive suppression. There are a number of reports of the restoration of menstrual cycling with weight gain or a decrease in exercise EE owing to a change in training or injury-induced rest (Drinkwater *et al.*, 1984; Prior & Vigna, 1985; Rosetta *et al.*, 1998; Warren, 1980). Intervention studies of amenorrhoeic athletes have also shown that amenorrhoea can be reversed by increasing EI and/or decreasing EE by reducing weekly training time (Dueck *et al.*, 1996; Kopp-Woodroffe *et al.*, 1999). Again, these observational studies are only suggestive as the researchers did not have good measures of the actual state of EB. That is they did not have good measures of either EI or EE. Further, these results do not rule out the possibility that exercise itself was a factor.

Prospective studies. A series of short prospective exercise studies by Williams *et al.* (1995) and Loucks and colleagues (Loucks *et al.*, 1998; Loucks & Thuma, 2003; Loucks, 2006) have demonstrated changes in reproductive hormones when EI is less than EE, i.e. when EB is negative. The studies have focused on LH (luteinizing hormone), a hormone secreted in a pulsatile fashion by the anterior pituitary. Luteinizing hormone along with FSH (follicle stimulating hormone), is responsible for the development of ovarian follicles. As the follicles grow and mature they release increasing quantities of oestrogen, which has a positive feedback effect on LH, until a surge of LH causes the dominant follicle to rupture and release an egg cell for fertilization. The process is dependent on optimal LH pulse frequency and amplitude, not only quantity. Control of LH pulsatility is dependent on GnRH (gonadotrophin-releasing hormone) pulses originating in the hypothalamus. It has been known since the work of Veldhuis *et al.* (1985) that reproductive dysfunction in amenorrhoeic athletes is associated with alterations in LH pulse frequency, but it was only recently that researchers like Williams and Loucks have been able to separate out the effects of exercise per se from that of energy availability.

Williams *et al.* (1995) demonstrated that an abrupt increase in the training volume of eumenorrhoeic runners disrupted LH pulse frequency, but only when EB was negative. The subjects were four eumenorrhoeic women with an average age of 28 ± 0.6 years (16.8 ± 0.3 years post-menarche), and average body mass index (BMI) and percentage body fat of 19.6 ± 0.9 and 17.3 ± 0.6, respectively. All ran 32–40 km per week, had been doing so for at least 5 years, and were asked to maintain the same level of training during the study. They were administered three different treatments of 7 days each administered in random order: (1) a diet designed to maintain EB, normal training during days 1–4, no training during days 5–7; (2) a diet designed to maintain EB, normal training during days 1–4, an increase (188%) in training distance run during days 5–7; (3) diet restricted to 60% of estimated need, normal training during days 1–4, an increase (188%) in training distance run during days 5–7. In comparison to the control conditions (treatment #1), the abrupt increase in training did not affect body weight or LH pulse parameters, except when EB was negative (treatment #3). In that treatment body weight declined (1.1 to 3.2 kg) as did LH pulse frequency, but not peak amplitude or mean serum LH. There was a non-significant tendency for T_3 to be lower in the negative EB treatment. Although the number of subjects studied was very small, the results suggest that LH pulse parameters are sensitive to short-term negative EB, at least when that negative EB is extreme (only 60% of estimated need).

Loucks *et al.* (1998) studied nine habitually sedentary women, 18–29 years of age. The women were young gynecologically (8.7 $\pm$ 1.1 years post-menarche), had no recent history of menstrual dysfunction and had menstrual cycles of normal length, and they had a mean percentage body fat of 26.6 $\pm$ 0.8%. The study used a 4-day exercise treatment (treadmill walking) designed to significantly increase EE, and was repeated under two different conditions: EB and food energy restriction. In comparison to the EB treatment, the energy-restricted treatment resulted in a decrease in LH pulse frequency (during waking hours), and an increase in pulse amplitude, but no change in mean serum LH. The energy-restricted treatment was clearly associated with negative EB as the women showed a loss of body weight (2.8%), and decreases in T_3, insulin and plasma glucose.

Using a similar experimental protocol, Loucks and Thuma (2003) demonstrated a dose response to negative EB, but not to an abrupt increase in exercise. They studied sedentary 21-year-old eumenarchial women (n = 29) using a 9-day protocol that controlled both diet and exercise. The women were 8–9 years post-menarche, and had body fat percentages of 24.9 to 26.2%. Subjects were monitored for days 1–3, and assigned an exercise treatment and specific diet on days 4–8. The exercise treatment was treadmill walking that effectively increased total EE by 33.5%. The specific diets included one designed to maintain EB and three designed to create energy deficits. All subjects were assigned the diet designed to maintain EB (control condition), and one of the three diets designed to create energy deficits. On the diet designed to provide 75% of estimated energy needs, LH parameters were similar to the control conditions. However, on the more energy-restricted diets (providing 58% and 42% of estimated need) LH pulse frequency decreased and amplitude increased proportionally to the amount of restriction. FSH was not affected by any of the treatments, and E_2 (24-hour mean) was only decreased by the most restrictive diet. This study, like that of Williams *et al.* (1995), effectively separated the effect of exercise on reproductive parameters from that of negative EB. Again, it was a very short-term study, and the two levels of energy restriction were quite severe.

Loucks (2006) went on to demonstrate that the effect of negative EB on LH pulsatility is greater in younger women. She used a protocol similar to that of the previous study (Loucks & Thuma, 2003) to compare habitually sedentary, regularly menstruating women of different ages: (1) "adolescents" with mean chronological and gynaecological ages of 20.5 and 6.9 years respectively, and a mean body fat percentage of 23.7%; (2) "adults" with mean chronological and gynaecological ages of 28.7 and 16 years respectively, and a mean body fat percentage of 26.8%. Under EB conditions there were no between-group differences in LH pulse parameters (frequency, amplitude and 24-hour mean). Under conditions of severe negative EB (EI = 42% of estimated energy need), LH pulse frequency and 24-hour mean were lower in adolescents, but unchanged in adults. The suppression of LH pulsatility in the adolescents was insufficient to reduce oestradiol in the mid-follicular phase, or induce anovulation, but did result in a decrease in the length of the luteal phase. There was no change in pulse amplitude in either group. Changes in body weight were not reported.

The finding of a greater effect of negative EB on younger women is in agreement with a number of studies (De Souza *et al.*, 1998, 2003; Laughlin & Yen, 1996) that have reported that amenorrhoea occurs more frequently in women of younger gynaecological ages. The lack of an effect of energy restriction on the adult women in the Loucks (2006) study conflicts with the finding of Williams *et al.* (1995). The women were of similar chronological and gynaecological ages, but interestingly the women in the Loucks (2006) study were habitually sedentary and had higher levels of body fat (26.8% versus 17.3%).

One long-term prospective exercise study (12–14 menstrual cycles) was done by Rogol *et al.* (1992). They failed to find an effect of exercise on menstrual cycle parameters or hormone levels, except for a decrease in the length of the luteal phase in the group training most intensively. The study was done with gynaecologically mature (mean gynaecological age of 17.8 years), previously sedentary, eumenorrhoeic women. There were no significant changes in body weight, although body fat decreased in the group training more intensively, and fat-free mass (FFM) increased in both groups. No measures of T_3 or other metabolic hormones that might be indicative of a hypometabolic state were reported.

In summary, these studies demonstrate that short-term acute increases in exercise do not affect LH pulse parameters except when dietary EI is restricted. Further, they demonstrate a more pronounced effect in women of younger gynaecological ages. The differences between the studies of Williams *et al.*

(1995) and Loucks (2006) are difficult to reconcile. The Rogol *et al.* (1992) study clearly demonstrates that longer-term increases in exercise do not necessarily lead to reproductive dysfunction. The assumption in that study being that EB was maintained over the long term.

Animal models. Intensive exercise has also been associated with reproductive suppression and its reversal in a non-human primate model (*Macaca fascicularis*) (Williams *et al.*, 2001a; Williams *et al.*, 2001b). Monkeys (n = 8) trained to run on a treadmill for up to 12 km/day, 7days/week, without an increased EI to compensate for the energy expended in running became amenorrhoeic in 7 to 24 months (mean = 14.3). The onset of amenorrhea was fairly abrupt (within two normal cycles) and not preceeded by a significant change in body weight (mean change <1%). When a subgroup (n = 4) was provided with supplemental food they increased their EI (131–181%) and body weight (3–11%) and re-established ovulatory cycles within 7 to 57 days. The correlation between the increase in EI and the resumption of ovulation in individual monkeys suggests that amenorrhea was the result of an energy shortage. However, it is not clear how the animals managed to sustain their body weight with the increase in exercise and lack of an increase in food intake for so long. Total EE was not measured. One can only assume that the monkeys down-regulated other aspects of EE, but could only sustain it for so long.

Recent work by DiMarco *et al.* (2007) with another animal model (rats) also supports the idea that reproductive function is suppressed by low energy availability, not exercise *per se*. DiMarco *et al.* (2007) demonstrated that energy-restricted rats suffered both a loss of weight and reproductive function (anoestrous) although their level of exercise (voluntary wheel running) did not differ from that of controls fed *ad libitum*.

Women in subsistence-based economies. The evidence presented above on the effects of negative EB on reproductive function is based largely on exercise-induced negative EB in which the levels of exercise can be characterized as strenuous to very strenuous. What about the situation of many women living in non-industrial societies who engage in physical work that is less intense, but perhaps of longer duration? An example might be women doing agricultural work for 4 to 6 hours a day. Would the same principles apply? They should apply because EB is based on the match between EI and expenditure at any and all levels. That is, regardless of the intensity of physical work, an EI less than EE will result in negative EB and the same physiological and behavioural outcomes.

Examples are limited, but conform to expectations. Rural Tamang women in Nepal who work in subsistence agriculture increase their subsistence work and EE during the monsoon season. This increase is accompanied by depressed salivary progesterone levels in women in negative EB as evidenced by weight loss

(Panter-Brick & Ellison, 1994). Lese horticulturalists in Zaire had depressed salivary progesterone levels in the pre-harvest season when food availability was low and body weight declining (Ellison *et al.*, 1989). Polish women farmers showed evidence of a loss of body fat and lowered salivary progesterone levels in response to a 2-month seasonal increase in farm work. The loss of body fat (based on triceps skinfold) suggests the women were in negative EB, even though body weight remained stable (Jasieńska & Ellison, 1998). Lastly, lactating Toba women in Argentina resumed ovulation after a period of sustained positive-energy balance (Valeggia & Ellison, 2004). This is especially interesting as the women were classified as well-nourished throughout lactation (mean pre-pregnancy and post-partum BMIs not less than 24.5).

Other explanations

Taken together the studies above provide strong evidence that the reproductive suppression seen with exercise is caused by a shortage of energy. They provide less support for three alternative hypotheses which have been proposed: (1) body fat hypothesis; (2) stress of exercise itself; (3) high-energy flux.

The association of body fat stores with reproductive function, or rather the hypothesis that adequate levels of body fat are necessary for normal reproductive function is associated with the work of Rose Frisch (Frisch & McArthur, 1974; Frisch 1984, 1987) and still appears in the thinking of some of the general public and clinicians (Loucks 2005; Warren & Perlroth, 2001). In the Frisch hypothesis, as it is widely known, fatness is a determinant of the onset and maintenance of menstruation. The hypothesis was based on the correlation between thinness and disrupted reproductive function in female athletes. Theoretically the hypothesis was based on an evolutionary argument, i.e. the idea that since reproduction was costly in energy terms, it would not be adaptive to begin the process when sufficient stores of energy were not available. This is an appealing argument, but the evidence to support it is missing in humans as well as in other mammals (Bronson & Manning, 1991). Recent prospective studies (Loucks & Thuma, 2003; Loucks, 2006) clearly demonstrate that changes in reproductive function (in this case LH pulsatility) occur more rapidly than measurable changes in body fat stores. Further, a number of studies have failed to find differences in body fatness between amenorrhoeic and eumenorrhoeic athletes (Kaiserauer *et al.*, 1989; Loucks & Horvath, 1984). Reviews by Bronson and Manning (1991), Loucks (2005), Sinning and Little (1987) and Rosetta (1993) argue against it. The work on non-human primates (Williams *et al.*, 2001a) also demonstrates that reproductive suppression can occur without a significant change in body weight (and hence presumably, body fat content).

It is noteworthy however, that women with lower levels of body fat seem to be more susceptible to reproductive dysfunction than do women with higher levels of body fat. For example, Alvaro *et al.* (1998) and Olson *et al.* (1995) reported that short-term fasts (3 days) resulted in more severe follicular phase abnormalities in leaner women than in the normal weight women. In a cross-sectional study of women classified as anorexic, Miller *et al.* (2004) found those who were amenorrhoeic had lower levels of body fat (16.7 versus 20.9%) than those with normal menstrual cycles, even though both groups had similar body weight and BMIs. Although Miller *et al.* (2004) attributed the differences in reproductive status to the differences in body fat percentage, age is a confounding factor. The amenorrhoeic women were younger in terms of both chronological (24.4 versus 28.3 years) and gynaecological age (10.9 versus 15.1 years), and as Loucks (2006) has argued, women with a gynaecological age of less than 14 years are more susceptible to reproductive suppression.

A second type of explanation for the reproductive suppression associated with exercise and/or higher than normal levels of physical activity is that it is caused by the stress of exercise itself, i.e. the stress of exercise beyond the energy cost of exercise. This idea that strenuous exercise itself could lead to reproductive dysfunction is probably traceable to observations that female athletes participating in strenuous activities had a higher incidence of menstrual dysfunction and amenorrhea than other women (Erdelyi, 1976; Speroff, 1980). Bullen *et al.*'s (1985) exercise intervention study demonstrated that the negative effects of exercise on reproductive function extended to non-athletes and lent support to the idea. In their study Bullen *et al.* assigned 28 habitually sedentary women with normal menstrual cycles to one of two different treatments: (1) exercise; (2) exercise plus an energy-restricted diet. The women were relatively young chronologically (22 ± 0.5 and 22 ± 0.8 years of age), as well as gynaecologically (10 ± 0.4 and 10 ± 0.8 years post-menarche). The exercise was an 8-week running programme that progressed from 4 to 10 miles a day. They found that the exercise treatment disrupted menstrual cyclicity in both groups, but to a greater extent in the exercise plus diet group. Although the assumption was that the disruptions were caused by exercise, women in both groups lost weight. Hence, a dietary energy deficit cannot be ruled out. Further research using salivary progesterone as a biomarker of menstrual function demonstrated ovarian suppression in recreational athletes (Ellison & Lager, 1986; Rosetta *et al.*, 1998), and in women farmers during the season of heavy agricultural work (Jasieńska & Ellison, 1998). These studies also suggested a direct effect of physical exercise itself on reproduction. However, since EB was not estimated with sufficient accuracy in these studies, the possibility of inadequate EI leading to negative EB cannot be ruled out.

In the above studies, the assumption has been that women with adequate food available and not consciously restricting their food intake will maintain EB when EE is increased. However, we now know that hunger mechanisms may not be sensitive enough to ensure adequate EI with the kind of increases in EE that accompany training, or seasonal work (Blundell *et al.*, 2003). Nor do they seem to be sensitive enough to provide adequate EI in individuals with chronically high EE, like endurance runners (Fudge *et al.*, 2006). Rather energy intake and expenditure appear to be only loosely coupled (Blundell & King, 1999). This is curious as we would expect natural selection to have favoured a tighter fit between energy intake and expenditure. Further, it appears that some individuals are better at compensating for increases in exercise-induced EE than others (Blundell *et al.*, 2003) and for some of them there is clearly a link between food behaviour and menstrual disturbances (Rosetta *et al.*, 2001). This kind of variability has not received much attention.

The issue of whether or not some kind of psychosocial stress associated with exercise may also play a role has been raised (see Williams, 2003). The work of Loucks (2006) and Williams *et al.* (1995), however, separate the energy cost of exercise from other factors associated with exercise by comparing women who exercise in EB with those who exercise the same amount but in negative EB. Their studies clearly support the idea that the energy cost of exercise, as opposed to other things associated with exercise, is causal.

Lastly, the idea that high-energy flux, i.e. high EE coupled with high EI, has been suggested as an explanation for suppression of progesterone profiles seen in well-nourished, weight-stable Polish farm women during the season of heavy agricultural work (Ellison, 2003). Since energy intake and expenditure were measured by recall, methods that lack the precision needed for the task, the possibility that the agricultural work tipped the women into negative EB cannot be ruled out. When total EE was measured using the doubly labeled method, the duration of amenorrhea in lactating Bangladeshi agricultural workers was found to be positively correlated with the total EE (Rosetta *et al.*, 2005). Further, the prospective studies which dramatically changed energy flux but without changing reproductive function (Loucks, 2006; Loucks & Thuma, 2003; Williams *et al.*, 1995) clearly demonstrate that a deficit of energy availability is a more likely explanation for the change in reproductive function than is a change in energy flux.

Mechanisms integrating energy metabolism and reproduction

In order for an energy deficit to impact reproductive function there has to be some kind of signal, or signals, as to the status of fuel availability at the cellular level that is transmitted to the GnRH pulse generator. The review by Krasnow and Steiner (2006) provides an overview of the host of molecules and physiological mechanisms that link metabolism to reproductive function. These include: leptin, insulin, metabolic fuels like glucose, thyroid hormones, growth hormone, insulin-like growth factor I, glucocorticoids, and gastrointestinal peptides like ghrelin, cholecystokinin and glucagon-like peptide I. As Krasnow and Steiner point out, it seems likely that there are a suite of molecules that link metabolism to reproductive function, and that the system exhibits redundancy. As examples, I will highlight two of the molecules that have been the focus of much recent research, leptin and ghrelin.

Leptin is a hormone secreted by adipocytes and found in concentrations that correlate with fat mass (Chan & Mantzoros, 2005). Serum levels decrease rapidly in response to energy deficits induced by short-term fasting in both rodents and humans (Ahima *et al.*, 1996; Chan *et al.*, 2003), and the reduction is proportionally much greater than the reduction in fat mass (Chan *et al.*, 2003). Hence, leptin is considered an ideal sensor of energy deficiency (Ahima, 2004). Serum levels are also low in individuals with chronic energy deficiency (Haspolat *et al.*, 2007; Welt *et al.*, 2004). In short-term fasting the administration of leptin (as r-metHuLeptin) normalized LH pulse parameters affected by fasting in normal weight men and women (Chan *et al.*, 2003, 2006). Leptin administration also normalized the level of reproductive hormones, follicular development and menstrual cycling in female athletes with amenorrhea presumably owing to chronic energy deficits (Welt *et al.*, 2004).

Ghrelin is a peptide synthesized primarily in the stomach that stimulates food intake and appetite (Wren *et al.*, 2001). It increases in response to energy deficits induced by food restriction with or without increased exercise (Kluge *et al.*, 2007; Leidy *et al.*, 2004, 2007), and thus acts to restore EB. De Souza *et al.* (2004) reported that elevated ghrelin levels distinguished athletes with exercise-induced amenorrhea from those with less severe menstrual disturbances. Interestingly, ghrelin also has a suppressive effect on LH secretion in rodents (Garcia *et al.*, 2007), as well as non-human primates (Vulliémoz *et al.*, 2004), and humans (Kluge *et al.*, 2007). These results suggest that leptin and ghrelin are key metabolic signals that serve to regulate both EB and reproductive function.

Unresolved issues

Our understanding of the linkages between EB and reproductive function have advanced enormously in the past 20 years, but there are still a number of unresolved issues.

One issue is the mechanism(s) of cellular-level sensing of energy availability during periods when the balance between energy intake and expenditure is tipped toward the negative, but not severely so, and periods of EB in a hypometabolic state. Most of the effort has been focused on understanding the acute response to diet and/or exercise-induced negative EB.

A second set of issues is related to the time course and magnitude of negative EB. Most of the prospective studies have been short (<10 days). Interestingly, De Souza (2003) has proposed that LPD is the result of short-term intermittent negative EB. Presumably, short term enough not to alter body weight in a measurable way, but long enough to provoke an endocrine response. How short, and how intermittent is not clear. The magnitude of energy imbalance necessary to generate the reproductive changes documented is also unclear. One short-term prospective study (Loucks, 2006) suggests it must be greater than 25%, but longer-term prospective studies would be needed to confirm that.

A third set of issues is methodological. The argument that energy deficits are responsible for the menstrual cycle dysfunctions associated with exercise depends on the concept of EB. Unfortunately, however, EB is difficult to measure in free-living humans with the precision necessary. In the studies reviewed above, the extent to which the components of EB (body weight, EI, EE) were measured or estimated varied considerably. And, this variation may have led to the diverse and sometimes conflicting results. At the whole-organism level changes in EB (positive and negative) can be measured in terms of changes in body weight. Although body weight is easily measured, it is a coarse measure and not capable of accurately distinguishing small changes in body mass from those of fluid shifts. Day-to-day fluctuation in healthy women can be as much as 2 kg (Robinson & Watson, 1965). Further it is clear that exercise–induced energy deficits can alter reproductive hormones (LH pulsatility) more rapidly than measurable changes in body weight.

The measurement of EI might seem simple enough, but it is not. The most common methods used are 24-hour dietary recalls, in which under-reporting is a significant problem (Buzzard, 1998), and 3–7 day weighed food records which require considerable subject cooperation, and can also lead to under-reporting (Buzzard, 1998), especially in weight-conscious women athletes (Edwards *et al.*, 1993; Myerson *et al.*, 1991; Rosetta *et al.*, 2001; Wilmore *et al.*, 1992). As regards EE, it is technically difficult to measure in free-living subjects. The only method capable of measuring EE in individuals with acceptable precision is the DLW (doubly labelled water) technique. Because it estimates EE over about 12 days, day-to-day differences in EB cannot be distinguished, nor can periods of short-term negative EB. Heart rate monitoring is another method, but can only provide acceptable estimates of EE for groups, not individuals. Activity records can also provide acceptable results for groups, but are not accurate enough for individuals and have a high subject burden. They do, however, have the potential to identify the behavioural changes that accompany negative EB. Other techniques, like the use of motion sensors and activity recalls, are not accurate enough.

Conclusions

The studies reviewed above generally support the idea that the changes in reproductive function associated with exercise in women are caused by energy deficits, not the stress of exercise itself. It also seems clear that there is considerable inter-individual variation in the response to exercise-induced energy deficits, gynaecological age being one. There remain significant gaps in our understanding, particularly of the molecular level linkages between energy metabolism and reproductive function, and the possible role of body composition.

References

Ahima, R. S. (2004). Body fat, leptin, and hypothalamic amenorrhea. *The New England Journal of Medicine*, 351, 959–62.

Ahima, R. S., Prabakaran, D., Mantzoros, C. (1996). Role of leptin in the neuroendocrine response to fasting. *Nature*, 382, 250–2.

Alvaro, R., Kimsey, L., Sebring, N. *et al.* (1998). Effects of fasting on neuroendocrine function and follicle development in lean women. *Journal of Clinical Endocrinology and Metabolism*, 83, 76–80.

Blundell, J. E. & King, N. A. (1999). Physical activity and regulation of food intake: current evidence. *Medicine and Science in Sports and Exercise*, 31, S573–83.

Blundell, J. E., Stubbs, R. J., Hughes, D. A., Whybrow, S. & King, N. A. (2003). Cross-talk between physical activity and appetite control: does physical activity stimulate appetite? *Proceedings of the Nutrition Society*, 62, 651–61.

Bronson, F. H. & Manning, J. M. (1991). The energetic regulation of ovulation: a realistic role for body fat. *Biology of Reproduction*, 44, 945–50.

Bullen, B. A., Skrinar, G. S., Beitins, I. Z. *et al.* (1985). Induction of menstrual disorders by strenuous exercise in untrained women. *The New England Journal of Medicine*, 312, 1349–53.

Buzzard, M. (1998). 24-hour dietary recall and food record methods. In *Nutritional Epidemiology*, Volume 30, 2nd edition. Oxford, UK: Walter Willett, pp. 50–73.

Chan, J. L., Heist, K., DePaoli, A. M., Veldhuis, J. D. & Mantzoros, C. S. (2003). The role of falling leptin levels in the neuroendocrine and metabolic adaptation to short-term starvation in healthy men. *The Journal of Clinical Investigation*, 111, 1409–21.

Chan, J. L. & Mantzoros, C. S. (2005). Role of leptin in energy deprivation states: normal human physiology and clinical implications for hypothalamic amenorrhea and anorexia nervosa. *Lancet*, 366, 74–85.

Chan, J. L., Matarese, G., Shetty, G. K. *et al.* (2006). Differential regulation of metabolic, neuroendocrine, and immune function by leptin in humans. *Proceedings of the National Academy of Sciences*, 103, 8481–6.

De Souza, M. J. (2003). Menstrual disturbances in athletes: a focus on luteal phase defects. *Medicine and Science in Sports and Exercise*, 35, 1553–63.

De Souza, M. J., Lee, D. K., VanHeest, J. L. *et al.* (2007). Severity of energy-related menstrual disturbances increases in proportion to indices of energy conservation in exercising women. *Fertility and Sterility*, 88, 971–5.

De Souza, M. J., Leidy, H. J., O'Donnell, E., Lasley, B. & Williams, N. I. (2004). Fasting ghrelin levels in physically active women: relationship with menstrual disturbances and metabolic hormones. *The Journal of Clinical Endocrinology and Metabolism*, 89, 3536–42.

De Souza, M. J., Miller, B. E., Loucks, A. B. *et al.* (1998). High frequency of luteal phase deficiency and anovulation in recreational women runners: Blunted elevation in follicle stimulating hormone observed during luteal follicular transition. *The Journal of Clinical Endocrinology and Metabolism*, 83 (12), 4220–32.

De Souza, M. J., Van Heest, J., Demers, L. M. & Lasley, B. L. (2003). Luteal phase deficiency in recreational runners: evidence for a hypometabolic state. *The Journal of Clinical Endocrinology and Metabolism*, 88, 337–46.

Deuster, P. A., Kyle, S. B. & Moser, P. B. (1986). Nutritional intakes and status of highly trained amenorrheic and eumenorrheic women runners. *Fertility and Sterility*, 46, 636–43.

Dimarco, N. M., Dart, L. & Sanborn, C. B. (2007). Modified activity–stress paradigm in an animal model of the female athlete triad. *Journal of Applied Physiology*, 103, 1469–78.

Drinkwater, B. L., Nilson, K. & Chesnut, C. H. (1984). Bone mineral content of amenorrheic and eumenorrheic athletes. *New England Journal of Medicine*, 311, 277–81.

Dueck, C. A., Matt, K. S., Manore, M. M. & Skinner, J. S. (1996). Treatment of athletic amenorrhea with a diet and training intervention program. *International Journal of Sport Nutrition*, 6, 24–40.

Edwards, J. E., Lindeman, A. K., Mikesky, A. E. & Stager, J. M. (1993). Energy balance in highly trained female endurance runners. *Medicine and Science in Sports and Exercise*, 25, 1398–1404.

Elia, M. (1997). Tissue distribution and energetics in weight loss and undernutrition. In *Physiology, Stress, and Malnutrition–Functional Correlates, Nutritional Intervention*, ed. J. Kinney & H. Tucker. Philadelphia, PA: Lippincott-Raven Publishers, pp. 383–411.

Ellison, P. T. (2003). Energetics and reproductive effort. *American Journal of Human Biology*, 15, 342–51.

Ellison, P. T. & Lager, C. (1986). Moderate recreational running is associated with lowered salivary progesterone profiles in women. *American Journal of Obstetrics and Gynecology*, 154, 100–3.

Ellison, P. T., Peacock, N. R. & Lager, C. (1989). Ecology and ovarian function among Lese women of the Ituri forest, Zaire. *American Journal of Physical Anthropology*, 78, 519–26.

Erdelyi, G. J. (1976). Effects of exercise on the menstrual cycle. *The Physician and Sports Medicine*, 4, 79–81.

Ferro-Luzzi, A., Scaccini, C., Taffese, S., Aberra, B. & Demeke, T. (1990). Seasonal energy deficiency in Ethiopian rural women. *European Journal of Clinical Nutrition*, 44, 7–18.

Frisch, R. E. (1984). Body fat, puberty and fertility. *Biological Reviews*, 59, 161–88.

Frisch, R. E. (1987). Body fat, menarche, fitness and fertility. *Human Reproduction*, 2, 521–33.

Frisch, R. E. & McArthur, J. W. (1974). Menstrual cycles: fatness as a determinant of minimum weight necessary for their maintenance or onset. *Science*, 185, 949–51.

Fudge, B. W., Westerterp, K. R., Kiplamai, F. K. *et al.* (2006). Evidence of negative energy balance using doubly labelled water in elite Kenyan endurance runners prior to competition. *British Journal of Nutrition*, 95, 59–66.

García, M. C., López, M., Alvarez, C. V. *et al.* (2007). Role of ghrelin in reproduction. *Reproduction*, 133, 531–40.

Haspolat, K., Ece, A., Gurkan, F. *et al.* (2007). Relationships between leptin, insulin, IGF-1 and IGFBP-3 in children with energy malnutrition. *Clinical Biochemistry*, 40, 201–5.

Jasieńska, G. & Ellison, P. T. (1998). Physical work causes suppression of ovarian function in women. *Proceedings of the Royal Society of London, Series B, Biological Sciences*, 265, 1847–51.

Kaiserauer, S., Snyder, A. C., Sleeper, M. & Zierath, J. (1989). Nutrition, physiological, and menstrual status of distance runners. *Medicine and Science in Sports and Exercise*, 21, 120–5.

Keys, A. (1950). Energy requirements of adults. *Journal of the American Medical Association*, 142, 333–8.

Kluge, M., Schüssler, P., Uhr, M., Yassouridis, A. & Steiger, A. (2007). Ghrelin suppresses secretion of luteinizing hormone in humans. *The Journal of Clinical Endocrinology and Metabolism*, 92, 3202–5.

Kopp-Woodroffe, S. A., Manore, M. M., Dueck, C. A., Skinner, J. S. & Matt, K. S. (1999). Energy and nutrient status of amenorrheic athletes participating in a diet and exercise training intervention program. *International Journal of Sport Nutrition*, 9, 70–88.

Krasnow, S. M. & Steiner, R. A. (2006). Physiological mechanisms integrating metabolism and reproduction. In *Knobil and Neill's Physiology of Reproduction*, 3rd edition. Burlington, MA: Jimmy Neill, pp. 2553–625.

Laughlin, G. A. & Yen, S. S. C. (1996). Nutritional, endocrine and metabolic aberrations in amenorrheic athletes. *Journal of Clinical Endocrinology and Metabolism*, 81, 4301–9.

Lebenstedt, M., Platte, P. & Pirke, K. M. (1999). Reduced resting metabolic rate in athletes with menstrual disorders. *Medicine and Science in Sports and Exercise*, 31, 1250–56.

Leidy, H. J., Dougherty, K. A., Frye, B. R., Duke, K. M. & Williams, N. I. (2007). Twenty-four hour ghrelin is elevated after calorie restriction and exercise training in non-obese women. *Obesity*, 15, 446–55.

Leidy, H. J., Gardner, J. K., Frye, B. R. *et al.* (2004). Circulating ghrelin is sensitive to changes in body weight during a diet and exercise program in normal-weight young women. *The Journal of Clinical Endocrinology and Metabolism*, 89, 2659–64.

Lopresti, J. S., Gray, D. & Nicoloff, J. T. (1991). Influence of fasting and re-feeding on 3, 3, 5' tri-iodothyronine metabolism in man. *Journal of Clinical Endocrinology and Metabolism*, 72, 130–6.

Loucks, A. B. (2005). Influence of energy availability on luteinizing hormone pulsatility and menstrual cyclicity. In *The Endocrine System in Sports and Exercise, The Encyclopaedia of Sports Medicine*, Volume 11, ed. W. J. Kraemer & A. D .Rogol. Oxford: Blackwell Publishing, pp. 232–50.

Loucks, A. B. (2006). The response of luteinizing hormone pulsatility to 5 days of low-energy availability disappears by 14 years of gynecological age. *Journal of Clinical Endocrinology and Metabolism*, 91, 3158–64.

Loucks, A. B. & Callister, R. (1993). Induction and prevention of low-T3 syndrome in exercising women. *American Journal of Physiology*, 264, R924–30.

Loucks, A. B. & Horvath, S. M. (1984). Exercise-induced stress responses of amenorrheic and eumenorrheic runners. *Journal of Clinical Endocrinology and Metabolism*, 59, 1109–20.

Loucks, A. B. & Thuma, J. R. (2003). Luteinizing hormone pulsatility is disrupted at a threshold of energy availability in regularly menstruating women. *Journal of Clinical Endocrinology and Metabolism*, 88, 297–311.

Loucks, A. B., Verdun, M. & Heath, E. M. (1998). Low-energy availability, not stress of exercise, alters LH pulsatility in exercising women. *Journal of Applied Physiology*, 84, 37–46.

Marcus, R., Cann, C., Madvig, P. *et al.* (1985). Menstrual function and bone mass in elite women distance runners: endocrine and metabolic features. *Annals of Internal Medicine*, 102, 158–63.

Miller, K. K., Grinspoon, S., Gleysteen, S. *et al.* (2004). Preservation of neuroendocrine control of reproductive function despite severe undernutrition. *The Journal of Clinical Endocrinology and Metabolism*, 89: 4434–8.

Myerson, M., Gutin, B., Warren, M. P. *et al.* (1991). Resting metabolic rate and energy balance in amenorrheic and eumenorrheic runners. *Medicine and Science in Sports and Exercise*, 23, 15–22.

Nelson, M. E., Fisher, E. C., Castos, P. D. *et al.* (1986). Diet and bone status in amenorrheic runners. *American Journal of Clinical Nutrition*, 43, 910–16.

Olson, B.R., Cartledge, T., Sebring, N., Defensor, R., Neiman, L. (1995). Short-term fasting affects luteinizing-hormone secretary dynamics but not reproductive function in normal-weight sedentary women. *The Journal of Clinical Endocrinology and Metabolism*, 80, 1187–93.

Panter-Brick, C. & Ellison, P. T. (1994). Seasonality of workloads and ovarian function in Nepali Women. *Annals of the New York Academy of Sciences*, 709, 234–5.

Prior, J. C. & Vigna, Y. (1985). Gonadal steroids in athletic women, contraception, complications and performance. *Sports Medicine*, 2, 287–95.

Robinson, M. F. & Watson, P. E. (1965). Day-to-day variations in body-weight of young women. *British Journal of Nutrition*, 19, 225–35.

Rogol, A. D., Weltman, A., Weltman, J. Y. *et al.* (1992). Durability of the reproductive axis in eumenorrheic women during 1 year of endurance training. *Journal of Applied Physiology*, 72, 1571–80.

Rosetta, L. (1993). Female reproductive dysfunction and intense physical training. *Oxford Reviews of Reproductive Biology*, 15, 113–41.

Rosetta, L., Conde Da Silva Fraga, E. & Mascie-Taylor, C. G. N. (2001). Relationship between self-reported food and fluid intake and menstrual disturbance in female recreational runners. *Annals of Human Biology*, 28: 444–54.

Rosetta, L., Harrison, G. A. & Read, G. F. (1998). Ovarian impairments of female recreational distance runners during a season of training. *Annals of Human Biology*, 25, 345–57.

Rosetta, L., Kurpad, A., Mascie-Taylor, C. G. N. & Shetty, P. S. (2005). Total energy expenditure (H2 18O), physical activity level, and milk output of lactating rural Bangladeshi tea workers and non-tea workers. *European Journal of Clinical Nutrition*, 59(5), 632–38.

Shetty, P. S. (1984). Adaptive changes in basal metabolic rate and lean body mass in chronic under nutrition. *Human Nutrition-Clinical Nutrition*, 38C, 443–51.

Sinning, W. E. & Little, K. D. (1987). Body composition and menstrual function in athletes. *Sports Medicine*, 4, 34–45.

Speroff, L. (1980). Can exercise cause problems in pregnancy and menstruation. *Contemporary Obstetrics and Gynecology*, 16, 57–70.

Tomten, S. E. & Høstmark, A. T. (2006). Energy balance in weight-stable athletes with and without menstrual disorders. *Scandinavian Journal of Medicine and Science in Sports*, 16, 127–33.

Valeggia, C. & Ellison, P. T. (2004). Lactational amenorrhoea in well-nourished Toba women of Formosa, Argentina. *Journal of Biosocial Science*, 36, 573–95.

Veldhuis, J. D., Evans, W. S., Demers, L. M. *et al.* (1985). Altered neuroendocrine regulation of gonadotrophin secretion in women distance runners. *Journal of Clinical Endocrinology and Metabolism*, 61, 557.

Vulliémoz, N. R., Xiao, E., Zhang, L. X. *et al.* (2004). Decrease in luteinizing hormone pulse frequency during a five-hour peripheral ghrelin infusion in the ovariectomized rhesus monkey. *Journal of Clinical Endocrinology and Metabolism*, 89, 5718–23.

Wade, G. N. & Jones, J. E. (2004). Neuroendocrinology of nutritional infertility. *American Journal of Physiology –Regulatory, Integrative and Comparative Physiology*, 287, R1277–96.

Warren, M. P. (1980). The effects of exercise on pubertal progression and reproductive function in girls. *Journal of Clinical Endocrinology and Metabolism*, 51, 1150–7.

Warren, M. P. & Perlroth, N. E. (2001). The effects of intense exercise on the female reproductive system. *Journal of Endocrinology*, 170, 3–11.

Welt, C. K., Chan, J. L., Bullen, J. *et al.* (2004). Recombinant human leptin in women with hypothalamic amenorrhea. *The New England Journal of Medicine*, 351, 987–97.

Williams, N. I. (2003). Lessons from experimental disruptions of the menstrual cycle in humans and monkeys. *Medicine and Science in Sports and Exercise*, 35, 1564–72.

Williams, N. I., Young, J. C., McArthur, J. W. *et al.* (1995). Strenuous exercise with caloric restriction: effect on luteinizing hormone secretion. *Medicine and Science in Sports and Exercise*, 27, 1390–8.

Williams, N. I., Caston-Balderrama, A. L., Helmreich, D. L. *et al.* (2001a). Longitudinal changes in reproductive hormones and menstrual cyclicity in cynomolgus monkeys during strenuous exercise training: abrupt transition to exercise-induced amenorrhea. *Endocrinology*, 142, 2381–9.

Williams, N. I., Helmreich, D. L., Parfitt, D. B., Caston-Balderrama, A. & Cameron, J. L. (2001b). Evidence for a causal role of low-energy availability in the induction of menstrual cycle disturbances during strenuous exercise training. *The Journal of Clinical Endocrinology and Metabolism*, 86, 5184–93.

Wilmore, J. H., Wambsgans, K. C., Brenner, M. *et al.* (1992). Is there energy conservation in amenorrheic compared with eumenorrheic distance runners? *Journal of Applied Physiology*, 72, 15–22.

Wren, A. M., Seal, L. J., Cohen, M. A. *et al.* (2001). Ghrelin enhances appetite and increases food intake in humans. *The Journal of Clinical Endocrinology and Metabolism*, 86, 5992–5.

Zanker, C. L. & Swaine, I. L. (1998). The relationship between serum oestradiol concentration and energy balance in young women distance runners. *International Journal of Sports Medicine*, 19, 104–8.

5 *Energetic cost of gestation and lactation in humans*

PRAKASH SHETTY

Introduction

The energy needs of humans are currently estimated from measures of energy expenditure of the individual. Until recently however, in the absence of sufficient data on habitual total energy expenditure of free-living humans, other approaches have enabled us to compute the energy requirements. While estimating the energy requirement of diverse demographic groups within the population, additional provision needs to be made for physiological processes characteristic of these groups. These include the additional provision of energy for optimal growth in infants and children and for the additional energy needs of pregnancy and lactation in women of reproductive age.

Energy requirement has been defined as 'the amount of food energy needed to balance energy expenditure in order to maintain body size, body composition and a level of necessary and desirable physical activity consistent with long-term good health' (Food and Agriculture Organization/World Health Organization/United Nations University – FAO/WHO/UNU, 2004). Energy balance is achieved when dietary intake is equal to the energy expenditure plus the energy cost of growth in infancy, childhood and pregnancy, or the energy cost to produce adequate milk during lactation. The recommended dietary intake of energy for a population group which should satisfy these requirements for the attainment and maintenance of optimal long-term good health, physiological functions and well-being is the mean energy requirement of the healthy and well-nourished individuals within this population group.

In this chapter the energy requirements of pregnancy and lactation in adult humans will be discussed separately and followed by a brief discussion on the implications for fertility.

Reproduction and Adaptation, eds. C. G. Nicholas Mascie-Taylor and Lyliane Rosetta.
Published by Cambridge University Press.

Energy requirements for human pregnancy

Women in the reproductive age range will require a dietary intake during pregnancy which will provide energy that will ensure the full-term delivery of a healthy, newborn infant of adequate size and appropriate body composition. For the purpose of arriving at estimates of energy requirements for pregnancy to ensure this birth outcome, the pregnant woman is expected to have a body weight and body composition and a physical activity level consistent with long-term good health and well-being, and to have become pregnant at an acceptable or normal body weight and with good nutritional status. This would imply that special considerations would necessarily apply to women who are under- or overweight or have body composition changes relative to normal standards when they enter pregnancy. On this basis the additional energy requirements of human pregnancy are those needed for adequate maternal weight gain to ensure the growth of the fetus, placenta and associated maternal tissues including stored energy. Dietary intake should also provide for the increased metabolic demands of pregnancy, in addition to the energy needed to maintain adequate maternal weight and body composition and physical activity consistent with the lifestyle, throughout the gestational period, as well as provide for sufficient energy stores for proper and adequate lactation post-partum.

An earlier International Consultation on energy and protein requirements (FAO/WHO/UNU, 1985) provided an excellent definition of the energy requirements of human pregnancy. This Consultation report states that the energy requirement of a pregnant woman is the level of energy intake from food that will balance her energy expenditure when the woman has a body size and composition and level of physical activity consistent with good health and that will allow for the maintenance of economically necessary and socially desirable physical activity, and further that the said energy requirement includes the energy needs associated with the deposition of tissues consistent with optimal pregnancy outcomes. Three basic principles underpin this definition of energy requirements for human pregnancy (Butte & King, 2005): (i) women should enter pregnancy with a body size and body composition consistent with long-term good health; (ii) pregnant women should gain weight at a rate and with a composition consistent with good health for herself and her infant; (iii) women during pregnancy should have energy intakes that will allow them to maintain economically necessary and socially desirable levels of physical activity during the duration of their pregnancy.

These recommendations would obviously be population specific to cater to the differences in body size and lifestyles. Lifestyles would dictate habitual activity patterns based on socio-economic and cultural factors which are specific to geographical regions and cultural attitudes related to pregnancy. Body sizes of normal weight for height women will also vary in different countries and these need to be considered, as does the variability within populations themselves. Despite the need to consider these factors it is important at the outset to clarify what one considers as the optimal gestational weight gain during pregnancy for an optimal pregnancy outcome. The optimal outcomes for the mother are defined by maternal mortality, complications of pregnancy, labour and delivery, and post-partum lactational performance and weight retention; while those outcomes defined in terms of the infant include fetal growth and birth weight, gestational duration, infant mortality and morbidity.

A World Health Organization (WHO) Collaborative study on maternal anthropometry and pregnancy outcomes (WHO, 1995) reviewed information on 110,000 births in 20 different countries to arrive at maternal anthropometric indicators which were predictive of fetal and maternal outcomes. The fetal outcomes considered were low birth weight (LBW), intrauterine growth retardation (IUGR) and pre-term birth, while the maternal outcomes examined were pre-eclampsia, post-partum haemorrhage and assisted delivery. The most important determinant of LBW and IUGR, but not of pre-term birth, was attained weight during pregnancy, i.e. pre-pregnancy weight plus the weight gained during pregnancy. Low birth weight was well predicted by maternal pre-pregnancy weight and attained weight at various periods during gestation. Pre-term delivery was also predicted by pre-pregnant maternal body weight. Women of short stature in developing countries were at increased risk of LBW, small for gestational age babies, and pre-term delivery, as well as obstetric complications. Based on this analysis, the study concluded that optimal maternal and fetal outcomes were associated with birth weights between 3.1 and 3.6 kg (mean 3.3 kg). They also concluded that birth weights over 3.0 kg were associated with maternal weight gains during gestation of between 10–14 kg and recommended that a mean weight gain of 12 kg was what was desired during pregnancy. The Institute of Medicine (IOM) in the USA recommends slightly higher maternal weight gains during gestation than those recommended by WHO (Institute of Medicine and Food and Nutrition Board, 1990). The estimation of the energy cost of human pregnancy is thus based on defining the desirable birth weight (mean of 3.3 kg) and the desirable gestational weight gain (mean 12 kg).

The energy cost of human pregnancy

The energy cost of pregnancy will include both the energy deposited in maternal and fetal tissues during the duration of pregnancy, and the increase in energy expenditure over and above that of the adult female at pre-pregnancy attributed

Table 5.1 *Cumulative energy cost of pregnancy based on the Hytten and Chamberlain theoretical model (1991).*

	Cumulative weight (g)	Cumulative energy (MJ)
Protein deposition	925	21.7
Fat deposition	3825	152.0
Contribution from increment in basal metabolism		149.4
Efficiency of energy utilization (taken as 0.90)		32.3
Total energy cost of pregnancy in MJ (kcal)		355.4 MJ (80,000 kcal)

to the increase in maintenance requirements and cost of physical activity with the weight gain. The increase in weight gain in pregnancy comprises the following: (i) products of conception which include the fetus, placenta and amniotic fluids; (ii) the increase in maternal tissues such as the uterus, breasts, blood, and other body fluids; and (iii) increase in maternal fat stores. This increase in tissue mass and body weight will contribute to increase the energy cost of maintenance and that of physical activity, which will in turn contribute to the energy cost of human pregnancy.

The earliest and careful estimation of the energy cost of human pregnancy in well-nourished women was carried out by Hytten & Chamberlain (1991). Their estimate of the total energy cost of pregnancy, summarized in Table 5.1, was based on a theoretical model with the following assumptions: (i) pre-pregnant body weight of between 60 and 65 kg; (ii) average gestational weight gain of 12.5 kg; and (iii) an average infant birth weight of 3.4 kg. This model and its derived energy costs of human pregnancy formed the basis for the international recommendations of energy requirements in 1985 (FAO/WHO/UNU, 1985). Better estimates of body composition with improved technology over the years and the use of the doubly labelled water (DLW) method to measure habitual total energy expenditure (TEE) in women during pregnancy has contributed to downward revision of these energy costs of pregnancy figures derived earlier by Hytten and Chamberlain (1991). An overestimate of the efficiency of energy utilization for fat and protein deposition also contributed to the slightly higher theoretical values derived by them.

More recently the availability of the DLW method to measure TEE and better body composition measurements in pregnant women has enabled more accurate estimates of the energy cost of human pregnancy for a gestational weight gain of 12 kg (FAO/WHO/UNU, 2004). The total energy cost of pregnancy in well-nourished women has been calculated based on the energy deposited from the estimated increase in protein and fat accretion, measured using more

Table 5.2 *Cumulative energy cost of pregnancy estimated from body composition changes and energy expenditure estimates by two different approaches in pregnant women with an average gestational weight gain of 12 kg.*

	Cumulative weight (gm)	Cumulative energy (MJ) estimated from BMR factorial approach	Cumulative energy (MJ) estimated from TEE measures
Protein deposition	597	14.1	14.1
Fat deposition	3741	144.8	144.8
Contribution from increment in basal metabolism		147.8	NA
Efficiency of energy utilization (taken as 0.90)		15.9	NA
Total energy expenditure (measured by DLW)		NA	161.3
Total energy cost of pregnancy in MJ (kcal)		322.6 MJ (77,100 kcal)	320.2 MJ (76,530 kcal)

BMR = Basal metabolic rate
TEE = Total energy expenditure
DLW = Doubly labelled water
NA = Not applicable

sophisticated body composition analysis. In addition, two approaches have been used for estimating the increase in energy expenditure. Energy expenditure was derived either factorially from the increment in basal metabolic rates (BMR) during pregnancy or from estimates of the increase in TEE using DLW data. The estimates of two approaches for TEE (Table 5.2) did not make much difference and the derived energy costs gave similar results of 322.6 and 320.2 MJs, i.e. 77,100 kcals and 76,530 kcals respectively; not too far from the slightly higher theoretical estimates (of 350.5 MJs or 80,000 kcals) of Hytten & Chamberlain (1991).

Energy requirements of human pregnancy

Based on these recent estimates and averaging them, the international recommendations for extra or additional energy needs of human pregnancy are 321 MJ (i.e. 77,000 kcals). The energy cost of pregnancy is not distributed equally throughout the duration of pregnancy as the deposition of protein occurs primarily in the second (20%) and third (80%) trimesters. It is assumed that the deposition of fat in the body also follows a similar pattern to the gestational weight gain of 11, 47 and 42% in the three successive trimesters (Institute of Medicine and Food and Nutrition Board, 1990). The increments in BMR were of the order of 5, 10 and 25% in the three trimesters. On this basis the recommendations for additional energy costs of pregnancy are 0.35 MJ (85 kcals) per day in the first trimester, 1.2 MJ (285 kcals) per day in the second trimester and 2.0 MJ (475 kcals) per day in the third trimester. Given that in many societies, women do not seek antenatal care during the first trimester, the practical recommendations for additional energy requirements and the dietary intakes for energy are that pregnant women increase their energy intake by 1.5 MJ (360 kcals) per day in the second trimester and 2.0 MJ (475 kcals) per day in the third trimester (FAO/WHO/UNU, 2004).

There has been much debate on the issue of metabolic adjustments and adaptation to meet energy requirements by adjusting basal metabolism or the thermic effect of feeding and the energy costs of activity, particularly among underweight or undernourished mothers and during marginal or seasonal variations in intake; and these have been previously reviewed (Prentice *et al.*, 1996).

Comparisons of energy intakes of women worldwide, both in affluent and poor countries, have recorded mean population energy needs ranging from as high as 520 MJ to as low as -30 MJ per pregnancy (Prentice & Goldberg, 2000). These energy costs were closely correlated with maternal energy status when analyzed both between and within populations. Hence this has prompted the suggestion that they represent functional adaptations that have been selected for their role in protecting fetal growth. It is important to recognize that although this metabolic plasticity represents a powerful mechanism for sustaining pregnancy under very marginal nutritional conditions, it must not be construed as a perfect mechanism that obviates the need for optimal nutritional care of pregnant women. The fetal weight represents up to 60% of total pregnancy weight gain in many pregnancies in poor and low- income societies compared to the norm of 25% in the well nourished and the well-to-do. Undoubtedly this indicates that the fetus is developing in the undernourished woman under suboptimal nutritional and physiologic conditions. This has long been recognized to have immediate consequences for the offspring in terms of increased perinatal mortality, and increased morbidity and mortality in infancy (Ashworth, 1998). Impaired fetal growth may also bear longer-term consequences in terms of adult susceptibility to non-communicable diseases (Barker, 2007).

The recommendations for energy requirements of human pregnancy will also have to take into account the practice in developed and industrialized societies of women reducing their physical activity levels during the third trimester, accounting for up to 0.6 MJ (145 kcal) per day based on studies in the Netherlands and Scotland (Prentice *et al.*, 1996). The requirements for energy also need to be adjusted for women who may be undernourished or underweight and those who are overweight or obese in pregnancy. Being underweight or overweight in pregnancy increases the risk of poor maternal and birth outcomes. Pre-pregnant weights of below 50 kg increase the risk of maternal complications, while pre-pregnant body weight of 45 kg or lower is associated with poor fetal outcomes. Short stature (less than 150 cm) is also a risk factor for poor maternal and fetal outcomes. This does not necessarily preclude that women in developing countries – very many of them short-statured and having pre-pregnancy bodyweights less than 50 kg – go through normal pregnancy, although the risk of poor birth outcomes such as low birth weight at full term, i.e. intrauterine growth retardation, is higher.

Maternal obesity increases the risk of maternal and fetal complications such as neural tube defects, spina bifida and congenital malformations. Pregnancy during adolescence is another special case where additional energy needs may be justified as 20 percent of the total growth in stature may occur during this period and will be compromised if these energy needs are not met. The risk of an adverse pregnancy outcome such as low birth weight and small-for-gestational age is higher among adolescent pregnant women.

Energy requirements of human lactation

The FAO/WHO/UNU consultation (1985) defined the energy requirements of human lactation as follows: 'The energy requirement of a lactating woman is the level of energy intake from food that will balance her energy expenditure when the woman has a body size and composition and breast milk production which is consistent with good health for herself and her child; and that will allow her for the maintenance of economically necessary and socially desirable physical activity.' Based on this definition the additional energy needs of lactation would be those needed to account for the energy cost of milk production, assuming that the rest of the energy needed would be unchanged, including those related to the resumption of normal physical activity patterns. Some of the additional energy needs of lactation would also be met by the additional stored energy during gestation and the consequent mobilization of tissues and weight loss (apart from loss of water) that occurs post-partum.

The ability to produce milk during lactation in humans is a robust physiological process which appears not to be influenced either by nutritional status or cultural settings. A review of the available data on human milk production concludes that the volume of milk produced at peak lactation is similar in mothers both from developed and developing societies, although the composition of the milk secreted may vary (Prentice *et al.*, 1986). However, the extent of exclusive breast feeding and the total duration of breast feeding is widely variable, despite the recommendations made by the World Health Organization (2001).

Table 5.3 *Milk production rates and the energy cost of milk production in exclusively breast feeding mothers in developed and developing countries.*

Post-partum	0–2 months	3–5 months	6–8 months	9–11 months
Milk production* (gm per day)				
• Industrialized countries	710	787	803	900
• Developing countries	714	784	776	
Energy cost of milk production** (MJ per day)				
• Industrialized countries	2.49	2.75	2.81	3.15
• Developing countries	2.50	2.74	2.72	

Compiled by Butte & King, 2005.
* Milk production data from Brown *et al.*, 1998.
** Energy cost of lactation based on milk production rates and using milk energy density of 2.8 KJ per gm with an energetic efficiency of synthesis of 0.80.

Energy cost of human lactation

Based on a comprehensive review of milk production rates in mothers from developing and developed countries (Brown *et al.*, 1998), both from exclusive breast feeding and partial breast feeding conditions and the energy content of the secreted milk, the energy cost of milk production has been computed. The assumptions made were that the milk energy density was equal to 2.8 kJ per gm of milk (based on several studies where proximate analysis was carried out on 24-hour representative milk samples) and that the energetic efficiency of human milk synthesis was assumed to be 0.80. The figure of 80% for efficiency was based on calculated biochemical efficiency, taking into consideration synthesis of lactose, protein and fat, as well as transfer of fat, and then adjusted downwards to account for the energy losses associated with digestion, absorption and transport of nutrients from food intake. Based on these assumptions and data sets of milk secretion from developing and developed countries, the energy cost of human lactation over a period of 24 months post-partum has been compiled (Butte & King, 2005) and is summarized in Table 5.3.

Energy requirements of human lactation

In much the same way as with human pregnancy the energy requirements of human lactation have been computed based on estimates of total energy expenditure by indirect calorimetry or the DLW method in lactating mothers. The energy requirements during lactation can be estimated and computed by adding the energy cost of milk production to the energy requirements of non-pregnant and non-lactating women of similar body size and composition. This needs of necessity to consider energy mobilized from tissue stores because energy is stored during pregnancy for the express purpose of supporting lactation post-partum in situations where individuals are energy replete. The computation of the energy cost of milk production can be made from measures of milk secretion, the energy density of milk and the energetic efficiency of human milk synthesis. Using this approach the total energy requirement of lactation would be equal to the total energy expenditure (TEE)(measured by DLW or estimated from the basal metabolic rate (BMR) and physical activity level) of a non-pregnant, non-lactating woman plus energy requirements of milk production (equal to daily milk production volume × energy density of milk × conversion efficiency) minus energy mobilization from tissue stores. The implications of using this approach assumes that lactating mothers have similar levels of physical activity as non-pregnant, non-lactating women despite exclusive breast feeding several times a day imposing sedentary behaviour patterns, but probably compensated for by the fact that carrying an infant imposes slightly greater energetic costs on other activities during the day.

The additional energy costs attributable to lactation over the first 6 months of exclusive breast feeding would amount to 2.8 MJ (675 kcal) per day. This value is arrived at from a mean milk output of 807 gm per day (see Table 5.4) multiplied by 2.8 kJ per gm (energy density of breast milk) into 0.8, i.e. efficiency of conversion. From 6 months onwards, infants are partially breast fed with the introduction of complementary feeds. With milk production on average dropping to 550 gm per day, the additional energy needs of lactation 6 months post-partum work out to 1.925 MJ (460 kcal) per day. It is however important to note that in developing countries women continue to breast feed for much longer durations, sometimes up to 24 months.

Recent studies from Bangladesh have shown that milk output at 12 months estimated using isotopic dilution techniques is well above the 550 gm (Rosetta *et al.*, 2005) suggested by the earlier review for the FAO/WHO/UNU Expert Consultation. In the Bangladeshi women the overall mean milk output at 703 gm per day was higher than the 550 gm per day high, even though most of the mothers were 12 months post-partum when the study was undertaken. The study by Rosetta and colleagues (Rosetta *et al.*, 2005) showed that mothers produced high quantities of milk even 12 months post-partum, an observation that supports earlier work from Papua New Guinea. In a longitudinal study of Papua New Guinean women over a 24 month period, Orr-Ewing and others (Orr-Ewing *et al.*, 1986) found that milk output rose from 601 gm per day at 1 month to a maximum of 901 gm per day at 9 months and fell to 501 gm per day at 24 months. This is an important area that needs further work, since it implies

Table 5.4 *Milk secretion rates and the energy cost of milk production in exclusively breast feeding mothers and the energy requirements of their infants.*

Months post-partum	Mean milk intake of infants[1] (gm/day)	Milk intake corrected for insensible water loss* (gm/day)	Gross energy secreted in milk (kJ/day)**	Energy cost of milk production (kJ/day)	Metabolizable energy intake in milk (kJ/day)***	Infant energy requirement based on TEE estimates by DLW method[1]
1 month	699	734	2055	2569	1946	1922
2 months	731	768	2149	2686	2035	2143
3 months	751	789	2208	2760	2091	2284
4 months	780	819	2293	2867	2172	2219
5 months	796	836	2340	2925	2216	2376
6 months	854	897	2511	3138	2378	2501
Mean of 6 months	769	807	2259	2824	2140	2241

[1] Milk volumes of well-nourished women with healthy babies and TEE estimates of infants using DLW method reported by Butte *et al.*, 2002.
TEE = Total energy expenditure; DLW = Doubly labelled water.
* Insensible water loss assumed to be equal to 5% of milk intake.
** Gross energy of milk measured by bomb calorimetry and found to be 2.8 kJ per gm.
*** Metabolizable energy based on proximate analysis of milk and found to be 5.3% lower than bomb calorimetry values.

extra energy needs at 6 months post-partum, probably up to 24 months after birth. With international agencies promoting both exclusive breast feeding up to 6 months and continued breast feeding as long as possible up to 24 months, providing the appropriate recommendations for energy requirements for this period when women may be back to their heavy work burden is crucial.

Estimates of energy mobilized from stored energy in well-nourished women have also been made. Post-partum loss of weight is highest in the first 3 months with a review of many studies suggesting that on average 0.8 kg body weight is lost per month from well-nourished women during lactation, compared to 0.1 kg in poorly nourished mothers (Butte & Hopkinson, 1998). Computed estimates are that they contribute to 0.72 MJ (170 kcal) per day in well-nourished mothers, which can be deducted from the recommended value of 2.8 MJ to a revised figure of 2.1 MJ (505 kcal) per day – an energy requirement closer to the recommended energy needed for lactation after 6 months post-partum. The recent international recommendations have considered this and recommend an additional energy intake of 2.1 MJ (505 kcal) per day for well-nourished mothers for the first 6 months of lactation.

As in the case of pregnancy, the role of probable metabolic adjustments by the human body to meet energy requirements during lactation have been examined with regard to changes in basal metabolism, thermic effect of feeding and by alterations in the energetic costs of activity. How important these metabolic adjustments are appears to be debatable although changes in physical activity patterns per se may play a more important role. Reductions are well documented for the first month post partum in developed societies, although their contributions may be much more significant in developing societies with any small reduction in the normally high levels of physical activity likely to contribute more to energy saving.

Energy requirements for human lactation: implications for fertility

The return of menstruation after the birth is delayed if a mother breast feeds her baby and the duration of post-partum amenorrhea is related to many factors such as breast feeding and the frequency and duration of suckling episodes (Rosetta, 1992). Breast feeding is therefore an important factor contributing to lactational amenorrhea. The duration of post-partum amenorrhea and the resumption of ovulation vary within individuals across pregnancies, between individuals, and between populations, and the duration can be as little as 2–3 months in Western societies or as long as 3 years in some hunter–gatherer societies in the Kalahari desert of Botswana and Namibia (Rosetta & Mascie-Taylor, 2009).

A WHO-sponsored multinational study in seven countries – five of them in developing countries, reported that the lowest median duration of lactational amenorrhea was 122 days and the longest was 282 days (WHO Task Force on Methods for the Natural Regulation of Fertility, 1998). The studies in Papua New Guinea (Orr-Ewing *et al.*, 1986) and Bangladesh (Rosetta *et al.*, 2005) underscore the importance of the additional energy requirements for lactation associated with long duration of breast feeding and their role in prolonging the infertile period. In Bangladesh the median duration of post-partum amenorrhea was reported to be 491 days (equivalent to over 70 weeks) with much longer duration of post-partum amenorrhoea found among tea garden workers (median 636 days, i.e. 91 weeks) compared with non-tea workers (375 days, i.e. 54 weeks). These recent observations of Rosetta and Mascie-Taylor (2009) are important for they show that our current estimates of post-partum amenorrhoea may be gross underestimates. More importantly they show that there is an interaction between levels of physical activity and continued breast feeding – the increased energy requirement of one compounded by the other to prolong the period of post-partum amenorrhoea and the time to return to fertility.

Conclusions

Estimates of human energy requirements and recommendations for the extra energy needed for proper gestation during pregnancy need to be population-specific, because of differences in body size, lifestyles and underlying nutritional status. Well-nourished women from economically developed societies have different energy needs during pregnancy as compared to women from low-income developing countries who in addition may be physically more active during much of pregnancy. Pregnancy energy needs of stunted or undernourished women may differ from those of overweight or obese individuals. Physical activity patterns that characterize lifestyles may also be different and change differently in pregnancy and are determined by socio-economic and cultural factors. Even within societies, high variability is observed in the rates of gestational weight gain and in energy expenditures and intakes of pregnant women.

During the first 6 months of lactation, the amount of breast milk produced daily appears to be similar among population groups of different cultural, socio-economic settings and even nutritional states although some variation in milk composition does occur and is related to maternal nutrition. Hence the main factors that influence energy needs during lactation are the duration of breast feeding and the extent of exclusive breast feeding, both of which differ considerably and hence highlight the need for population-specific recommendations.

The continued breast feeding after 6 months, sometimes up to 24 months, in women in developing societies increases the energy needs in lactation, especially if they are associated with physically very active lifestyles. Studies now suggest that our current estimates of milk output after 6 months may be underestimates as women seem to produce good quantities of milk for much longer periods. These interactions between energy needs of physical activity add to the energy needs of lactation and probably contribute to influencing the duration of post-partum amenorrhoea and return to fertility.

It is important that pregnancy and lactation are considered as reproductive processes in continuum and hence their energetics, studied ideally from the beginning of pregnancy, contributing to appropriate weight gain during gestation and the attainment of adequate reserves to be mobilized to support proper and adequate lactation and milk output for whatever period lactation continues.

References

Ashworth, A. (1998). Effects of intrauterine growth retardation on mortality and morbidity in infants and young children. *European Journal of Clinical Nutrition*, 52, S34–42.

Barker, D.J. (2007). The origins of the developmental origins theory. *Journal of Internal Medicine*, 261, 412–17.

Brown, K., Dewey, K.G. & Allen, L. (1998). *Complementary Feeding of Young Children in Developing Countries: A Review of Current Scientific Knowledge.* Geneva: World Health Organization.

Butte, N.F. & Hopkinson, J.M. (1998). Body composition changes during lactation are highly variable among women. *Journal of Nutrition*, 128, S381–5.

Butte, N.F. & King, J.C. (2005). Energy requirements during pregnancy and lactation. *Public Health Nutrition*, 8, 1010–27.

FAO/WHO/UNU (2004). Human energy requirements. *Report of Joint FAO/WHO/UNU Expert Consultation, Rome, October 2001. Food and Nutrition Technical Report Series 1.* Rome: Food and Agriculture Organization.

FAO/WHO/UNU (1985). Energy and protein requirements. *Report of Joint FAO/WHO/UNU Expert Consultation, 1981. WHO Technical Report series 724.* Geneva: World Health Organization.

Hytten, F.E. & Chamberlain, G. (1991). *Clinical Physiology in Obstetrics.* Oxford: Blackwell Scientific Publications.

Institute of Medicine and Food and Nutrition Board (1990). *Nutrition during Pregnancy.* Washington, DC: National Academy Press.

Orr-Ewing. A., Heywood, P. & Coward, W.A. (1986). Longitudinal measurements of breast milk output by a 2H2O tracer technique in rural Papua New Guinean women. *Human Nutrition: Clinical Nutrition*, 40, 451–67.

Prentice, A. & Goldberg G.R. (2000). Energy adaptations in human pregnancy: limits and long-term consequences. *American Journal of Clinical Nutrition*, 71, 1226S–32.

Prentice, A., Paul A., Black, A., Cole, T. & Whitehead, R. (1986). Cross-cultural differences in lactational performance In: *Human Lactation 2: Maternal and Environmental Factors* (Editors: Hamosh & Goldman). New York, NY: Plenum Press.

Prentice, A.M., Spaaij, C.J.K., Goldberg, G.R. *et al.* (1996). Energy requirements of pregnant and lactating women. *European Journal of Clinical Nutrition*, 50, S82–111.

Rosetta, L. (1992). Aetiological approach of female reproductive physiology in lactational amenorrhoea. *Journal of Biosocial Science*, 24, 301–15.

Rosetta, L., Kurpad, A., Mascie-Taylor, C.G.N. & Shetty, P.S. (2005). Total energy expenditure ($H_2^{18}O$), physical activity level and milk output of lactating rural Bangladeshi tea workers and non-tea workers. *European Journal of Clinical Nutrition*, 59, 632–8.

Rosetta, L. & Mascie-Taylor, C.G.N. (2009). Factors in the regulation of fertility in deprived populations. *Annals of Human Biology*, 36, 642–52.

World Health Organization (1995). Maternal anthropometry and pregnancy outcomes – A WHO Collaborative Study. *Bulletin of the World Health Organization*, 73, S1–69.

World Health Organization Task Force on Methods for the Natural Regulation of Fertility (1998). The World Health Organization multinational study of breast-feeding and lactational amenorrhea. II. Factors associated with the length of amenorrhea. *Fertility & Sterility*, 70, 461–71.

World Health Organization (2001). Expert consultation on the optimal duration of exclusive breastfeeding. *Conclusions and Recommendations*. Geneva: World Health Organization.

6 *Adaptive maternal, placental and fetal responses to nutritional extremes in the pregnant adolescent: lessons from sheep*

JACQUELINE M. WALLACE

Introduction

Inadequate fetal nutrient supply and the resulting intrauterine growth restriction (IUGR) and premature delivery continue to cause unacceptably high rates of infant mortality and morbidity throughout the world. Indeed in the relatively affluent UK, recent statistics reveal that the incidence of low birth weight (<2500 g) has increased from 67 per 1000 births in 1989 to 78 per 1000 births in 2006 (Fabian Society, 2006). While these figures may in part reflect both the changing ethnic makeup of UK society and the increased availability of assisted conception procedures and hence multiple births, the trend is worrying as both premature delivery and low birth weight are associated with a lifetime legacy of health issues. For the extremely small and premature baby there is an increased risk of cerebral palsy, autism, visual and aural impairment, and of experiencing developmental problems such as low IQ, poor cognitive function and learning difficulties with their obvious social, ethical and economic costs (Hack & Merkatz, 1995). In addition, there is compelling evidence from a large number of epidemiological studies that low birth weight, even within the normal range, is a major risk factor for the subsequent development of metabolic syndrome and its co-morbidities, particularly when the infant is born into a calorie-rich environment (Barker, 1998, 2006). It is axiomatic that reducing the incidence of low birth weight is a major research priority with the potential to impact immediate survival and lifelong health of the individual.

The role of nutrition

The nutritional status and dietary intake of the mother plays an important and potentially modifiable role in optimising pregnancy outcome. Nutritional

Reproduction and Adaptation, eds. C. G. Nicholas Mascie-Taylor and Lyliane Rosetta. Published by Cambridge University Press.

risk factors for low birth weight and premature delivery include short inter-pregnancy intervals, low pre-pregnancy weights, insufficient gestational weight gain, multi-fetal pregnancies and a young maternal age (reviewed in King, 2003; Luther *et al.*, 2005a). At the other end of the nutritional spectrum a high maternal body mass index (BMI) and excessive pregnancy weight gain are variously associated with an increased incidence of stillbirth, premature delivery, IUGR or fetal macrosomia (Cnattingius *et al.*, 1998; Castro & Avina, 2002). While appropriate maternal nutrition is imperative for ensuring a favourable pregnancy outcome in all these vulnerable groups, it is particularly important in young adolescent mothers.

Pregnancy during adolescent life (<19 years of age) accounts for one-fifth of all births worldwide and is associated with a particularly high risk of miscarriage, premature delivery, low birth weight, neonatal and infant mortality, and maternal death. These negative pregnancy outcomes are most prevalent in very young girls who are gynaecologically immature and have yet to complete their own body growth (reviewed in Wallace *et al.*, 2006a). As suboptimal dietary intakes are commonplace in the general adolescent population, many young adolescents may be in danger of becoming pregnant with poor nutrient stores and/or subsequently experiencing inadequate gestational weight gains. In support, insufficient pregnancy weight gains (a proxy indicator of maternal undernutrition) in adolescent mothers have been associated with low birth weight in several studies (Hediger *et al.*, 1989; Scholl *et al.*, 1991; Stevens-Simon *et al.*, 1993). In contrast, data from the Camden Adolescent Pregnancy and Nutrition Project (based in New Jersey, one of the poorest cities in the United States) suggest that continued maternal growth occurs in approximately 50% of their pregnant adolescent population. This continued maternal growth as measured by sequential changes in knee height is associated with larger pregnancy weight gains and increased fat stores, but in spite of this the babies are smaller than those born to non-growing adolescents and mature women (Scholl *et al.*, 1997). These latter effects are attributed to a competition for nutrients between the maternal body and her gravid uterus and is unique to the adolescent growth period.

Clearly the relationship between nutritional status, gestational dietary intake and pregnancy outcome is complex when pregnancy coincides with the continued or incomplete growth of the mother. As ethical boundaries limit direct investigations of mother and fetus, highly controlled animal models have been developed to examine the role of maternal nutrition in mediating pregnancy outcome in the young, but still growing adolescent. The pregnant sheep is the species of choice. The dam is similar in size and adiposity to the human mother and pregnancies can be limited to a single fetus, which accumulates fat prenatally and is of equivalent weight and maturity at birth. Furthermore, sheep have a relatively long gestation, similar ontogeny for all major organ systems, and when required, the fetal circulation can be catheterised to directly measure fetal nutrient uptakes and metabolism in utero.

Adolescent sheep paradigms

The basic paradigms are unique to my laboratory and involve assisted conception procedures to establish singleton pregnancies to a single sire in adolescent ewes of equivalent age, live weight and adiposity at conception. This approach controls for the main peri-conceptual factors known to influence feto–placental growth and maximises the genetic homogeneity of the resulting fetuses (Wallace *et al.*, 1996). Nutritional treatments typically commence immediately after embryo transfer and two contrasting nutritional perturbations have been developed.

The first (and to date most extensively studied) model involves overnourishing the adolescent dam to promote rapid maternal growth, which mimics pregnancy in the 50% of adolescent girls who continue to grow significantly while pregnant. In the second model, adolescent dams are prevented from growing during pregnancy by relative underfeeding (low intake, $\sim 0.70 \times$ maintenance). The control cohort for both models involves a moderate dietary intake calculated to maintain normal maternal adiposity throughout gestation and hence to meet the estimated nutrient requirements for optimum conceptus growth. In practice this requires step-wise increases in maternal intake of control dams during the final third of gestation.

Contrasting routes to adverse pregnancy outcome in young adolescents

The key characteristics of pregnancy outcome at term in the overnourished and undernourished models relative to optimally nourished controls are detailed in Figure 6.1. High dietary intakes to promote rapid and continued maternal growth throughout gestation are associated with an increased incidence of miscarriage or stillbirth in late pregnancy. The remaining surviving fetuses are on average 30–40% smaller and spontaneously delivered 3 to 4 days earlier than control fetuses, with viable lambs being born as early as day 135 of gestation (term = 145 days). Moreover the quantity and quality of colostrum produced by the dams immediately after parturition is markedly reduced and fails to meet minimum requirements for over 50% of lambs born. Consequently neonatal survival of the most perturbed lambs can be compromised, in spite of

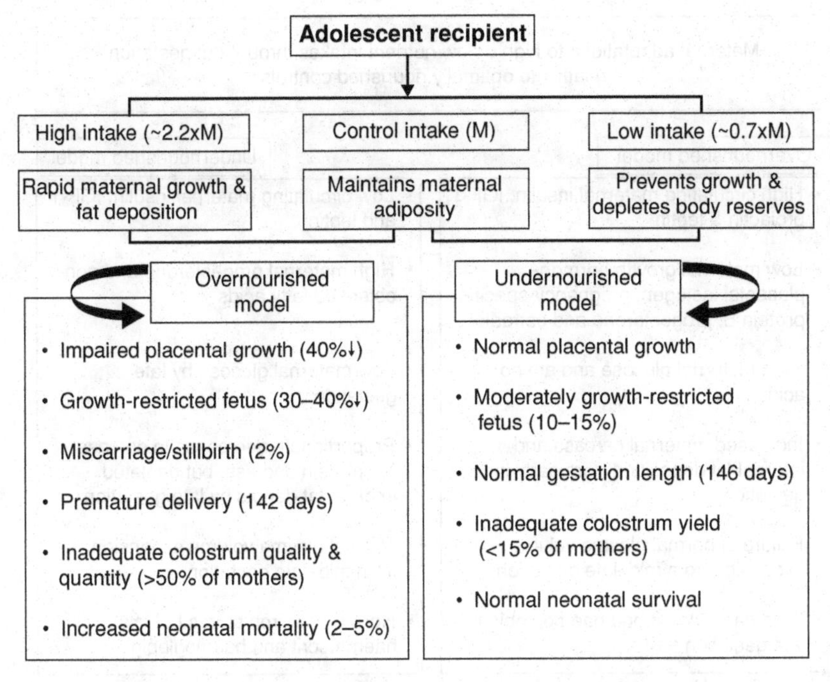

Figure 6.1. Key features of pregnancy outcome in ovine adolescent paradigms. Adolescent ewes of equivalent age, weight and adiposity were implanted with a single embryo derived from a single sire on day 4 post-oestrus. Thereafter ewes were offered either a moderate or control intake to maintain maternal adiposity throughout gestation, a high nutrient intake to promote rapid maternal growth and increasing adiposity (Overnourished model) or a low nutrient intake to prevent further growth and deplete nutrient reserves (Undernourished model). Data summarised from Wallace *et al.*, 2001, 2004a, 2006a; Aitken *et al.*, 2007.

human intervention and colostrum supplementation. As detailed below, placental growth restriction is central to all these adverse effects in the overnourished paradigm. In contrast in the undernourished adolescents, placental mass and gestation length are equivalent to control pregnancies and, to date, no incidences of miscarriage or neonatal death have been recorded. The fetuses are on average 10–15% smaller than controls and although the quantity and quality of colostrum is reduced, it meets the minimum requirement in more than 85% of cases. Thus, while dietary intakes at both ends of the nutritional spectrum negatively influence fetal growth in these ovine adolescent paradigms, it is clear that the underlying mechanisms differ. Moreover, it is the overnourished model which most closely replicates the human with respect to the key adverse pregnancy outcomes, namely an increased risk of miscarriage, preterm delivery, low birth weight and neonatal mortality.

Maternal adaptations to high or low nutrient intakes throughout gestation relative to optimally nourished controls

Overnourished model	Undernourished model
• High circulating maternal insulin, IGF-1, prolactin & leptin	• Low circulating maternal insulin, IGF-1 and leptin
• Low maternal growth hormone, placental lactogen, pregnancy-specific protein B, progesterone and estradiol	• High maternal progesterone and non-esterified fatty acids
• High maternal glucose and amino acids	• Low maternal glucose by late gestation
• Increased maternal carcass and internal fat deposition from early in gestation	• Proportionate decreases in carcass fat, protein and ash, but depleted internal fat depots by late gestation
• Failure of normal plasma volume expansion from mid–late gestation	• Normal plasma volume expansion from mid–late gestation
• High haematocrit and haemoglobin in late gestation	• Normal liver iron stores but low haematocrit and haemoglobin

Figure 6.2. Maternal adaptations to variations in gestational dietary intake. Relative to optimally nourished control adolescents, ewes were offered a high (overnourished) or low (undernourished) nutrient intake throughout gestation. Original references detailed in text.

Maternal adaptations to nutritional extremes in young adolescents

Pregnancy per se is associated with major changes in metabolism and cardiovascular function and impacts on virtually every organ in the body. Many of the physiological adaptations to pregnancy and the resulting partitioning of nutrients between the dam and her gravid uterus are under the control of endocrine hormones of maternal, placental and fetal origin. The maternal and placental endocrine responses to diverse gestational intakes and the associated changes in nutrient availability, maternal body composition and blood volume are summarised in Figure 6.2. Relative to optimally nourished control dams, high nutrient intakes are associated with elevated insulin and insulin like growth factor 1 (IGF-1) concentrations from early in gestation, providing a sustained anabolic stimulus to maternal tissue deposition. The overfed dams are also insulin resistant and circulating maternal glucose concentrations are similarly high throughout gestation (Wallace *et al.*, 1997b, 1999). Cross-sectional assessments of body composition post-mortem reveal an increase in maternal carcass fat and internal fat depots as early as day 50 of gestation with dams becoming increasingly obese as pregnancy progresses (Wallace *et al.*, 2004a). Maternal leptin concentrations reflect this increase in adiposity (Thomas *et al.*, 2001), which in turn may ultimately compromise the supply of blood to the gravid uterus. Maternal cardiac output during pregnancy normally increases by ~70% in the sheep and results in a redistribution of the percentage of cardiac output going to the various organs, particularly the gravid uterus (Rosenfeld, 1977). Clearly both cardiac output and the partitioning of blood to the maternal versus gravid uterine tissues may be influenced by the increasing adiposity of the dams. Although this aspect has not been directly tested it is noteworthy that the normal blood volume expansion of pregnancy is significantly impaired in overnourished dams when expressed relative to maternal body mass, and that maternal haematocrit and haemoglobin levels are elevated (Luther *et al.*, 2005a). In contrast relative underfeeding to maintain maternal weight at conception is associated with reduced circulating insulin, IGF-1 and leptin concentrations relative to optimally fed control animals. By late gestation circulating maternal glucose concentrations are reduced in the undernourished dams, non-esterified fatty acid concentrations are high and internal fat depots are depleted (Luther *et al.*, 2007a). Moreover, while relative blood volume is equivalent in underfed versus control dams, haematocrit and haemoglobin concentrations are reduced. Thus it appears that the low availability of nutrients in the maternal circulation is the primary cause of the modest reduction in growth during late gestation in undernourished adolescents. This is in complete contrast to the scenario in overnourished adolescents, where the placenta is the major limitation to fetal nutrient supply.

Placental adaptations to nutritional extremes in young adolescents

In spite of the ready availability of nutrients in the overnourished (high intake) dam the fetus is severely growth-restricted at term. Serial autopsies at key stages of pregnancy reveal that placental mass per se is not significantly reduced relative to control pregnancies until the final third of gestation (Figure 6.3) and by late gestation, and at term, the weight of the placenta is highly correlated with that of the fetus. By late gestation, uterine and umbilical blood flows, uteroplacental glucose and oxygen consumption and lactate production, and placental glucose transport are all reduced in absolute terms by approximately 35–40%. This is strikingly similar to the magnitude of the reduction in placental

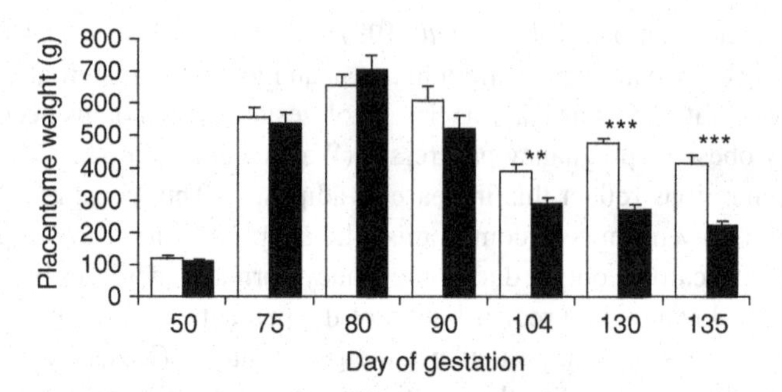

Figure 6.3. Total placentome mass in adolescent ewes receiving a control (□) or high (■) nutrient intake throughout gestation (mean ± sem). Data obtained at seven time points in five discrete studies (Wallace *et al.*, 1996, 2002, 2004b, 2006b; Redmer *et al.*, 2009). All pregnancies were derived by singleton embryo transfer to a single sire and all aspects of the experimental design were identical between studies.

and fetal weight and accordingly when expressed on a fetal or placental weight-specific basis all these indices are equivalent in the growth- restricted compared with the control pregnancies (Wallace *et al.*, 2002, 2003). Thus it is the small size of the placenta rather than altered nutrient uptake, metabolism and transport which mediates the slowing of fetal growth in the final third of gestation in rapidly growing adolescents. So what goes wrong in these pregnancies and when?

Early nutritional switch-over studies indicated that the placental and fetal growth trajectories are most sensitive to high maternal intakes during the second two-thirds of gestation. Reducing maternal intake from a high to a control level at day 50 of pregnancy stimulated placental growth and enhanced pregnancy outcome. In contrast increasing maternal intake at this time inhibited placental development and fetal growth to the same degree as in continuously overnourished pregnancies (Wallace *et al.*, 1999). More recent data suggest that it is the second third of gestation which is the crucial period of placental adaptation. When dams are overnourished for the first two-thirds of gestation and then have their intakes decreased for the final third, placental and fetal weights at autopsy in late gestation are equivalent to continuously overnourished animals (J.M. Wallace, unpublished). Serial sampling of the maternal circulation reveals that a number of placental hormones are attenuated in the putatively growth-restricted pregnancies of overnourished dams relative to controls. Reductions in progesterone, placental lactogen (oPL) and pregnancy-specific protein-B concentrations (PSPB) are detected during the first two-thirds of gestation. Additionally there is a relative delay in the onset of both oPL and PSPB secretion indicative of reduced trophoblast cell migration (Wallace *et al.*, 1997a, b; Lea *et al.*, 2007). Moreover at Day 80 of gestation (the apex of placental growth, Figure 6.2) the placentae of rapidly growing dams exhibit less proliferative activity in the fetal trophectoderm, increased protein level of bax (the pro-apoptosis gene product of the bcl-2 family) and reduced mRNA expression of several angiogenic growth factors and their receptors (Lea *et al.*, 2005; Redmer *et al.*, 2005).

These nutritionally mediated changes in indices of proliferation, apoptosis and angiogenesis occur before differences in gross placental mass become apparent but indicate these placentae are already on a different developmental trajectory. Indeed recent measurements of placental vascularity obtained post-mortem have detected reduced placental vascular development (namely, capillary area density and area per capillary) at Day 50 of gestation in the fetal component of the placenta from overnourished dams (Redmer *et al.*, 2009). Similarly at mid-gestation (Day 88) in vivo assessment of uterine blood flow using perivascular flow probes reveals a 42% reduction in uterine blood flow relative to control pregnancies and a strong positive correlation between uterine blood flow at this early stage and fetal weight some 50 days later (Wallace *et al.*, 2008). These data suggest that early measurement of uterine blood flow may help to identify at-risk pregnancies. Indeed recent reviews of the human literature suggest that uterine artery Doppler screening can predict subsequent fetal growth restriction as early as the first trimester and that the risk of poor pregnancy outcome is particularly high if abnormal Doppler indices persist into the second trimester (Papageorghiou & Leslie, 2007). It follows that therapies that target placental blood flow may be used to ameliorate fetal growth restriction in putatively compromised pregnancies.

The mechanisms programming these haemodynamic effects have not been fully elucidated but my working hypothesis is that the nutritionally induced suppression of both major sex steroids (oestrogen, progesterone) in the overnourished dams may be inhibiting uteroplacental angiogenesis, pregnancy associated plasma volume expansion and blood flow.

In the undernourished adolescent paradigm, placental proliferation, gross morphology and mass of the placenta were not significantly perturbed during mid–late gestation or following spontaneous delivery at term (Aitken *et al.*, 2007; Luther *et al.*, 2007b). Nevertheless capillary area density in the maternal component of the placentome was reduced by 20% at both mid and late gestation, suggesting that maternal nutrient restriction impacted on the vascular development of the placenta independent of changes in placental growth per se. Moreover the negative effect on capillary density cannot be reversed by subsequent re-alimentation to control intakes, emphasising that the developmental trajectory of the vasculature in the maternal caruncle is established in the first half of pregnancy, during the period of rapid cellular proliferation. Preliminary assessments of uterine blood flow in vivo in contemporaneous undernourished compared with control intake dams suggest that this reduction in capillary development is associated with a reduction in uterine blood flow of approximately 25% between mid and late gestation (average daily flow = 454 ± 49 and 340 ± 28 ml/min in control (n = 7) and underfed (n = 9) groups respectively, $P < 0.07$, J.M. Wallace, J.S. Milne and R.P. Aitken, unpublished). It remains to be established whether this modest reduction in uterine blood flow is of sufficient magnitude to play a role in mediating the previously reported reduction in maternal and hence fetal nutrient supply. The mechanism whereby reduced maternal nutrient intake compromised placental vascular development and uterine blood flow is unknown but may in part be secondary to maternal anaemia. The progressive induction of chronic maternal anaemia in adult sheep (50% decrease in haematocrit) by repetitive exchange transfusions prevents the normal gestational increase in uterine blood flow and is associated with a 40% reduction in fetal weight in late pregnancy (Mostello *et al.*, 1991). Although less severe, observations in undernourished adolescent dams indicate that low dietary intakes are associated with a 26% decrease ($P < 0.001$) in maternal haematocrit relative to optimally nourished controls at Day 130 of gestation (26 ± 0.5 versus 35 ± 0.4%, respectively, n = 16 per group, J.M. Wallace and R.P. Aitken, unpublished). Similarly in human adolescents, anaemia during mid and late pregnancy is associated with an increased risk of pre-term delivery, low birth weight, and maternal and infant mortality (Loto *et al.*, 2004; Menacker *et al.*, 2004; Conde-Agudelo *et al.*, 2005).

Fetal adaptations to nutritional extremes in young adolescents

One of the main advantages of sheep paradigms of prenatal growth restriction is the ability to access and sample the fetal and placental circulations. Accordingly using this approach in the overnourished paradigm (i.e. most perturbed fetuses) we have been able to quantify the absolute umbilical (fetal) nutrient uptakes of essential nutrients, namely glucose, oxygen and amino acids from the uterine (maternal) circulation. All three nutrient types are significantly reduced in the growth-restricted fetus but are equivalent to the normally growing control fetuses when expressed on a fetal weight-specific basis (Wallace *et al.*, 2002, 2003, 2005). Paradoxically in spite of increased fetal glucose extraction to compensate partially for reduced glucose supply, fetal glucose and insulin concentrations in the growth-restricted fetus remain low.

The adaptive mechanisms that allow a growth-restricted fetus to maintain normal fetal weight-specific nutrient uptakes in a substrate-deficient environment have recently been examined. Thus, fetal sensitivity to insulin and glucose during fetal hyperinsulinemic– euglycemic and hyperglycaemic–euinsulinemic clamps were investigated (Wallace *et al.*, 2007). Independently increasing fetal plasma insulin or glucose concentrations over a wide but physiological range increased total fetal glucose utilisation rate in both groups to a similar degree, indicating an equivalent capacity with respect to the insulin-regulated pathways of glucose uptake and metabolism, at least in the short term. Although absolute fetal glucose utilisation was attenuated throughout both studies in the growth-restricted group, when expressed per kilogram fetus, glucose utilisation was similar to the normally growing control group. Such normal bodyweight specific metabolic responses to these short-term increases in plasma insulin and glucose are indicative of maintained or up-regulated mechanisms of insulin action and/or glucose uptake/utilisation capacity in response to chronic hypoglycemia. This adaptive response allows the fetus to preserve essential metabolic functions (i.e. oxidative metabolism) at the expense of its growth, which in this paradigm progressively slows during the final third of gestation as placental size and hence total nutrient supply become limiting. While such adaptive mechanisms may have a beneficial effect for ensuring fetal survival, albeit at the expense of growth, they may also have implications beyond the fetal period. Thus, if these maintained mechanisms of insulin action persist, the prenatally growth-restricted fetuses may be predisposed to increased fat deposition when exposed to high glycemic diets during post-natal life. These short-term fetal metabolic studies, each carried out in the course of a single day clearly demonstrate that the growth-restricted fetus can respond appropriately to acute changes in both insulin and glucose supply. However, it is arguably more important to establish whether the fetus can adapt to the re-introduction of specific nutrients and hormones over the longer term by increasing protein accretion and growth. Clinically even small improvements in fetal weight, as little as 100 to 200 g, are likely to increase gestation length and ultimately fetal survival (Costeloe *et al.*, 2000; Tommiska *et al.*, 2003).

Clearly there are a number of points along the fetal nutrient supply chain which can be targeted in an attempt to improve essential nutrient availability. As the placenta is the primary limitation to fetal growth in the overnourished dams, initial studies focused on manipulating the placental growth trajectory by supplementing maternal hormones with putative roles in nutrient partitioning between the mother and her gravid uterus. Progesterone supplementation of overnourished dams during early pregnancy (day 5 to 55) increased fetal growth by 30% above that recorded for untreated overfed dams. However this increase in fetal weight was not mediated by changes in placental weight which was equivalent at term. Although these results do not preclude the possibility that progesterone may have impacted on placental vascularity, blood flow or nutrient transfer capacity, the available evidence suggests a direct effect on the differentiation of the inner cell mass is more likely (Wallace *et al.*, 2003).

In contrast, maternal growth hormone (GH) supplementation throughout the period of rapid placental proliferation (day 35 to 80) enhanced uteroplacental mass as assessed at mid-gestation (Wallace *et al.*, 2004b). However, a direct effect of GH on placenta growth has been ruled out as a subsequent study revealed that exogenous GH had a major effect on maternal metabolism, particularly when administered between mid- and late gestation (day 95 to 125). The resulting three-fold increase in maternal glucose availability stimulated a 21% increase in fetal growth relative to untreated pregnancies (Wallace *et al.*, 2006b). Although this indirect way of increasing nutrient supply to the fetus was encouraging, the therapeutic potential is limited by the fact that maternal GH treatment had a major effect on fetal body composition, namely the fetuses were more than 40% fatter (per kg fetus) than the contemporaneous controls. While this increase in lipid deposition would initially be of benefit as an energy source during the early neonatal period, it is highly likely that such a major increase in fetal adiposity would be undesirable in the longer term. Current studies are investigating the therapeutic potential of more physiological increases in fetal glucose supply to the growth-restricted fetus via direct glucose infusion into the mother.

Fetal nutrient uptakes and metabolism have not been directly studied in the undernourished adolescent paradigm. Nevertheless data obtained at necropsy in late gestation suggest that the more modestly growth-restricted fetuses from undernourished dams have a different phenotype to that of the more perturbed growth-restricted fetuses from the overnourished dams. When both cohorts of fetuses are compared with the normally growing fetuses from the optimally nourished control dams, there is evidence of brain sparing. Although growth of the brain is somewhat restricted, its growth is preserved relative to the rest of the body and particularly the abdominal organs, resulting in an increased brain: liver weight ratio (Wallace *et al.*, 2002; Luther *et al.*, 2007a). By late gestation, growth-restricted fetuses from overnourished dams have a higher relative fetal weight-specific peri-renal fat mass and carcass fat content compared with controls (Matsuzaki *et al.*, 2006). In addition they have higher plasma cholesterol and LDL concentrations, all indicative of a 'fat' phenotype. In contrast, in the more modestly growth restricted fetuses from the underfed adolescents, relative perirenal fat mass was reduced while relative carcass ash content (an index of bone mass) was increased, suggesting that skeletal growth was largely preserved while lipid stores were depleted, all indicative of a 'thin' phenotype (Luther *et al.*, 2007a). In spite of this difference in fetal body composition, fetal plasma lipid concentrations were equivalent to the normally growing control fetuses. Relative to controls both cohorts of growth-restricted fetuses were hypoglycaemic and hypoinsulinaemic but only the more perturbed fetuses from the overnourished dams have significant reductions in plasma

IGF-1 (Wallace *et al.*, 2000; Luther *et al.*, 2007a). We are currently investigating the expression of key energy balance regulatory genes in the hypothalamus of these contrasting fetal phenotypes. Intriguingly early data suggest that these genes are present from early in gestation and are sensitive to extremes in glucose supply at mid-gestation (Adam *et al.*, 2008).

Postnatal consequences of in utero adaptation

The contrasting maternal, placental and fetal adaptations to the wide variations in maternal nutrient supply detailed in the preceding sections differentially impact on birth weight. Preliminary examination of the early postnatal phenotype of lambs born to over- and undernourished mothers relative to controls suggests that these in utero adaptations also influence postnatal growth and metabolism. The offspring of overnourished dams have the lowest birth weight but the highest fractional growth rate to weaning, confirming that they have, as previously suggested, the metabolic capacity to benefit from an improved nutrient supply once released from the constraining in utero environment (Aitken *et al.*, 2007). Relative to controls the low birth weight offspring of overnourished dams have higher total plasma and LDL cholesterol at birth and during lactation, but not beyond. In contrast evidence of altered glucose handling (higher fasting glucose and glucose area under the curve following a glucose tolerance test) was detected at 6 months of age and beyond.

Whether this alteration in metabolism programs an obese phenotype is unclear but in a separate cohort low birth weight offspring had equivalent adiposity but at reduced stature in adulthood (Milne *et al.*, 2007). The intermediate birth weight offspring of undernourished dams have identical fractional growth rates to the normally growing controls and are similar in all metabolic aspects studied to date.

Concluding remarks

Here and elsewhere in this book the risks associated with becoming pregnant during adolescent life are obvious. Policy initiatives aimed at reducing the incidence of pregnancy in the young adolescent population have largely failed in the UK and accordingly these girls require appropriate prenatal care to optimise pregnancy outcome and prevent the perpetuation of health inequalities across the generations. Similarly, in many developing countries early marriage soon after menarche is commonplace and hence pregnancy during adolescent life is normal (e.g. 143 pregnancies per 1000 girls in sub-Saharan Africa (Treffers,

2003) and less stigmatised than in the developed world. Nevertheless, for the majority of these girls, low gestational intakes are likely to be a major issue and the risks associated with pregnancy during adolescence will be exacerbated by a poor standard of prenatal care.

To date, our studies in adolescent sheep have largely focused on manipulating calorie intake and hence growth immediately after conception. Using this approach it is clear that when gynaecological age, maternal weight and adiposity at conception are controlled, gestational dietary intake at both ends of the nutritional spectrum is a powerful determinant of fetal growth in young adolescents. However, adolescent pregnancies are often unplanned and girls become pregnant from diverse nutritional backgrounds, which may interact with subsequent gestational intake to influence pregnancy outcome. We recently examined this aspect in sheep with either a good or poor body mass index (BMI) at conception and then exposed to high, control or low gestational intakes as previously. Initial BMI at conception did influence fetal nutrient supply and birth weight but high gestational intakes were the predominant negative effect.

In addition the incidence of fetal growth restriction and early delivery was highest in adolescents with poor initial BMI and high gestational weight gains (Aitken *et al.*, 2008). Formulating appropriate dietary advice for young adolescent girls is likely to be complex and one size is unlikely to fit all. Nevertheless biomarkers of growth and nutritional status at the time of conception and the use of ultrasound to diagnose early deficiencies in uteroplacental growth or umbilical blood flow may prove beneficial in identifying human adolescents at risk of poor pregnancy outcome. Similarly, and irrespective of maternal age, pregnancies complicated by severe intrauterine growth restriction may benefit from therapies aimed at improving fetal nutrient availability by increasing uterine blood flow and/or transplacental nutrient supply.

References

Adam, C.L., Findlay, P.A., Chanet, A. *et al.* (2008). Expression of energy balance regulatory genes in the developing ovine fetal hypothalamus at mid-gestation and influence of hyperglycemia. *American Journal of Physiology, Regulatory Integrative and Comparative Physiology*, 294(6), R1895– 900.

Aitken, R.P., Milne, J.S. & Wallace, J.M. (2007). Wide variations in gestational dietary intake differentially impact on pregnancy outcome in young adolescents and influence early postnatal offspring phenotype. *Pediatric Research*, 62, 385(Abstract).

Aitken, R.P., Milne, J.S. & Wallace, J.M. (2008). Effect of weight and adiposity at conception, and gestational dietary intake on pregnancy outcome in young adolescent sheep. *Proceedings Physiological Society*, 11, PC36.

Barker, D.J.P. (1998). *Mothers, Babies and Health in Later Life*, 2nd Edn. Edinburgh: Churchill Livingstone.

Barker, D.J.P. (2006). Adult consequences of fetal growth restriction. *Clinical Obstetrics and Gynecology*, 49, 270–83.

Castro, L.C. & Avina, R.L. (2002). Maternal obesity and pregnancy outcomes. *Current Opinions in Obstetrics and Gynecology*, 14, 601–6.

Cnattingius, S., Bergstrom, R., Lipworth, L. & Kramer, M.S. (1998). Prepregnancy weight and the risk of adverse pregnancy outcomes. *New England Journal of Medicine*, 338, 147–52.

Conde-Agudelo, A., Belizan, J.M. & Lammers, C. (2005). Maternal–perinatal morbidity and mortality associated with adolescent pregnancy in Latin America: Cross-sectional study. *American Journal of Obstetrics and Gynecology*, 192, 342–9.

Costeloe, K., Hennessy, E.M., Gibson, A.T., Marlow, N. & Wilkinson, A.R. (2000). The EPICure Study: Outcomes to discharge from hospital for infants born at the threshold of viability. *Pediatrics*, 106, 659–79.

Hack, M. & Merkatz, I.R. (1995). Preterm delivery and low birth weight – a dire legacy. *New England Journal of Medicine*, 333, 1772–4.

Hediger, M.L., Scholl, T.O., Belsky, D.H., Ances, I.G. & Salmon, R.W. (1989). Patterns of weight gain in adolescent pregnancy: effects on birth weight and preterm delivery. *Obstetrics and Gynecology*, 74, 6–12.

King, J.C. (2003). The risk of maternal nutritional depletion and poor outcomes increases in early or closely spaced pregnancies. *Journal of Nutrition*, 133, 1723S–36.

Lea, R.G., Hannah, L.T., Redmer, D.A. *et al.* (2005). Developmental indices of nutritionally-induced placental growth restriction in the adolescent sheep. *Paediatric Research*, 57, 599–604.

Lea, R.G., Wooding, P., Stewart, I. *et al.* (2007). The expression of ovine placental lactogen, StAR and progesterone-associated steroidogenic enzymes in placentae of over-nourished growing adolescent ewes. *Reproduction*, 133, 785–96.

Loto, O.M., Ezechi, O.C., Kalu, B.K. *et al.* (2004). Poor obstetric performance of teenagers: is it age- or quality-of-care-related? *Journal of Obstetrics and Gynecology*, 24, 395–8.

Luther, J.S., Aitken, R.P., Milne, J.S. *et al.* (2005a). Maternal and fetal micronutrient status in rapidly growing pregnant adolescent sheep. *Paediatric Research*, 58, 1113.

Luther, J.S., Redmer, D.A., Reynolds, L.P. & Wallace, J.M. (2005b). Nutritional paradigms of ovine fetal growth restriction: Implications for human pregnancy. *Human Fertility*, 8, 179–87.

Luther, J.S., Aitken, R.P., Milne, J.S. *et al.* (2007a). Maternal and fetal growth, body composition, endocrinology and metabolic status in undernourished adolescent sheep. *Biology of Reproduction*, 77, 343–50.

Luther, J.S., Aitken, R.P., Milne, J.S. *et al.* (2007b). Placental growth, angiogenic gene expression and vascular development in undernourished adolescent sheep. *Biology of Reproduction*, 77, 351–7.

Matsuzaki, M., Milne, J.S., Aitken, R.P., & Wallace, J.M. (2006). Overnourishing pregnant adolescent ewes preserves perirenal fat deposition in their growth-restricted fetuses. *Reproduction, Fertility and Development*, 18, 357–64.

Menacker, F., Martin, J.A., MacDorman, M.F. & Ventura, S.J. (2004). Births to 10–14 year-old mothers, 1990–2002: trends and health outcomes. *National Vital Statistics Report*, 53, 1–18.

Milne, J.S., Aitken, R.P., Green, L. & Wallace, J.M. (2007). Nutritionally mediated prenatal growth restriction and postnatal hypothalamic–pituitary–adrenal function in female sheep. *Pediatric Research*, 62, 389 (Abstract).

Mostello, D., Chalk, C. & Khoury, J. (1991). Chronic anemia in pregnant ewes: maternal and fetal effects. *American Journal of Physiology*, 261, R1075–83.

Narrowing the Gap. The Final Report of the Fabian Commission on Life Chances and Child Poverty (2006). Ed. Tom Hampson, London: Fabian Society.

Papageorghiou, A.T. & Leslie, K. (2007). Uterine artery Doppler in the prediction of adverse pregnancy outcome. *Current Opinion in Obstetrics and Gynecology*, 19, 103–9.

Redmer, D.A., Aitken, R.P., Milne, J.S., Reynolds, F. & Wallace, J.M. (2005). Influence of maternal nutrition on mRNA expression of placental angiogenic factors and their receptors at mid-gestation in adolescent sheep. *Biology of Reproduction*, 72, 1004–9.

Redmer, D.A., Luther, J.S., Milne, J.S. *et al.* (2009) Fetoplacental growth and vascular development in overnourished adolescent sheep at day 50, 90 and 130 of gestation. *Reproduction*, 137, 749–57.

Rosenfeld, C.R. (1977). Distribution of cardiac output in ovine pregnancy. *American Journal of Physiology*, 232, H231–5.

Scholl, T.O., Hediger, M.L., Khoo, C.S., A. & Rawson, N.L. (1991). Maternal weight gain, diet and infant birth weight: correlations during adolescent pregnancy. *Journal of Clinical Epidemiology*, 44, 423–8.

Scholl, T.O., Hediger, M.L. & Schall, J.I. (1997). Maternal growth and fetal growth: pregnancy course and outcome in the Camden study. *Annals New York Academy of Science*, 81, 292–301.

Stevens-Simon, C., McAnarney, E.R. & Roghmann, K.J. (1993). Adolescent gestational weight gain and birth weight. *Pediatrics*, 92, 805–9.

Thomas, L., Wallace, J.M., Aitken, R.P. *et al.* (2001). Circulating leptin concentrations during ovine pregnancy in relation to maternal body composition and pregnancy outcome. *Journal of Endocrinology*, 169, 465–76.

Tommiska, V., Heinonen, K., Kero, P. *et al.* (2003). A national two-year follow-up study of extremely low birthweight infants born in 1996–1997. *Archives of Disease in Childhood*, 88, SI29–34.

Treffers, P.E. (2003). Teenage pregnancy, a worldwide problem. *Ned Tijdschr Geneeskd*, 22, 2320–5.

Wallace, J.M., Aitken, R.P. & Cheyne, M.A. (1996). Nutrient partitioning and fetal growth in the rapidly growing adolescent ewe. *Journal of Reproduction and Fertility*, 107, 183–90.

Wallace, J.M., Aitken, R.P., Cheyne, M.A. & Humblot, P. (1997a). Pregnancy-specific protein B and progesterone concentrations in relation to nutritional regime, placental mass and pregnancy outcome in growing adolescent ewes carrying singleton fetuses. *Journal of Reproduction and Fertility*, 109, 53–8.

Wallace, J.M., Aitken, R.P., Milne, J.S. & Hay, W.W. Jr. (2004a). Nutritionally mediated placental growth restriction in the growing adolescent: consequences for the fetus. *Biology of Reproduction*, 71, 1055–62.

Wallace, J.M., Bourke, D.A., Aitken, R.P. & Cruickshank, M.A. (1999). Switching maternal dietary intake at the end of the first trimester has profound effects on placental development and fetal growth in adolescent ewes carrying singleton fetuses. *Biology of Reproduction*, 61, 101–10.

Wallace, J.M., Bourke, D.A., Aitken, R.P. *et al.* (2000). Relationship between nutritionally mediated placental growth restriction and fetal growth, body composition and endocrine status during late gestation in adolescent sheep. *Placenta*, 21, 100–8.

Wallace, J.M., Bourke, D.A., Aitken, R.P., Leitch, N. & Hay, W.W. Jr. (2002). Blood flows and nutrient uptakes in growth- restricted pregnancies induced by overnourishing adolescent sheep. *American Journal of Physiology – Regulatory, Integrative and Comparative Physiology*, 282, R1027–36.

Wallace, J.M., Bourke, D.A., Aitken, R.P., Milne, J.S. & Hay, W.W. Jr. (2003). Placental glucose transport in growth-restricted pregnancies induced by overnourishing adolescent sheep. *Journal of Physiology*, 547, 85–94.

Wallace, J.M., Bourke, D.A., Da Silva, P. & Aitken, R.P. (2001). Nutrient partitioning during adolescent pregnancy. *Reproduction*, 122, 347–57.

Wallace, J.M., Bourke, D.A., Da Silva, P. & Aitken, R.P. (2003). Influence of progesterone supplementation during the first third of pregnancy on fetal and placental growth in overnourished adolescent ewes. *Reproduction*, 126, 481–7.

Wallace, J.M., Da Silva, P., Aitken, R.P. & Cruickshank, M.A. (1997b). Maternal endocrine status in relation to pregnancy outcome in rapidly growing adolescent sheep. *Journal of Endocrinology*, 155, 359–68.

Wallace, J.M., Luther, J.S., Milne, J.S. *et al.* (2006a). Nutritional modulation of adolescent pregnancy outcomes. *Placenta*, 27, S61–8.

Wallace, J.M., Matsuzaki, M., Milne, J.S. & Wallace, J.M. (2006b). Late but not early gestational maternal growth hormone treatment increases fetal adiposity in overnourished adolescent sheep. *Biology of Reproduction*, 75, 231–39.

Wallace, J.M., Milne, J.S. & Aitken, R.P. (2004b). Maternal growth hormone treatment from Day 35 to 80 of gestation alters nutrient partitioning in favour of uteroplacental growth in overnourished adolescent sheep. *Biology of Reproduction*, 70, 1277–85.

Wallace, J.M., Milne, J.S., Aitken, R.P. & Hay, W.W. Jr. (2007). Sensitivity to metabolic signals in late gestation growth restricted fetuses from rapidly growing adolescent sheep. *American Journal of Physiology – Endocrinology and Metabolism*, 293, E1233–41.

Wallace, J.M., Milne, J.S., Matsuzaki, M. & Aitken, R.P. (2008). Attenuated uterine blood flow is an early defect in growth-restricted pregnancies induced by overnourishing adolescent dams. *Placenta*, 29, 718–724.

Wallace, J.M., Regnault, T.R.H., Limesand, S.W., Hay, W.W. Jr. & Anthony, R.V. (2005). Investigating the causes of low birth weight in contrasting ovine paradigms. *Journal of Physiology*, 565, 19–26.

7 *Growth and sexual maturation in human and non-human primates: a brief review*

PHYLLIS C. LEE

Introduction

"... adult size can be reached through many pathways involving timing differences and rate differences."

(Leigh, 2001, p. 236)

If there are indeed many pathways to the adult endpoint during growth, what becomes of interest is the nature of those different pathways and, more significantly, the consequences of differences in pathways for early survival, for the timing of the onset of reproductive activity, and for subsequent lifespan. This chapter explores some of the strategies for growth observed in primates, and outlines potential consequences of these strategies. I do this in the context of an abundance of detailed literature on growth rates, growth constraints and models of rates of growth for humans (see for example Bogin, 1999) and non-human primates (e.g. Leigh, 1996). Models of growth can provide explanations for primate life history variation (Leigh, 1995; Dirks & Bowman, 2007) and, in addition, potentially elucidate the variation that exists at the level of the individual and population between rapid and late maturation (e.g. Wilson *et al.*, 1983; Bercovitch & Berard, 1993; Setchell *et al.*, 2001; Altmann & Alberts, 2005).

Physical maturation is obviously a consequence of growth, but the timing of physiological or hormonal reproductive onset, rather than simply attained size, determines much of the subsequent reproductive success of the individual. Early reproduction, combined with high survival, provides significant advantages in terms of lifetime reproductive success. And the selective advantage of early reproduction has led cercopithecoids into an unusual life history paradigm among the non-human primates where females typically reproduce well before they reach adult size (Wilner & Martin, 1985; Martin *et al.*, 1994; Setchell *et al.*, 2001; Dirks & Bowman, 2007).

Reproductive maturation among non-human primates has recently received attention in terms of hormonal growth factors acting as controlling mechanisms

Reproduction and Adaptation, eds. C. G. Nicholas Mascie-Taylor and Lyliane Rosetta. Published by Cambridge University Press.

along with other influences on hypothalamic pulse generators (Ulijaszek, 2002; Bernstein *et al.*, 2008). Puberty is initiated by a diurnal increase in pulsatile gonadotrophin-releasing hormone (GnRH) release, and the activities of GnRH neurons are controlled by neuronal and glial networks under the influence of specific genetically determined transcription factors (Heger *et al.*, 2008). Other studies also find major genetic determinants of puberty, at least for its delay in female macaques (Ojeda *et al.*, 2006). While selective breeding studies on ungulates have long demonstrated an underlying genetic basis for aspects of age at maturation (e.g. Patterson *et al.*, 2002; Rosendo *et al.*, 2007), the many associations between growth rates, attained size, body composition, insulin-like growth-factors (IGFs) and other growth hormones, and the onset of reproductive function remain to be explored (Wilson, 1997, 1998), and in addition sex differences are poorly understood.

Growth to reproductive maturation among the primates is of evolutionary interest since one key transition in human evolution is argued to be growth prolongation with regard to reproductive onset (Leigh & Park, 1998; Bogin, 1999; Kelley, 2002). Here, I briefly explore trends in reproductive maturation among non-human primates, meaning the physical traits associated with the onset of hormonal function, specifically in relation to patterns and processes of early growth. My focus is generally on the anthropoid primates, by contrast with the strepsirrhines, who can partition gestational investment into several infants per reproductive event and have pre- and post-natal growth strategies that are distinct from those of higher primates (Kappeler & Heymann, 1996; Lee & Kappeler, 2003). I also attempt to place these "primate-specific" strategies into a more general mammalian context where data exist, using phylogenetically specific relationships.

In addition to evolved strategies within phylogenies, the long-term consequences of individual experiences of growth variation on survival and reproductive success are becoming better understood. A variety of studies – from humans (e.g. Barker, 2001; Painter *et al.*, 2008) to rats (Fowden *et al.*, 2006) – have demonstrated reproductive, morbidity and mortality consequences for adults, not all of which are negative (e.g. Helle *et al.*, 2005), as a result of growth perturbations in utero. Thus a perspective on growth from conception to reproductive onset should provide insights into trans-generational phenotypic plasticity, individual adaptive plastic responses, and facultative maternal investment responses to local environmental variability.

Phases of growth

Growth represents a series of response functions from the initial formation of the zygote to a terminal size of the adult which can be described by a

Table 7.1 *Definitions of growth phases in relation to different developmental periods (e.g. Periera & Leigh, 2002; Setchell & Lee, 2004).*

Growth phase	Energy source	Costs
Gestational	Growth sustained through maternal investment via placenta	Relatively inexpensive energetically for primates; potentially expensive for mammals with precocial offspring; potential for genomic conflict in rates of fetal growth
Infancy	Rapid growth sustained primarily through maternal care allocation in lactation, protection, thermoregulation and transport	Relatively expensive at peak lactation; during the period of peak growth, rates of investment are relative to maternal size and, for primates, as a result of carrying costs
Juvenile	Slower growth, sustained through independent feeding	Growth costs reduced, costs of independent locomotion increase, risks of care-independent mortality potentially high
Adolescent	Costly growth spurt, slowing as reproductive maturation and adult size are approached, sustained through independent feeding	High growth costs, combined with costs of competition and social integration. Risks of mortality caused by growth, requiring energy that is unable to be sustained in an uncertain environment. Risks of poor longevity and/or reproductive outcome if stunted

variety of mathematical equations (Roche & Sun, 2003). Whatever the (hotly debated) shape of the appropriate growth function at each stage, there are a number of distinct and discontinuous phases to primate growth (Table 7.1; Bogin, 1999; Lee, 1999; Leigh *et al.*, 2003). The initial division between prenatal and postnatal growth represents growth that is fuelled by different means and has distinct mechanisms to buffer against environmental insults, and the consequences of rapid and slow growth in each phase may vary over the long and short term.

Throughout the investment period, primate mothers partition their allocation of energy to infants differently between prenatal and postnatal phases of growth. In the prenatal phase, there is a potential trade-off between gestation length and fetal growth rates and, for primates, an interaction with fetal brain growth. These interactions were first described as size-dependent *Maternal Energy* strategies (see Martin, 1996) and illustrate correlations of neonate mass, gestation length, maternal mass and maternal basal metabolic rate (BMR); maternal energy turnover needs to be able to sustain the specific and higher costs of rapid anthropoid brain growth over those for somatic tissue growth (Martin, 1996;

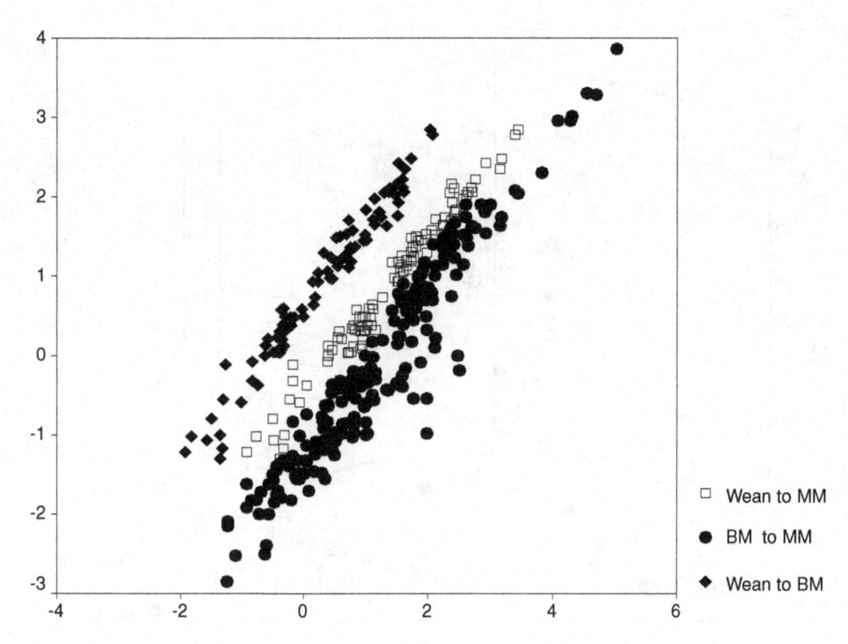

Figure 7.1. Absolute mass relations for 243 mammalian species (94 primates, 29 ungulates, 28 pinnipeds) comparing infant mass at birth (BM) to maternal Mass (MM); Infant mass at weaning (Wean) to MM; and Wean to BM, showing the consistencies in the allometric relationships.

Leigh, 2004). As a result of these additional costs, the exponent of the allometric relationships between maternal mass and birth mass is higher than that for other traits (Lee, 1996; Martin, 1996). Across a variety of primate and non-primate mammalian species (ungulates and pinnipeds), the overall allometric relationships appear to be relatively constant (Figure 7.1), while residuals from phylogenetically corrected regressions can be used to illustrate those species with specifically faster or slower growth than expected for their body mass within taxonomic groups.

As discussed many years ago (Martin & MacLarnon, 1985), one mechanism by which these differential residuals come about is the distinction between altricial and precocial species among mammals. The existence of these two major strategies of investment suggests that there are at least two specific modes of energy allocation during gestation, which are to some extent unrelated to postnatal growth. Relatively altricial species, born immature after a short gestation for body mass, do not all grow rapidly (catch-up to some taxa norm) in the postnatal period, while some precocial species which are born highly developed after a relatively long gestation for body mass can actually grow relatively rapidly. Among the primates, the results of an exploration of these

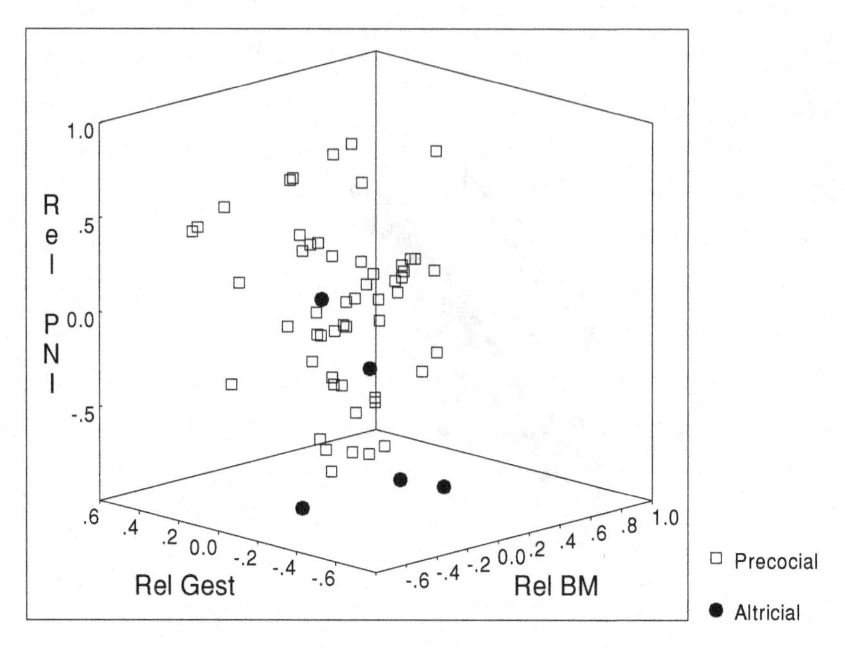

Figure 7.2. Relations between relative birth mass (Rel BM), relative duration of gestation (Rel Gest) and the relative duration of postnatal investment (Rel PNI) for primates. Precocial and altricial species are clearly distinct in gestation, but overlap in both birth mass and duration of postnatal investment.

patterns confirm this variation; species with longer durations of investment in gestation are not necessarily those with the largest, most rapidly developing offspring nor do they represent the smallest, least developed offspring in a dichotomous picture (Figure 7.2). In addition, relative gestational times are unassociated with either postnatal growth or the duration of that growth. At this stage, there are too few data for most mammals on postnatal growth rates to weaning to be able to demonstrate any systematic distinctions between altricial and precocial infants.

An alternative or potentially interacting factor acting to produce a prenatal–postnatal growth discontinuity is that of a "*Risky Environment*" strategies (Ross, 1998; Lee, 1999). While originally applied to the post-weaning juvenile growth phase by Janson and van Schaik (1993), there are marked differences in the associations between mass-controlled residuals of early growth when comparing primates with high, as opposed to low, levels of environmental predictability (e.g. Ross, 1988). Those species born into unpredictable and highly variable environments tend to be relatively larger than expected for a relatively short gestation, while those born into predictable environments have a relatively long

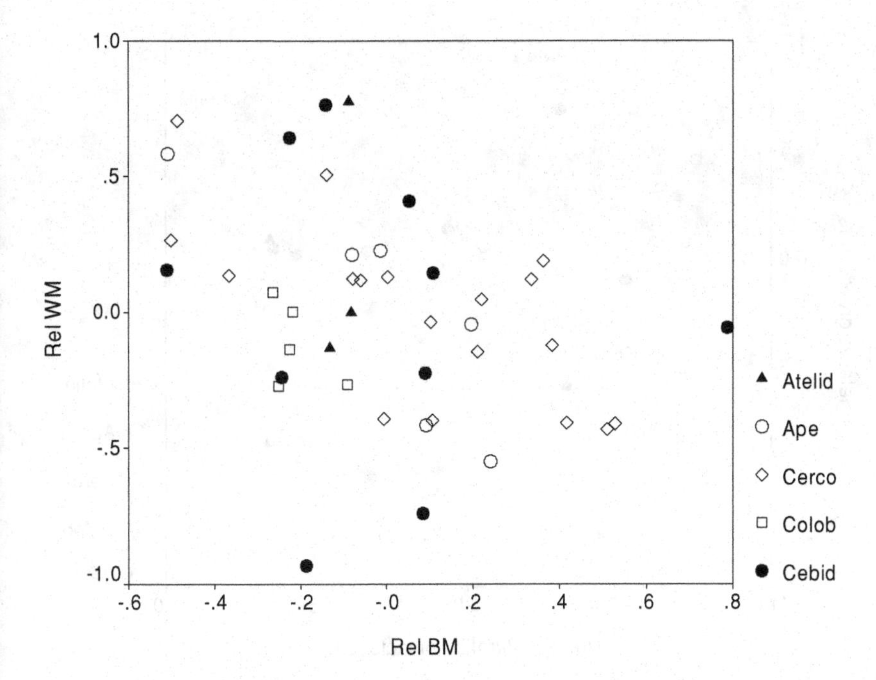

Figure 7.3. Relative wean mass (Rel WM) plotted against relative birth mass (Rel BM) for major groups of anthropoid primates using phylogenetically specific regressions.

gestation but relatively low birth mass (e.g. Lee, 1999). Predictability of energy availability to mothers, which is an indicator of environmental risk, appears to underlie at least some of the size-corrected prenatal maternal investment strategies.

By contrast, after birth the relationships between growth to weaning, maternal mass and duration of lactation as proxies for investment become more consistent for primates (Figure 7.3), with an additional, individual capacity for facultative adjustment to local environments. For the major anthropoid groups, postnatal growth can be divided between "quick" (energy expensive/time cheap) and "slow" energy cheap/time expensive strategies within and between species. Both strategies focus on attaining a metabolic weaning mass (Lee *et al.*, 1991), but the trade-off can be sensitive to the local environment, producing marked variation that is both inter- and intra-specific.

Previous analyses found a relationship between predation risk and relative birth and weaning mass (Lee, 1999). Species experiencing a generally low risk of predation tend to give birth to relatively smaller neonates, but wean infants at a relatively larger mass. Those species experiencing a high risk of predation

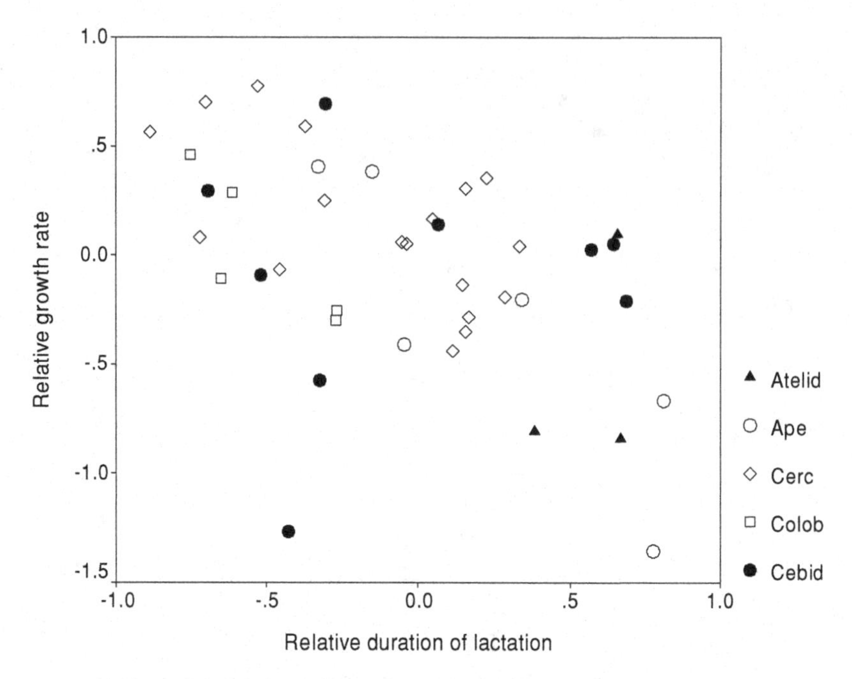

Figure 7.4. Relative duration of lactation plotted against relative growth rate (both derived from phylogenetically specific residuals) over the period of lactation for anthropoid primates.

give birth to infants with a relatively larger mass at birth, possibly in order to minimise the risks of predation early in life; but these same groups wean at a relatively smaller mass, which minimises the investment in an infant with a higher risk of care-independent mortality.

Overall, postnatal growth rates to weaning show a generally negative association with the duration of lactation – longer lactations are associated with slower rates of growth (Figure 7.4). There are marked differences within populations in maternal strategies to attain a weaning mass through growth and in how mothers deliver energy to their infants. This variation is illustrated by plotting the mass attained at weaning for maximum and minimum duration of lactation for a number of well-studied species with relatively constant weaning mass (Figure 7.5; and see Lee, 1999).

As noted by several authors, the growth trajectory during the juvenile phase prior to any pubertal growth spurt takes a different shape from that of growth in infancy (Janson & van Schaik, 1993; Bogin, 1999; Leigh, 2004). Juvenile growth occurs while the individual is foraging for itself, and thus the evolved, inter-specific rates of growth appear to have been strongly selected in the

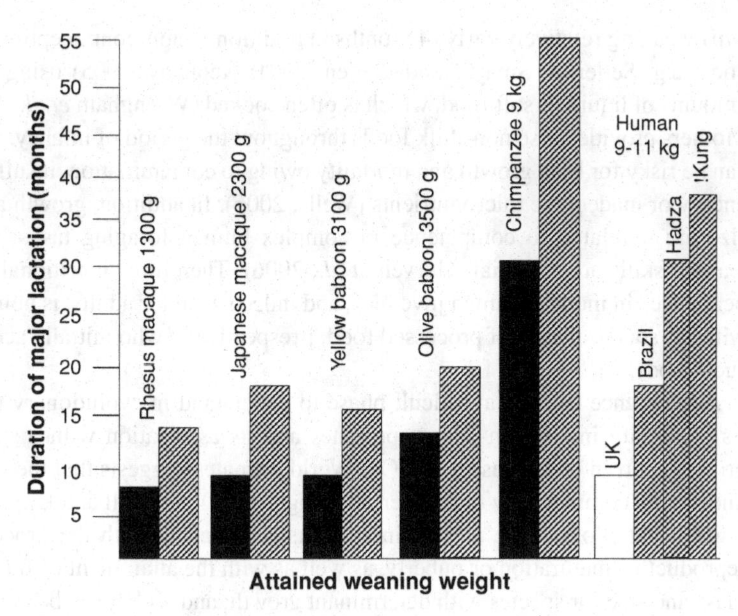

Figure 7.5. Minimum and maximum duration of major lactation (prior to resumption of menses) for several anthropoid primates in relation to attained weaning weight.

context of risky environments – either in terms of energy availability or in terms of predation risk.

Bogin's (1999) designation of an intermediate growth phase, that of *childhood* between the juvenile and adolescent growth periods and sustained by post-weaning provisioning, as unique to humans remains to be explored in detail by comparison to the anthropoids. It is worth noting however that for many higher primates it is not whether mothers or juveniles themselves energetically support growth (i.e. Bogin's justification of the special nature of childhood), but rather mortality in the absence of maternal support that may be more interesting as evidence of a common intervening life phase. For example, weaned yearling vervets have reduced survival after the birth of their sibling, despite being able to forage efficiently (Lee, 1983), while juvenile orangutans maintain a period of maternal dependence while both foraging for themselves and suckling for up to eight years in order to learn modes and areas of food exploitation (van Noordwijk & van Schaik, 2004). The linkage between mortality and evolution of life history traits (e.g. Janson & van Schaik, 1993) suggests that considering how survival varies as a result of being embedded in a social context for the acquisition of feeding skills and avoiding predation might be a better predictor of the timing and duration of distinctive life phases than is a simple perspective on who provides the calories for growth. It has long been known that humans *start* weaning relatively early (4 months in traditional, non-contracepting societies: e.g. Sellen & Smay, 2000; Sellen, 2001; Kennedy, 2005) using small amounts of liquid or soft food, which is often cooked (Wrangham *et al.*, 1999). Mothers provide these non-milk foods throughout the period of infancy, which can be risky for both growth and mortality owing to contamination, insufficient energy or inadequate micronutrients (Wells, 2006). In addition, growth and/or size are unrelated to competence in complex human foraging tasks; rather learned skills are essential (Gurven *et al.*, 2006). Therefore "nutritional independence" in modern human juveniles, and indeed even for adults, is bound up with the social sharing of processed food, irrespective of who initially acquires such foods.

Adolescence itself is a difficult phase to understand in evolutionary terms. Its generality in the non-human primates and its association with a specific growth spurt in body mass in the Old World primates suggests that there is an underlying commonality to this period (Leigh, 1996). Setchell and Lee (2004) argued that adolescence was defined by its relationship with the process of reproductive maturation or puberty, as well as with the attainment of full adult mass and size for species with determinant growth, and with those behavioural characteristics facilitating reproductive capacity and reproductive competition. It is thus a phase when behavioural, social and physical competences are achieved, and it may be associated with an increase in the risks of mortality.

The costs of a growth spurt and of behavioural maturation, of pre-reproductive dispersal, and of the onset of reproductive competition are marked for most primate males and indeed many other mammalian males. By contrast, females have a reduced growth velocity, a shorter duration of rapid growth, and while their mortality risks in adolescence are poorly known, it is likely that mortality does not increase markedly until first reproduction. Sex differences in mass growth are illustrated by data from Setchell *et al.* (2001) for one of the most dimorphic primates, the mandrill (Figure 7.6).

The adolescent growth spurt in primates has been extensively investigated by Leigh and colleagues (Leigh, 1996; 2001; Leigh & Park, 1998). They have developed a series of hypotheses to explain variation in the timing and extent of growth spurts associated with both environmental predictability and richness (for example: low-energy leaf versus high-energy fruit diets), with age-specific mortality hazards, and with the extent of sexual selection for large male size. These evolutionary models explain differences between species, but do not easily explain why growth spurts exist and why they are so specific to anthropoid primates. Indeed, Gurven and Walker (2006) suggest that the growth spurt observed in humans is distinctively human, and associated with the costs of provisioning juveniles. Were juveniles to prolong the rates of growth needed to attain adult size, they would be too expensive for their foraging mothers

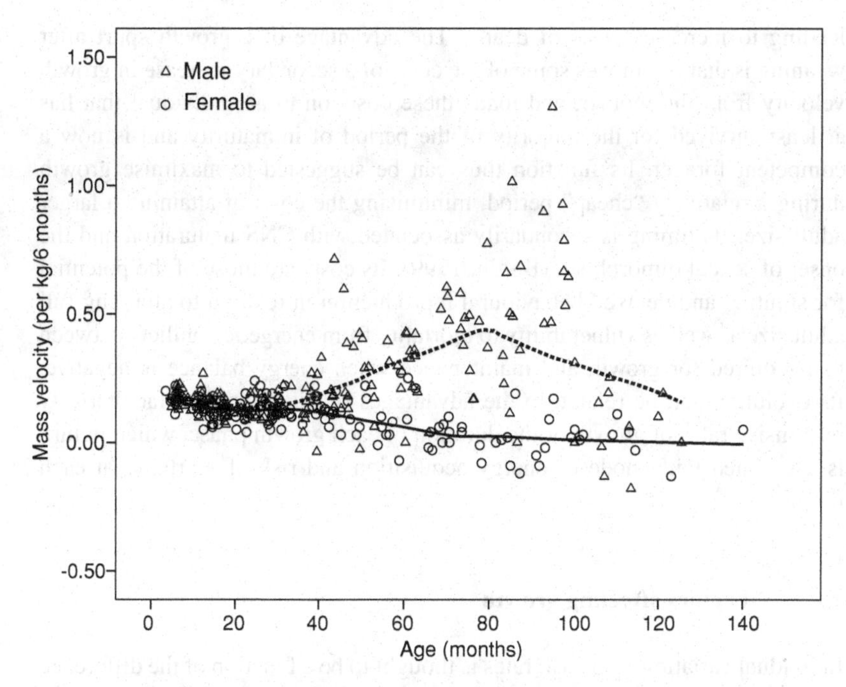

Figure 7.6. Mass velocity for male and female mandrills showing peak velocity for females as they approach sexual maturity at 36 months, and that for males, which is shifted to 72–96 months (after Setchell *et al.*, 2001).

to sustain since they are unable to forage sufficiently to contribute to their own energy balance when rapid growth is required. These authors also suggest that the anthropoid growth spurt is restricted to mass growth, while humans exhibit spurts in both mass and stature. In part, this observation may simply be the result of a human preoccupation with our bipedal posture and thus with long-bone and back growth. In primates as in other quadrupeds, crown–rump (back-length) growth rates may be constrained in their rate of growth by the mechanics of weight-bearing, and relatively few data on limb growth in primates as yet exist (but see Setchell *et al.*, 2001). Some highly dimorphic mammals achieve their considerable dimorphism in both mass and height by faster growth, a longer growth period, or both, in one sex by comparison to the other (e.g. red deer, Clutton-Brock *et al.*, 1982; elephants, Lee & Moss, 1995), while other species have a clear growth spurt but only in the dimorphic sex (e.g. pinnipeds: McLaren, 1993). Why then do primates, including humans, show this accelerated mass-growth velocity prior to and during reproductive maturity, when any food limitations could impact on both the capacity to attain the large adult size necessary for reproduction and on current energy balance leading to increased risks of death? The advantage of a growth spurt after weaning is that it removes some of the costs of a secondary increase in growth velocity from the mother, and loads these costs on to an individual that has at least survived for the majority of the period of immaturity and is now a competent forager. Its function thus can be suggested to maximise growth during a relatively "cheap" period, minimising the costs of attaining a larger adult size. Its timing is secondarily associated with CNS maturation and the onset of sexual dimorphism (Bogin, 1999). Its costs are those of the potential for stunting and delayed behavioural reproduction in relation to attaining full adult size as well as vulnerability to mortality from energetic conflicts between that required for growth and maintenance when energy balance is negative. Its evolution can be related to the advantages of differential or facultatively responsive rates of growth or the duration of each growth phase, which in turn is associated with mode of energy acquisition and risk of mortality at each phase.

Factors affecting growth

Individual variation in growth rates is thought to be a function of the difference between energy acquisition and energy expenditure, although there may be limits to the potential for rate variation owing to the needs of tissue-specific development and metabolism (see Mangel & Munch, 2005). The key factors influencing pre-weaning growth rates are the maternal ability to allocate care based on her physical condition, her energy balance, health and/or disease status and her dominance rank or status. High levels of seasonality can cause weight loss and gain in a number of wild species (e.g. chimpanzees, Pusey *et al.*, 2003; Japanese macaques, Kurita *et al.*, 2002) and dominant individuals are consistently heavier than subordinates when dominance confers a feeding advantage (Altmann *et al.*, 1993; Pusey *et al.*, 2003). High maternal dominance is further translated into a feeding and survival enhancement for offspring (Altmann, 1991; Setchell *et al.*, 2002; Charpentier *et al.*, 2008). Recently, Wilson & Kinkead (2008) found a gene–environment interaction such that sexual maturation was delayed in those subordinate macaque females with an allele associated with sensitivity to psycho-social stressors. Puberty delay was also associated with reduced growth hormone (GH) and leptin secretion during the juvenile phase of these females. Interestingly, the authors found no differences in the growth velocity from birth to puberty between early and late maturers, suggesting some sensitivity to hormone production prior to the point at which growth velocity "takes off" (e.g. Bogin, 1999). The complexity of the gene–social context interaction is further noted by Charpentier *et al.* (2008), who found that hybrid baboons and those of higher rank with many allies had earlier menarche.

The second issue affecting early growth is that of offspring "need" for investment; these needs can be a function of relative size for age, absolute growth rates, extraction efficiency and nutrient availability as weanlings start sampling foods. A third factor is that of enhanced cost to mothers of sustaining the growth of sons over that of daughters, and their need to meet these additional costs in order to minimise infant male mortality. As noted above, were primate mothers to support the higher rates and prolonged growth necessary to achieve sexual dimorphism through lactation alone, it might be simply too costly in the absence of a growth spurt (see Gurven & Walker, 2006). Mothers are engaged in a constant trade-off between current and future reproductive events; they thus need to balance the costs of a son in the current event against their own and their offspring's potential reproductive advantages in the long term. This trade-off is illustrated among captive olive baboons by the mothers of heavy infants where the infants attain weaning mass through rapid early growth, and thus mothers incur a shorter delay until the resumption of cycling (Garcia *et al.*, 2006), and by rhesus macaques where mothers do not invest as much in early infant growth when they have the potential to re-conceive relatively early in their infant's life (Bowman & Lee, 1995). Both of these strategies are alternatives in situations of high infant survival – invest in early growth and the early subsequent resumption of reproduction, or invest in maintenance of maternal condition with low infant growth and again early return to reproduction. In ecological and social contexts where there could be mortality consequences to infants owing to low growth rates or small attained size, or to mothers through maternal depletion, such strategies can be suggested to be part of a continuum of maternal responses to local energy and mortality conditions at each reproductive event (Lee, 1984; Lee *et al.*, 1991; Hauser & Fairbanks, 1988; Fairbanks & McGuire, 1996).

Despite the lower degree of dimorphism in infant primates, by comparison to many sexually dimorphic mammals, sons are still slightly heavier and grow faster in almost all species with adult dimorphism (Garcia *et al.*, 2009). So, some head-start on the attainment of larger male adult mass or size is initiated in infancy. Males thus cost slightly more in terms of allocation of milk for growth, and in addition the costs of carrying them, which is a function of mass, will be greater (e.g. Altmann & Samuels, 1992; Watson *et al.*, 2008). In the highly sexually dimorphic mandrills, high-ranking mothers have infants who are heavier for their age than do low-ranking mothers, and especially compared with the low-ranking mothers of sons (Setchell *et al.*, 2001).

A final influence on rates and processes of growth is that of mortality risk. In the model of Janson and van Schaik (1993), risk of mortality predicts variance in growth rates among juveniles of different primate species. High early mortality risk associates with rapid growth, while low risk and low growth rates go together. As discussed above, risk of mortality appears to affect growth rates differently in gestation, in the infant phase, and in the juvenile period. In addition, the suggested inter-specific relationships between mortality and growth rates might also hold at a proximate level within species, but vary between mothers and over each reproductive event, as outlined above. However, despite all the uncertainties and lack of data available to test such predictions at this time, it seems likely that care-independent mortality could be potentially as influential on rates of growth during each separate period of immaturity as is the quality and quantity of care allocated by mothers.

Reproductive maturation and growth

If we start with the classical presumption that the adult life phase is one that prioritises reproduction over somatic investment, then the attainment of adulthood should be associated with a transition in costs and benefit trade-offs. Survival remains a consideration throughout each growth phase, but hazards reduce considerably as the adult phase is approached. Therefore mortality risks need to be incorporated into these trade-offs. The timing of reproductive maturation might be a constant – a life history invariant reached at some specific proportion of body mass, or some specific proportion of the lifespan (Stearns, 1992; Charnov, 1993). In either case, we would expect some relationship between growth and reproductive maturation within species as theoretically modelled by Kozlowski and Weiner (1997) and empirically suggested by a number of studies (Nieuwenhuijsen *et al.*, 1987; Bercovitch & Berard, 1993). However, whether these observed relationships are causal or simply correlated are as yet empirically untested.

Physiological reproductive maturation, as discussed above, is related to the onset of pulsatile release of reproductive hormones, which are thought to be strongly under functional genetic controls (Ojeda *et al.*, 2006). Physical maturation, however, is associated with energy availability and expenditure, affecting mass growth, statural growth and body composition (fat to lean mass ratio; e.g. Frisch, 1987), all of which are in some way related to the onset of puberty (see Bogin, 1999). Historically, age at puberty has been used as an indicator of the nutritional status of a population, especially in developing countries (e.g. Chowdhury *et al.*, 2000) or in comparisons of populations among primates (e.g. Ross, 1998), and the "secular trend" towards a declining age at puberty among girls was assumed to be related to improvements in nutrition, health and growth over the past century (e.g. Wyshak & Frisch, 1982). The evidence for such a trend is now questionable (Juul *et al.*, 2006; Euling *et al.*, 2008), although there is a strong association between rapid early growth, high fat mass and/or high BMI and early female menarche across a variety of human populations (e.g. Cooper *et al.*, 1996; dos Santos Silva *et al.*, 2002). Adaptive phenotypic plasticity in hormone production as adults (e.g. Núñez-de la Mora *et al.*, 2007), and epigenetic signals via inter-uterine hormonal programming (e.g. Fowden *et al.*, 2006) may underlie some of these population differences in reproductive onset. The causal factors again remain to be teased apart in an increasingly complex interaction between fetal environment, genetics, and individual experiences and behaviour.

Among anthropoid primates, as discussed above, the growth spurt frees the mother from the costs of sustaining rapid and prolonged growth. During the juvenile period, individuals have low luteinising hormone (LH) and low T (Setchell, 2003; Gesquiere *et al.*, 2005), but there is some increase in IGF secretion (Bernstein *et al.*, 2008). This period is associated with maturation of the hormonal controls on puberty via the central nervous system. As individuals age, or alternatively increase in mass and stature during the growth spurt, and attain adolescence, we see an LH rise, and peak IGF secretion (Ulijaszek, 2002; Bernstein *et al.*, 2008). Of course, the sexes differ in their development in this stage such that for males, the testes develop with the capacity for sperm production (high T), while for females, the onset of ovulatory cycles is initiated (Setchell, 2003; Gesquiere *et al.*, 2005). In addition, adult dentition erupts.

In women, a first birth is only rarely associated with menarche or the actual onset of fecundity after a peri-menarchial period of subfertility, in contrast to some non-human primates. In both humans and non-humans, however, younger primipares suffer reduced infant survival, prolonged interbirth intervals and low infant growth rates (humans: Kramer, 2008; baboons: Bercovitch & Strum, 1993; Johnson, 2006; rhesus: Wilson *et al.*, 1988; Gomendio, 1989; marmosets: Smucny *et al.*, 2004). Thus a discontinuity exists between menarche/puberty, fertility and the start of reproduction in female anthropoids, which can be exacerbated by contexts of social competition and a lack of social support.

This period of onset – from puberty to first successful reproduction – is considerably prolonged in most anthropoid primates, lasting from one to up to five years (Setchell & Lee, 2004). It can be even longer for some males, where there is a mismatch between the timing of physiological as opposed to social maturation (e.g. orangutans: Utami-Atmoko & van Hooff, 2004). The adult phase is marked by regular (cyclical) production of reproductive hormones, full physiological reproductive competence and, for females, adult stature is typically attained at some point during this period.

Questions that are unexplored here are associations between social status, early growth and reproductive onset. As noted above, among some anthropoids, high rank tends to be associated with higher fecundity (reviewed in Garcia *et al.*, 2006) and enhanced offspring growth (Setchell *et al.*, 2001), as well as earlier maturation (Bercovitch & Berard, 1993; Bercovitch & Strum, 1993). Low status may or may not be a stressor for primates, depending on the stability and persistence of hierarchies. Thus, rank or social effects on female maturation probably depend on the nature and stability of female–female competitive interactions. For humans, psycho-social stressors have known effects on adult reproductive competence but their effects on offspring growth and reproductive onset appear to be mediated by attachment or parenting quality, with failure-to-thrive infants associated with depressed mothers (e.g. Skuse *et al.*, 1995) and early puberty in girls associated with harsh maternal control (Belsky *et al.*, 2007).

A final intriguing question is that of the consequences of the timing of reproductive maturation: is there an association between early reproductive onset and early reproductive senescence or a reduced lifespan, or the opposite with high-quality individuals out-performing others throughout their reproductive careers (e.g. Hamel *et al.*, 2009).

Conclusions

Growth in each of the different and, as I have argued, discontinuous phases of early development – gestation, infancy, juvenile and adolescent periods – has specific implications for both individual and maternal energetics, and for future reproduction and survival. Thus sexual maturation represents the endpoint of a series of complex processes, as well as the starting point for a variety of new problems and processes. Achieving sexual maturation reflects a trade-off between somatic growth and reproductive tissue functioning for females; for males behavioural competence may be more of a constraint, at least on the timing of their reproductive onset. What controls the timing of physical and physiological maturation depends on the energy available for growth, on the nature of intra-sexual competition and on age-specific survival probabilities.

A vast number of interesting questions remain to be addressed with regards to reproductive maturation among the primates more generally. We have better information from humans than from most other primate species, and even here controversies over cause and consequence persist. Thus the universality or specificity of these patterns are, as yet, poorly understood. Does rapid early growth reduce age at first reproduction and hence increase lifetime reproductive success? By contrast does an early age at reproduction increase vulnerability to mortality? And finally, is rapid early growth and early onset of reproduction associated with specific body composition as opposed to body mass or size?

The endpoint of sexual maturation has specific consequences for reproductive success, for behavioural competence and integration, and for understanding elements of primate life history evolution. It is a subject that remains relatively poorly studied and understood but of vital theoretical and practical importance.

Acknowledgements

For advice, collaboration, collegiality and inspiration in thinking about growth and infancy – thanks to Pat Bateson, Louise Barrett, Robin Dunbar, Cecile Garcia, Robert Hinde, John Lycett, Bob Martin, Nick Mascie-Taylor, Cynthia Moss, Lyliane Rosetta, Jo Setchell and Kelly Stewart.

References

Altmann, S. A. (1991). Diets of yearling female primates (*Papio cynocephalus*) predict lifetime fitness. *Proceedings of the National Academy of Sciences of the United States of America*, 88, 420–3.

Altmann, J. & Alberts, S. C. (2005). Growth rates in a wild primate population: ecological influences and maternal effects. *Behavioral Ecology and Sociobiology*, 57, 490–501.

Altmann, J. & Samuels, A. (1992). Costs of maternal care: infant carrying in baboons. *Behavioral Ecology and Sociobiology*, 29, 391–8.

Altmann, J., Schoeller, D., Altmann, S. A., Murithi, P. & Sapolsky, R. M. (1993). Body size and fatness of free-living baboons reflect food availability and activity levels. *American Journal of Primatology*, 30, 149–61.

Barker, D. J. (2001). The malnourished baby and infant. *British Medical Bulletin*, 60, 69–88.

Belsky, J., Steinberg, L. D., Houts, R. M. *et al.* (2007). Family-rearing antecedents of pubertal timing. *Child Development*, 78, 1302–21.

Bercovitch, F. B. & Berard, J. D. (1993). Life history costs and consequences of rapid reproductive maturation in female rhesus macaques. *Behavioral Ecology and Sociobiology*, 32, 103–9.

Bercovitch, F. B. & Strum, S. C. (1993). Dominance rank, resource availability, and reproductive maturation in female savanna baboons. *Behavioral Ecology and Sociobiology*, 33, 313–18.

Bernstein, R. M., Leigh, S. R., Donovan, S. M. & Monaco, M. H. (2008). Hormonal correlates of ontogeny in baboons (*Papio hamadryas anubis*) and mangabeys (*Cercocebus atys*). *American Journal of Physical Anthropology*, 136, 156–68.

Bogin, B. (1999). *Patterns of Human Growth* (2nd ed.). Cambridge: Cambridge University Press.

Bowman, J. E. & Lee, P. C. (1995). Growth and threshold-weaning weights among captive rhesus macaques. *American Journal of Physical Anthropology*, 96, 159–75.

Charnov, E. L. (1993). *Life History Invariants*. Oxford: Oxford University Press.

Charpentier, M. J. E., Tung, J., Altmann, J. & Alberts, S. C. (2008). Age at maturity in wild baboons: genetic, environmental and demographic influences. *Molecular Ecology*, 17, 2026–40.

Chowdhury, S., Shahabuddin, A. K., Seal, A. J. *et al.* (2000). Nutritional status and age at menarche in a rural area of Bangladesh. *Annals of Human Biology*, 27, 249–56.

Clutton-Brock, T. H., Albon, S. & Guinness, R. (1982). *The Red Deer: The Ecology of Two Sexes*. Princeton: Princeton University Press.

Cooper, G., Kuh, D., Egger, P., Wadsworth, M. & Barker, D. J. P. (1996). Childhood growth and age at menarche. *British Journal of Obstetrics and Gynecology*, 103, 814–17.

Dirks, W. & Bowman, J. E. (2007). Life history theory and dental development in four species of catarrhine primates. *Journal of Human Evolution*, 55, 309–20.

dos Santos Silva, I., De Stavola, B. L., Mann, V. *et al.* (2002). Prenatal factors, childhood growth trajectories and age at menarche. *International Journal of Epidemiology*, 31, 405–12.

Euling, S. Y., Herman-Giddens, M. E., Lee, P. A. *et al.* (2008). Examination of US puberty-timing data from 1940 to 1994 for secular trends: panel findings. *Pediatrics*, 121 (Suppl. 3), S172–91.

Fairbanks, L. A. & McGuire, M. T. (1996). Maternal condition and the quality of maternal care in vervet monkeys. *Behaviour*, 132, 733–54.

Fowden, A. L., Giussani, D. A. & Forhead, A. J. (2006). Intrauterine programming of physiological systems: causes and consequences. *Physiology*, 21, 29–37.

Frisch, R. E. (1987). Body fat, menarche, fitness and fertility. *Human Reproduction*, 2, 521–33.

Garcia, C., Lee, P. C. & Rosetta, L. (2006). Dominance and reproductive rates in captive female olive baboons, *Papio anubis*. *American Journal of Physical Anthropology*, 131, 64–72.

Garcia, C., Lee, P. C. & Rosetta, L. (2009). Growth in colony-living anubis baboon infants and its relationship with maternal activity budgets and reproductive status. *American Journal of Physical Anthropology*, 138, 123–35.

Gesquiere, L. R., Altmann, J., Khan, M. Z. *et al.* (2005). Coming of age: steroid hormones of wild immature baboons (*Papio cynocephalus*). *American Journal of Primatology*, 67, 83–100.

Gomendio, M. (1989). Differences in fertility and suckling patterns between primiparous and multiparous rhesus mothers (*Macaca mulatta*). *Journal of Reproduction and Fertility*, 87, 529–42.

Gurven, M. & Walker, R. (2006). Energetic demand of multiple dependents and the evolution of slow human growth. *Proceedings of the Royal Society B*, 273, 835–41.

Gurven, M., Kaplan, H. & Gutierrez, M. (2006). How long does it take to become a proficient hunter? Implications for the evolution of extended development and long life span. *Journal of Human Evolution*, 51, 454–70.

Hauser, M. D. & Fairbanks, L. A. (1988). Mother–offspring conflict in vervet monkeys: variation in response to ecological conditions. *Animal Behaviour*, 36, 802–13.

Hamel, S., Côté, S. D., Gaillard, J. M. & Festa-Bianchet, M. (2009). Individual variation in reproductive costs of reproduction: high-quality females always do better. *Journal of Animal Ecology*, 78, 143–51.

Heger, S., Körner, A., Meigen, C. *et al.* (2008). Impact of weight status on the onset and parameters of puberty: analysis of three representative cohorts from central Europe. *Journal of Pediatric Endocrinology & Metabolism*, 21, 865–77.

Helle, S., Lummaa, V. & Jokela, J. (2005). Are reproductive and somatic senescence coupled in humans? Late, but not early, reproduction correlated with longevity in historical Sami women. *Proceedings of the Royal Society B*, 272, 29–37.

Janson, C. H. & van Schaik, C. P. (1993). Ecological risk aversion in juvenile primates: slow and steady wins the race. In *Juvenile Primates: Life History, Development, and Behavior*, ed. M. E. Pereira and L.A. Fairbanks, pp. 57–74. Oxford: Oxford University Press.

Johnson, S. E. (2006). Maternal characteristics and offspring growth in chacma baboons: A life-history perspective. In *Reproduction and Fitness in Baboons: Behavioral, Ecological, and Life History Perspectives*, ed. L. Swedell and S. R. Leigh, pp. 177–97. New York, NY: Springer.

Juul, A., Teilmann, G., Scheike, T., *et al.* (2006). Pubertal development in Danish children: comparison of recent European and US data. *International Journal of Andrology*, 29, 247–55.

Kappeler, P. M. & Heymann, E. (1996). Nonconvergence in the evolution of primate life history and socio-ecology. *Biological Journal of the Linnean Society*, 59, 297–326.

Kelley, J. (2002). Life-history evolution in Miocene and extant apes. In *Human Evolution Through Developmental Change*, ed. N. Minugh-Purvis and K. J. McNamara, pp. 223–48. Baltimore, MD: Johns Hopkins University Press.

Kennedy, G. E. (2005). From the ape's dilemma to the weanling's dilemma: early weaning and its evolutionary context. *Journal of Human Evolution*, 48, 123–45.

Kozlowski, J. & Weiner, J. (1997). Interspecific allometries are by-products of body-size optimization. *American Naturalist*, 149, 352–80.

Kramer, K.L. (2008). Early sexual maturity among Pumé foragers of Venezuela: fitness implications of teen motherhood. *American Journal of Physical Anthropology*, 136, 338–50.

Kurita, H., Shimomura, T. & Fujita, T. (2002). Temporal variation in Japanese macaque bodily mass. *International Journal of Primatology*, 23, 411–28.

Lee, P. C. (1983). Effects of parturition on the mother's relationship with older offspring. In *Primate Social Relationships: An Integrated Approach*, ed. R. A. Hinde, pp. 134–9. Oxford: Blackwell Scientific.

Lee, P. C. (1984). Ecological constraints on the social development of vervet monkeys. *Behaviour*, 91, 245–62.

Lee, P. C. (1996). The meaning of weaning: growth, lactation, and life history. *Evolutionary Anthropology*, 5, 87–96.

Lee, P. C. (1999). Comparative ecology of postnatal growth and weaning among haplorhine primates. In *Comparative Primate Socioecology*, ed. P. C. Lee, pp. 111–39. Cambridge: Cambridge University Press.
Lee, P. C. & Kappeler, P. M. (2003). Socio-ecological correlates of phenotypic plasticity in primate life history. In: *Primate Life Histories*, ed. P. M. Kappeler and M. Pereira, pp. 41–65. Chicago, IL: University of Chicago Press.
Lee, P. C., Majluf, P. & Gordon, I. J. (1991). Growth, weaning and maternal investment from a comparative perspective. *Journal of Zoology*, 225, 99–114.
Lee, P. C. & Moss, C. J. (1995). Statural growth in known-age African elephants (*Loxodonta africana*). *Journal of Zoology*, 236, 29–41.
Leigh, S. R. (1995). Socioecology and the ontogeny of sexual size dimorphism in anthropoid primates. *American Journal of Physical Anthropology*, 97, 339–56.
Leigh, S. R. (1996). Evolution of human growth spurts. *American Journal of Physical Anthropology*, 101, 455–74.
Leigh, S. R. (2001). Evolution of human growth. *Evolutionary Anthropology* 10, 223–36.
Leigh, S. R. (2004). Brain growth, life history, and cognition in primate and human evolution. *American Journal of Primatology*, 62, 139–64.
Leigh, S. R. & Park, P. B. (1998). Evolution of human growth prolongation. *American Journal of Physical Anthropology*, 107, 331–50.
Leigh, S. R., Shahb, N. F. & Buchanan, L. S. (2003). Ontogeny and phylogeny in papionin primates. *Journal of Human Evolution*, 45, 285–316.
Mangel, M. & Munch, S. B. (2005). A life-history perspective on short- and long-term consequences of compensatory growth. *American Naturalist*, 166, E155–76.
Martin, R. D. (1996). Scaling of the mammalian brain: the maternal energy hypothesis. *News in Physiological Sciences*, 11, 149–56.
Martin, R. D. & MacLarnon, A. M. (1985). Gestation period, neonatal size, and maternal investment in placental mammals. *Nature*, 313, 220–3.
Martin, R., Willner, L. & Dettling, A. (1994). The evolution of sexual size dimorphism in primates. In *The Differences Between the Sexes*, ed. R. Short and E. Balabar, pp. 159–200. Cambridge: Cambridge University Press.
McLaren, I. A. (1993). Growth in pinnipeds. *Biological Reviews of the Cambridge Philosophical Society*, 68, 1–79.
Nieuwenhuijsen, K., Bonke-Jansen, M., de Neef, K. J., Van Der Werff ten Bosch, J. J. & Slob, A. K. (1987). Physiological aspects of puberty in group-living stumptail monkeys (*Macaca arctoides*). *Physiology & Behavior*, 41, 37–45.
Núñez-de la Mora, A., Chatterton, R. T., Choudhury, O. A., Napolitano, D. A. & Bentley, G. R. (2007). Childhood conditions influence adult progesterone levels. *PLoS Medicine*, 4, 813–21.
Ojeda, S. R., Lomniczi, A., Mastroniardi, C. *et al.* (2006). Minireview: the neuroendocrine regulation of puberty: is the time ripe for a systems biology approach? *Endocrinology*, 147, 1166–74.
Painter, R. C., Osmond, C., Gluckman, P. *et al.* (2008). Transgenerational effects of prenatal exposure to the Dutch famine on neonatal adiposity and health in later life. *British Journal of Obstetrics and Gynaecology*, 115, 1243–9.

Patterson, J. L., Ball, R. O., Willis, H. J., Aherne, F. X. & Foxcroft, G. R. (2002). The effect of lean growth rate on puberty attainment in gilts. *Journal of Animal Science*, 80, 1299–310.

Periera, M. E. & Leigh, S. R. (2002). Modes of primate development. In *Primate Life Histories*, ed. P.M. Kappeler and M. Pereira, pp. 149–79. Chicago, IL: University of Chicago Press.

Pusey, A. E., Oehlert, G. W., Williams, J. M. & Goodall, J. (2003). Influence of ecological and social factors on body mass of wild chimpanzees. *International Journal of Primatology*, 26, 3–31.

Roche, A. & Sun, S. (2003). *Human Growth: Assessment and Interpretation*. Cambridge: Cambridge University Press.

Rosendo, A., Druet, T., Gogué, J., Canario, L. & Bidanel, J. P. (2007). Correlated responses for litter traits to six generations of selection for ovulation rate or prenatal survival in French Large White pigs. *Journal of Animal Science*, 85, 1615–24.

Ross, C. (1988). The intrinsic rate of natural increase and reproductive effort in primates. *Journal of Zoology*, 214, 199–219.

Ross, C. (1998). Primate life histories. *Evolutionary Anthropology*, 6, 54–63.

Sellen, D. W. (2001). Comparison of infant feeding patterns reported for non-industrial populations with current recommendations. *Journal of Nutrition*, 131, 2707–15.

Sellen, D. & Smay, D. (2000). Relationship between subsistence and age at weaning in "preindustrial" societies. *Human Nature*, 12, 47–87.

Setchell, J. M. (2003). Behavioural development in male mandrills (*Mandrillus sphinx*): puberty to adulthood. *Behaviour*, 140, 1053–89.

Setchell, J. & Lee, P. C. (2004). Development and sexual selection in primates. In *Sexual Selection in Primates: Causes, Mechanisms and Consequences*, ed. P. M. Kappeler and C. P. van Schaik, pp. 175–95. Cambridge: Cambridge University Press.

Setchell, J., Lee, P. C., Wickings, E. & Dixson, A. (2001). Growth and ontogeny of sexual size dimorphism in the mandrill (*Mandrillus sphinx*). *American Journal of Physical Anthropology*, 115, 349–60.

Setchell, J., Lee, P. C., Wickings, E. & Dixson, A. (2002). Reproductive parameters in female mandrills (*Mandrillus sphinx*). *International Journal of Primatology*, 23, 51–68.

Skuse, D., Wolke, D., Reilly, S. & Chan, I. (1995). Failure to thrive in human infants: the significance of maternal well-being and behaviour. In *Motherhood in Human and Nonhuman Primates*, ed. C. R. Pryce, R. D. Marin and D. Skuse, pp. 162–70, Basel: Karger.

Smucny, D. A., Abbott, D. H., Mansfield, K. G. *et al.* (2004). Reproductive output, maternal age, and survivorship in captive common marmoset females (*Callithrix jacchus*). *American Journal of Primatology*, 64, 107–21.

Stearns, S. C. (1992). *The Evolution of Life Histories*. Oxford: Oxford University Press.

Ulijaszek, S. J. (2002). Serum insulin-like growth factor-I, insulin-like growth factor binding protein-3, and the pubertal growth spurt in the female rhesus monkey. *American Journal of Physical Anthropology*, 118, 77–85.

Utami-Atmoko, S. S. & van Hooff, J. A. R. A. M. (2004). Alternative male reproductive strategies: male bimaturism in orangutans. In *Sexual Selection in Primates: New and Comparative Perspectives*, ed. P. M. Kappeler and C. P. van Schaik, pp. 196–207. Cambridge: Cambridge University Press.

van Noordwijk, M. A. & van Schaik, C. P. (2004). Development of ecological competence in Sumatran orangutans. *American Journal of Physical Anthropology*, 127, 79–94.

Watson, J. C., Payne, R. C., Chamberlain, A. T., Jones, R. K. & Sellers, W. I. J. (2008). The energetic costs of load-carrying and the evolution of bipedalism. *Human Evolution*, 54, 675–83.

Wells, J. C. K. (2006). The role of cultural factors in human breastfeeding: adaptive behaviour or biopower? *Human Ecology*, Special Issue No. 14, 39–47.

Wilner, L. A. & Martin, R. D. (1985). Some basic principles of mammalian sexual dimorphism. In *Human Sexual Dimorphism*, ed. J. Ghesquiere, R. D. Martin and F. Newcombe, pp. 1–42. London: Taylor and Francis.

Wilson, M. E. (1997). Administration of IGF-I affects the GH axis and adolescent growth in normal monkeys. *Journal of Endocrinology*, 153, 327–35.

Wilson, M. E. (1998). Premature elevation in serum insulin-like growth factor-I advances first ovulation in rhesus monkeys. *Journal of Endocrinology*, 158, 247–57.

Wilson, M. E. & Kinkead, B. (2008). Gene–environment interactions, not neonatal growth hormone deficiency, time puberty in female rhesus monkeys. *Biology of Reproduction*, 78, 736–43.

Wilson, M. E., Walker, M. L. & Gordon, T. P. (1983). Consequences of first pregnancy in rhesus monkeys. *American Journal of Physical Anthropology*, 61, 103–10.

Wilson, M. E., Walker, M. L.,Pope, N. S. & Gordon, T. P. (1988). Prolonged lactational infertility in adolescent rhesus monkeys. *Biology of Reproduction*, 38, 163–74.

Wrangham, R., Jones, J., Laden, G., Pilbeam, D. & Conklin-Brittain, N. (1999). The raw and the stolen: cooking and the ecology of human origins. *Current Anthropology*, 40, 567–94.

Wyshak, G. & Frisch, R. E. (1982). Evidence of a secular trend in age of menarche. *New England Journal of Medicine*, 306, 1033–5.

8 *The evolution of post-reproductive life: adaptationist scenarios*

LYNNETTE LEIDY SIEVERT

Introduction

In adaptationist scenarios, traits are selected for if they are positively correlated with reproductive success (Fisher, 1930; Price, 1970). It seems counterintuitive, then, that natural selection would select for a dampening, or complete cessation, of reproductive ability before the end of the somatic lifespan. If the production of offspring is a means by which favorable traits can be passed to future generations, then the continued production of offspring – even the costliest of offspring – would seem to be beneficial. Why was there a phylogenetic shift in female reproductive strategy away from continued egg production in fish, amphibians, and reptiles, to a finite number of eggs that slowly dwindles across the lifespan to the point, in some mammalian species, of follicular exhaustion and low levels of ovarian hormones prior to death? In other words, why is there a menopause?

This chapter reviews the evidence for two categories of adaptationist scenarios. In the first, menopause and post-reproductive life are the direct products of natural selection. In the second, menopause and post-reproductive life are the indirect by-products of natural selection for other traits. A similar differentiation between menopause as either adaptation or epiphenomenon has already been made (Peccei, 2001), but the argument presented here takes a slightly different tack by emphasizing the byproduct scenario as also adaptationist. In the byproduct scenarios presented here, mammalian patterns of early oogenesis (egg production) and lifelong atresia (ovarian follicle loss) are the reproductive strategies that underwent positive selection. Negative selection acted to conserve the patterns of oogenesis and atresia among all mammals by selecting against mutations that would bring about change. When coupled with a lengthened lifespan, the byproducts of these conserved strategies are menopause and post-reproductive life. This chapter does not address why lifespans lengthened in species such as humans, killer whales, and short-finned pilot whales (Foote,

Reproduction and Adaptation, eds. C. G. Nicholas Mascie-Taylor and Lyliane Rosetta.
Published by Cambridge University Press.

2008). Instead, this chapter asks why the ability to produce offspring was not extended to the end of a longer life.

Female menopause and post-reproductive life

Menopause is clinically and epidemiologically defined as the cessation of menses caused by the loss of ovarian follicular activity (WHO, 1996). It is a universal event experienced by all women who live beyond 60 years of age (Sievert, 2006). After menopause, women can spend more than a third of their life in post-reproductive years. Jeanne Louise Calment, who died in 1997 at the age 122, may have spent 72 years in post-reproductive life! It is the potentially great length of post-reproductive life, rather than the event of menopause per se, that is unique to human females.

Not all mammals menstruate, therefore the definition of menopause must be broadened for cross-species comparisons. In a survey of post-reproductive life among 42 mammals, Cohen (2004) defined menopause as "the irreversible loss of the physiological capacity to produce offspring due to intrinsic biological factors" (p. 734). The definition of post-reproductive life can also be broadened to mean the length of life that exceeds one species-specific inter-birth interval plus two standard deviations of inter-birth interval (Cohen, 2004; Pavelka & Fedigan, 1999).

Varying lengths of post-reproductive life have been observed among an assortment of mammals. Western lowland gorillas have a maximum longevity of 52 years but few give birth past the age of 37 (Atsalis & Margulis, 2006). A study carried out among gorillas in American zoos of fecal progestogen concentrations, daily estrus status, and weekly activity patterns demonstrated that, among 22 females aged 32 or older, five (23%) were not cycling (Margulis *et al.*, 2007). Among chimpanzees in the Mahale mountains of Tanzania, 25 middle aged (18–33 years) females were observed until death. Six were post-reproductive for 5 years or longer, with a mean age at death of 40.2 years (Nishida *et al.*, 2003). Among free-ranging Japanese macaques, all females ceased reproduction after age 25 but lived only an average of 2.1 years past their final birth. Only 3% lived to the age of 26 (Pavelka & Fedigan, 1999). East African elephants, with a maximum lifespan of 60 years, are often cited as an example of animals with post-reproductive life; however, age-specific fecundity (the probability of giving birth to a female calf) is still above 15% for females aged 50–54 (Croze *et al.*, 1981). Short-finned pilot whales and killer whales are much better examples of species demonstrating post-reproductive life (Foote, 2008). Short-finned pilot whales demonstrate an average post-reproductive lifespan of 14 years (Marsh & Kasuya, 1986). Female killer whales have a maximum longevity of at least 80 and perhaps 90 years, but post-reproductive life is estimated to begin at a mean age of 40.1 years (Olesiuk *et al.*, 1990). See Cohen (2004) for more data on post-reproductive lifespan in mammal species.

Menopause is a mammalian trait

In all vertebrates, primordial germ (stem) cells give rise to *oogonia* (undeveloped eggs) through mitosis (Baker, 1986). A single layer of granulosa cells surrounds each oogonium, and meiosis begins in the nucleus of what is now called an *oocyte* (Guraya, 1998). During meiosis, the full complement of genetic material present in the oocyte is halved, but this process does not happen all at once. Meiosis begins in the oocyte and then stops (meiotic arrest) until ovulation occurs (Crisp, 1992). An oocyte encircled by flattened granulosa cells forms a *primordial follicle*. Primordial follicles do not increase in size, but they live the longest – potentially for 50 years or longer depending on the species. According to Rothchild (2003), the common organization of the follicle in all vertebrates "must mean that it was a specific trait of the common ancestor of all vertebrates – and probably of all chordates" (p. 338).

In most mammals, the primordial follicle develops into a primary and then secondary (pre-antral) follicle. A layer of theca cells surrounds the granulosa cells. Development continues as the follicle becomes an antral (vesicular or Graafian) follicle through the proliferation of granulosa cells and a vesicle that fills with follicular fluid (Figure 8.1). The largest antral follicles are called pre-ovulatory follicles.[1] Antral follicles produce the hormones necessary for communication within the hypothalamic–pituitary–ovarian axis (Rajkovic *et al.*, 2006). In mammals, massive numbers of oogonia and follicles in all stages of development are lost through atresia, "the process by which an ovarian follicle loses its integrity and the egg is lost by means other than ovulation" (Crisp, 1992). Guraya (1985) notes that all follicles appear destined to become atretic, but a few are rescued at a critical point in development through "a complicated web of interactions between the oocyte, granulosa cells, thecal cells, pituitary and hypothalamus" (Rajkovic *et al.*, 2006). Only a few oocytes will ovulate (about 400 in humans) and only a few follicles will become corpus lutea (progesterone-producing glands) (Rothchild, 2003).

[1] Follicular growth is somewhat different in monotremes and non-mammalian vertebrates because of the accumulation of yolk within the oocyte (Rothchild, 2003).

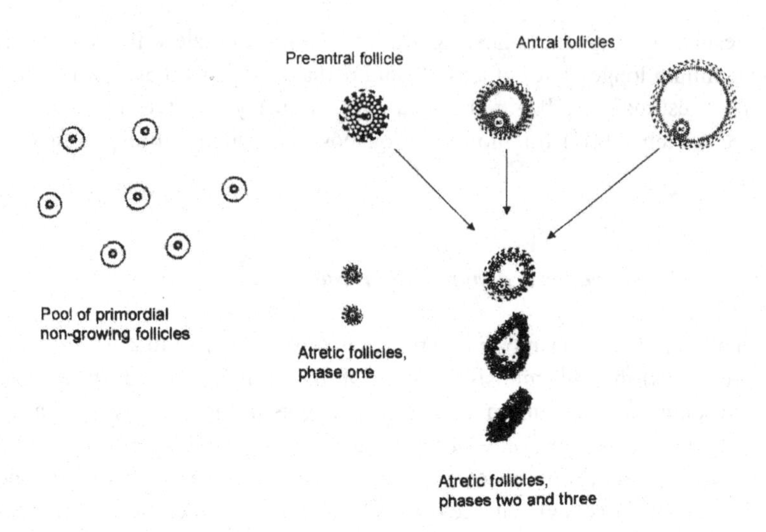

Figure 8.1. Follicles in various stages of development (after Peters & McNatty, 1980, p. 15). Phases one, two, and three refer to stages of loss in Figure 8.2.

Mammals and birds produce a single large population of oogonia only once (Figure 8.2).[2] In birds, oogenesis ends with hatching (Guraya, 1989). In mammals, oogenesis ceases before or just after birth (Peters & McNatty, 1980). Oogonia, oocytes, and follicles in all stages of development are lost through the process of atresia from the prenatal period, through the pre-reproductive and reproductive periods, into the early post-reproductive lifespan. It appears that at least 34 species of mammals experience the complete exhaustion of follicles prior to death (Cohen, 2004). In mammals, humans in particular, the loss of follicles can result in menopause and post-reproductive life. Post-reproductive life is not an avian characteristic (Ricklefs *et al.*, 2003; Ricklefs, 2008).

In contrast to the bird and mammalian pattern, fish, amphibians, and possibly all reptiles have stem cells which persist throughout the female's life. New oogonia can be produced during each breeding season through mitosis. Atresia of oogonia or follicles is less common in fish and amphibians compared to mammals; almost all oogonia may be ovulated as mature oocytes (Habibi & Andreu-Vieyra, 2007; Rothchild, 2003).[3]

[2] Cyclostomes (hagfish and lampreys) and elasmobranchs (rays, sharks, and skates) also follow this pattern of egg production. Possible exceptions to this pattern in mammals include some prosimians (Baker, 1986; Byskov & Hoyer, 1988; Comfort, 1979; Greenwald & Terranova, 1988) and mice (Johnson *et al.*, 2004), in which limited oogenesis may continue.

[3] Guraya (1986) gives a somewhat different perspective, however, noting that "follicles (or oocytes) in various stages of growth in the fish ovary are lost through atresia during any time of the year" (p. 177). Differences in perspective may result from differences among fish. Some

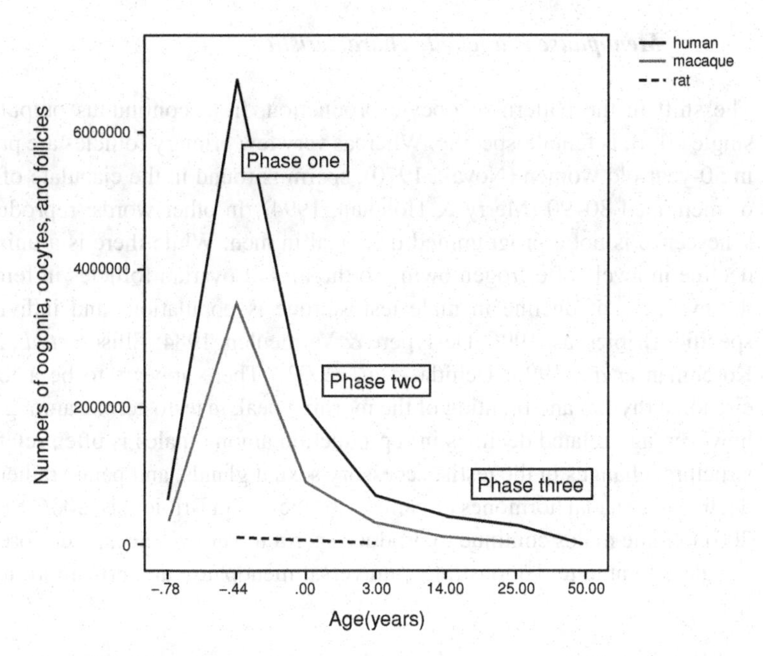

Figure 8.2. Rise and decline in numbers of oogonia, oocytes, and follicles in human, rhesus macaque, and rat, during the prenatal period, at birth, age 3 (approximate life expectancy for rat and age at menarche for rhesus macaque), age 14 (approximate age at menarche for humans – midway between current and historic estimates), age 25 (approximate age at menopause for macaques), and age 50 (approximate age at menopause for humans). (Counts from Baker, 1986; Hansen *et al.*, 2008; Ryan 1981.)

The timing of menopause is determined by: (1) the number of eggs formed in the female ovary during fetal development; (2) the rate of loss of those eggs across the lifespan through the processes of ovulation and atresia; and (3) the threshold number of follicles needed to maintain reproductive function because, in addition to nurturing the oocyte, follicles produce the ovarian hormones critical to the maintenance of fertile cycles (Sievert, 2006). Menopause is a mammalian characteristic because of the pattern of oogenesis (all at once, just once) and the process of atresia across the lifespan to the point of the exhaustion of follicular reserves.

do not demonstrate much follicle loss through atresia (Thorsen *et al.*, 2006), while others do (Bromley *et al.*, 2000; Murua *et al.*, 2003; Murua and Saborido-Rey, 2003). Also, because fish eggs are yolk-rich, atresia may play a role in resource recovery for fish when environmental conditions become unfavorable, rather than the editing or quality-control role suggested here for mammalian atresia (although see Hardardottir *et al.*, 2003 for a mention of quality control in fish).

Menopause is a female characteristic

The shift in the pattern of oocyte production, from continuous output to a single effort, is female-specific. Whereas very few primary follicles are present in 50-year-old women (Novak, 1970), sperm is found in the ejaculate of 48% of men aged 80–90 (Merry & Holehan, 1994). In other words, reproductive senescence is not a programmed universal in men. While there is a universal decline in levels of estrogen owing to the loss of ovarian follicles in females, the evidence of decline in male testosterone is population- and individual-specific (Bribiescas, 1996; Deslypere & Vermeulen, 1984; Ellison *et al.*, 2002; Korenman *et al.*, 1990; Uchida *et al.*, 2002.) There appears to be a loss of circadian rhythm and blunting of the morning peak in testosterone among men; however, age-related declines in reproduction among males is often related to structural changes in the testis, accessory sexual glands, and penis, rather than decline in gonadal hormones or gametes (reviewed in Bribiescas, 2006; Sievert, 2001). Some males continue to produce sperm to very old ages, therefore only females are able to demonstrate a universal menopause and post-reproductive life.

The evolution of menopause: adaptive scenarios

The process of continuous oogenesis worked well for millions of years in fish, amphibians, and reptiles, and continuous spermatogenesis functions pretty well in mammalian males. Why, then, did the pattern of egg production change so that mammalian females make all of their eggs up front, lose them through the process of atresia, and potentially run out of viable oocytes prior to the end of the somatic lifespan? Many have argued that human menopause and/or post-reproductive life is an adaptation, a direct product of natural selection.

Adaptive scenarios apply Darwinian evolution to explain the appearance or persistence of particular traits. For natural selection to occur, three conditions must be met: (1) variation in the trait of interest among individuals in a population, (2) heritability of the variation, and (3) correlation between the variation and reproductive success (Langdon, 2005). Menopause satisfies each of these three conditions.

First, there is species-level and population-level variation in the length of the reproductive span. Across human populations, mean or median ages at menopause vary by about 10 years, from 42 to 52.6 years (Beyene, 1986; DeLille Henderson *et al.*, 2008; Reynolds & Obermeyer, 2005; Thomas *et al.*, 2001). Part of this variation can be explained by methodological inconsistencies (Sievert & Hautaniemi, 2003); however, application of the same methods also yields meaningful differences in the median age at menopause across populations. Within populations, there is substantial variation among women in age at menopause. This is phenotypic variation upon which natural selection can act, particularly if premature ovarian failure is included in the continuum (Peccei, 1995). For example, recalled ages at natural menopause range from 36 to 56 years in Asuncion, Paraguay (n = 132) and 28 to 56 years in Puebla, Mexico (n = 268) (Sievert, 2006).

Second, a number of studies have measured heritability of age at menopause. Estimates of heritability range from 63% among British twins (Snieder *et al.*, 1998), to estimates of 72–87% for Dutch singleton sisters and twins (de Bruin *et al.*, 2001), 31–53% for Australian twins (Treloar *et al.*, 1998), and 40–50% for US mother–daughter pairs (Peccei, 1999).

The third condition, the correlation between the timing of menopause and reproductive success, is harder to measure (Shanley *et al.*, 2007), but has become the subject of five adaptationist scenarios. These scenarios can be summarized as aging eggs, aging mothers, the grandmother hypothesis, energy conservation, and the prevention of reproductive cancers.

Aging eggs

Menopause may have been selected for in order to ensure that old eggs are not fertilized (Hrdy, 1999; O'Rourke & Ellison, 1993; Pollard, 1994). As noted above for mammals, oogenesis and the first division of meiosis occur when the female is still a fetus. In humans, pre-ovulatory meiotic arrest can last for 15 to 50 years while oocytes wait in the ovary. This wait time is related to an age-related increase in risk of chromosomal abnormalities (Kohn, 1978; Martin, 2008) and fetal loss (Forbes, 1997; Holman *et al.*, 2000; Wood, 1994). Older oocytes may accumulate injuries "because selection pressure cannot act to maintain vigor of surviving cells in a nongrowing population" and because oocytes can be exposed "within their leaky follicle walls" to damaging toxins (Gosden & Faddy, 1994:271).

Also, as women age, the ovarian reserve declines so that there are fewer primordial follicles to choose from and, thus, an increased risk of chromosomal abnormalities in the oocytes that are selected to develop and ovulate (Freeman *et al.*, 2000; Kline & Levin, 1992; Kline *et al.*, 2000). Menopause may function to prevent the ovulation and fertilization of abnormal oocytes.

Aging mothers

Menopause ensures that mothers are young enough to survive pregnancy, delivery, and the infancy of their offspring (Pollard, 1994; Peccei, 1995; Shanley *et al.*, 2007). The demands of motherhood are heightened by the extreme altriciality of human infants (Williams, 1957), and, as shown among the Ache foragers of eastern Paraguay, offspring survivorship is low when mothers die in the first 5 years of the child's life (Hill & Hurtado, 1991).[4]

With menopause, women cease childbearing early enough to allow the last child to remain dependent, perhaps for 15 to 16 years (Lancaster & Lancaster, 1983). This adaptationist scenario implies that maternal death, more than paternal death, threatens the survival of the youngest offspring owing to differential parental investment (Gaulin, 1980; Kuhle, 2007).

From a different perspective, menopause protects older women from the demands of pregnancy and childbirth. With increasing age, women are at greater risk of complications from pregnancy (reviewed in Sievert, 2001) which, untreated, would result in death. Using data from pre-industrial Quebec, Pavard *et al.* (2008) explain menopause with models that include an increase in maternal mortality at older ages. However, using data from The Gambia, Shanley *et al.* (2007) concluded that "increasing mortality risk of giving birth at older ages was not sufficient on its own to select for menopause."

The grandmother hypothesis

Many authors have looked beyond aging eggs and aging mothers to the extended family and argued that menopause and post-reproductive aging were selected for by the evolutionary benefits gained through grandmothering (Voland *et al.*, 2005). The general idea is that, during human evolution, females gained more, in terms of inclusive fitness, by investing in their grandchildren than they would have gained by continuing to produce children of their own (Alexander, 1974; Dawkins, 1976; Hamilton, 1966; Williams, 1957). Postmenopausal grandmothers were selected for because they served as reservoirs of knowledge, provided toddler care and attention, produced surplus calories (Hrdy, 1999; O'Connell *et al.*, 1999),[5] and may have been the first birth attendants when hominins moved from solo to assisted births. Hawkes *et al.* (1997) argue that if a grandmother fed a weaned but still dependent grandchild, then her daughter could have more children more quickly. The fecundity of the daughter benefits the reproductive success of the grandmother.

In a test of the grandmother hypothesis, using data from historic Canadian and Finnish farming communities, Lahdenperä *et al.* (2004) showed that

[4] However, there is a very low probability of a woman dying within the first 5 years after a birth until she is very old. Shanley *et al.* (2007) found a similarly low probability of being orphaned using 1950–1975 data from The Gambia.

[5] This is not true among the Ache of Paraguay (Kaplan *et al.*, 2000).

post-reproductive women gained two extra grandchildren for every 10 years that they lived beyond age 50. In Finland, both sons and daughters who had a living, post-menopausal mother had children sooner, at shorter intervals, and more of those children reached adulthood (Lahdenperä *et al.*, 2004). Other studies have also demonstrated an increase in reproductive success among offspring because of grandmothers in historical Tokugawa, Japan (Jamison *et al.*, 2002), rural Gambia (Sear *et al.*, 2000, 2002; Shanley *et al.*, 2007), and historic Germany (Voland & Beise, 2005).

However, some of these same studies have shown that only maternal grandmothers have a positive effect on the survival of grandchildren (Jamison *et al.*, 2002; Voland & Beise, 2002; Voland *et al.*, 2005), whereas paternal grandmothers and grandfathers are associated with an increased risk of infant death. Using data from Atenas, Costa Rica, Madrigal & Meléndez-Obando (2008) showed that the longevity of maternal grandmothers was negatively correlated with number of grandchildren. Among non-human species, Packer *et al.* (1998) found no evidence among lions or baboons for contributions of grandmothers to grandchild fitness. Finally, although a significant proportion of contemporary women in hunting and gathering societies do live beyond menopause (Hawkes, 2007), the lifespan was much shorter for earlier hominins, not just because of infant and child mortality but because of the harshness of everyday life. In this case, the potential for a grandmother effect is unlikely to have had much of an impact.

Energy conservation

Roberta Hall (2004) suggests that high metabolic costs are associated with reproduction, therefore menopause evolved as an opportunity to save energy. Just as menstrual bleeding may save the energy required to maintain an enriched endometrium (Strassman, 1996), menopause may have evolved so that energy spent in menstrual cycling could otherwise be used to support offspring survival.

Prevention of reproductive cancers

Risk factors for breast cancer include early menarche, delayed childbirth, and the low parity patterns characteristic of contemporary Western societies. Each of these factors increases a woman's exposure to periodic elevations in levels of estrogen which, over time, increase the risk of breast cancer (Kelsey *et al.*, 1993; Eaton & Eaton, 1999). Without menopause, women would continue to be exposed to high, cyclic levels of estrogen into the seventh or eighth decades of life. Menopause is, therefore, adaptive within the context of contemporary industrialized societies because it lowers the risk of breast cancer and other estrogen-dependent cancers which can occur before the end of the fertile period and hence directly reduce reproductive success.

The evolution of menopause: byproduct scenarios

In contrast to the adaptive scenarios reviewed above, menopause can also be understood to be the byproduct of (1) a conserved mammalian pattern of oogenesis and follicular atresia coupled with (2) a lengthened lifespan. This is true not just for humans but for killer whales, short-finned pilot whales, and other mammals as well (Cohen, 2004; Foote, 2008). Instead of focusing on menopause and post-reproductive life as direct products of natural selection, the following adaptationist scenarios use evidence from cross-species studies of ovarian physiology to construct menopause and post-reproductive life as neutral byproducts of natural selection.

In mammals, many more oogonia are produced shortly before, or at the time of birth, than will ever develop and ovulate. Almost all oogonia, oocytes, and follicles are lost through the degenerative process of atresia. The loss occurs in three phases – prior to birth, prior to puberty, and prior to menopause (Figure 8.2). Authors who present log-transformed counts of follicles in relation to age have argued for acceleration in the rate of atresia as menopause approaches (Gosden & Faddy, 1994; Gougeon *et al.*, 1994; Hansen *et al.*, 2008; Richardson *et al.*, 1987); however, when untransformed data are used, as in Figure 8.2, one can see that fewer follicles are lost with age (Leidy *et al.*, 1998). Each phase of oocyte and follicle loss is detailed below. The outcome, or byproduct, of this pattern of oocyte production and loss is the eventual depletion of all viable ovarian follicles in some long-lived species, resulting in menopause and post-reproductive life.

Phase one: the rapid pre- and neonatal decline

As Figure 8.2 illustrates, millions of oogonia are produced by mitosis at the beginning of human female life. Production begins during the eleventh or twelfth week and continues until seven million oogonia and oocytes are present in the ovaries by the fifth month in utero (Baker, 1986; Crisp, 1992). Across species, the peak number of oogonia and oocytes is roughly correlated with the length of the lifespan, e.g. 100,000 in the rat, more than 3 million in the macaque, and 7 million in the human female (Figure 8.2, Baker, 1986). The numbers of oogonia produced are high, not unlike the numbers of oogonia seen in fishes (Martin, 1985); however, mammalian oocytes are not spawned and oocyte reserves are depleted primarily through atresia, not ovulation. While mammals may appear to have evolved an inefficient pattern of oocyte overproduction and loss, once the reproductive system was in place, the cost of producing many eggs was not much higher than the cost of producing a few (Ghiselin, 1987). Why was there a selection for this pattern of oogenesis and atresia?

In humans, during the fifth month in utero, and for the remainder of the pregnancy, oocytes are lost through atresia at a rate of about 1 million per month (Rothchild, 2003) until the neonate is born with approximately 2 million primordial follicles (Baker, 1986). This pattern, characteristic of all mammals, may have been selected for as a reproductive filtering mechanism (Wasser & Place, 2001) to reduce the number of eggs with developmental errors, such as naked oocytes (oocytes without surrounding granulosa cells, Crisp, 1992) or excessive chromosomal abnormalities (Baker, 1986). During the early phase of oocyte production, the transition from mitosis to meiosis takes place (Rothchild 2003). The first stage of meiosis involves duplicating and halving the number of chromosomes within the oocyte. Crossing over also occurs, which is when chromosomes exchange homologous pieces (Baker, 1986). This exchange increases the risk of chromosomal mutations. During the prenatal period, oogonia that have not become part of a primordial follicle and entered meiosis will die via apoptosis (programmed cell death) (Parker & Schimmer, 2006).

Atresia, the shrinking and loss of early oocytes within small follicles, can be described as an editing function that was selected for as a means to increase the likelihood of producing viable offspring from the primordial follicles that remain (Guraya, 1985). A substantial initial production of oocytes allows for a large margin of error in the establishment of the pool of viable follicles. Except in the case of environmental stress (Habibi & Andreu-Vieyra, 2007), fish and amphibians don't generally "clean" the pool of developing oogonia and primary oocytes (see note 3). But mammals are K-selected creatures by comparison – K-selected species invest more heavily in fewer offspring, and each offspring has a relatively high probability of surviving to adulthood. It makes sense that an animal which ovulates fewer oocytes, and invests more in offspring through lactation, more carefully removes abnormalities early in the process of reproduction.

Phase two: the less rapid postnatal pre-reproductive decline

From birth to the onset of puberty, the loss of follicles in the human ovary slows to a rate of 12,000 per month (Rothchild, 2003) and the female enters puberty with approximately 400,000 follicles (Baker, 1986). Because ovulation does not occur during childhood, all follicles that enter the growth phase during this period are destined to degenerate (Peters & McNatty, 1980; Rajkovic *et al.*, 2006). Atresia during this phase is not caused by the shrinking of the oocyte, as in phase one, but caused by the degeneration of follicles surrounding the oocyte. More specifically, granulosa cells of most early antral follicles undergo apoptosis, which results in follicle death (Rajkovic *et al.*, 2006). As noted above, atresia can happen during any stage of follicle development (Figure 8.1). Some of this loss may be caused by defects that arise when granulosa cells start to proliferate, or by problems that occur when oocytes start to grow (Rothchild, 2003).

From an adaptationist point of view, pre-pubertal follicular development and loss may be part of a physiological trade-off that benefits early fertility at the expense of later reproductive decline. Waves of developing and degenerating follicles during pre-reproductive life may help to initiate regular cycles. As levels of FSH increase, more follicles reach late antral size and are capable of hormone production. Repetitions of cyclic ovarian follicular growth and atresia may eventually contribute to ovulatory cycles (Vihko & Apter, 1981). Alternatively, atretic follicles can become steroid-producing interstitial tissue in the pre-pubertal ovary. This tissue may contribute to the development and maturation of a dominant follicle, or to ovarian communication with the hypothalamus and pituitary (Baker, 1986; Guraya, 1985).

When a physiological trait has beneficial effects early in the lifespan, but deleterious effects later in the lifespan, this is known as antagonistic pleiotropy (Wood *et al.*, 2001). Waves of follicle development and degeneration may be necessary prior to puberty to prime the hypothalamic–pituitary–ovarian axis, and to ready the female for ovulation during the reproductive period. The cost, however, is a loss of follicles, contributing to the eventual exhaustion of ovarian reserves, menopause, and post-reproductive life in long-lived species.

Phase three: the least rapid decline during the reproductive period

From puberty to menopause, approximately once per month, the human female may ovulate one oocyte. In addition, she will lose follicles through the process of atresia at an average rate of approximately 700 per month across the reproductive span (Rothchild, 2003). The rate of loss varies across species (Peters & McNatty, 1980), but it appears that the follicular depletion rate among chimpanzees is almost identical to that of humans (Jones *et al.*, 2007). During this third phase of follicle loss, as in phase two, the follicles rather than the oocytes degenerate. Although atresia occurs in follicles at all stages of development, it occurs more frequently in the advanced stages of follicular growth (Guraya, 1985; Rolaki *et al.*, 2005). Follicle loss continues until few primary and early vesicular follicles remain in the postmenopausal ovaries (Mossman & Duke, 1973).

One adaptationist scenario for a process of atresia that depletes ovarian reserves could be described as: many produced to serve a chosen few (Crisp, 1992). An ovarian follicle serves two functions: it can nurture an oocyte to maturity and ovulation, and/or produce the hormones needed for successful reproduction.

During each ovulatory cycle, a group of primordial follicles begins to grow, with as many as 50 reaching a diameter of 1 mm or more in the follicular (early) phase of the cycle. Three or fewer follicles grow to a diameter of 8 mm by midcycle. The follicle destined to ovulate grows to 20 mm in diameter or more (Hunter, 2003). As these follicles mature (Figure 8.1), they produce hormones such as estrogen and inhibin. These hormones help to maintain regular cycles at early adult ages – another example of antagonistic pleiotropy (Wood *et al.*, 2001). The development of many ovarian follicles maintains regular cycles; this is the benefit. However, only one follicle will be rescued from atresia for ovulation, and the loss of non-ovulatory follicles can be viewed as a deleterious trait at later ages when older women run out of the ovarian follicles needed for fertility.

The same hormonal changes that induce ovulation (e.g. the surge in FSH) may also induce atresia among the non-ovulatory follicles (Guraya, 1986), perhaps to prevent the full maturation of too many follicles (Rolaki *et al.*, 2005). Although talking about fish, Guraya (1986) observes that follicular atresia limits the number of eggs that require maternal investment in vitellogenesis (formation of yolk), maturation, and ovulation. In humans, the degeneration of non-ovulatory follicles results in singleton births. In other mammals, the general pattern of oogenesis and atresia can allow for flexibility in litter size in relation to environmental conditions so that, for example, a deer mouse can produce two to eight offspring per litter in any given year. In birds which are determinant layers (who lay a fixed number of eggs) the number of eggs is controlled by either the number of follicles maturing, or by follicular atresia and yolk resorption when the clutch is complete (Sadleir, 1973).

A pool of developing follicles also provides atretic follicles that can contribute interstitial cells to the ovary. These cells may secrete steroid hormones needed for the growth, differentiation, and maturation of normal follicles (Guraya, 1985). Atretic follicles can also be luteinized to produce progesterone (Byskov, 1979). For example, in the rabbit there is a quantitative relationship between the number of corpora lutea and the ability of the female to retain eggs within the uterus (Adams, 1966). In the African elephant, as many as 40 corpora lutea may be present to help maintain a pregnancy (Hogarth, 1976). In the porcupine, the accessory corpora lutea function exactly as the corpus luteum, growing during early pregnancy and remaining as accessory glands for more than one year (Mossman & Duke, 1973). In humans there is no physiological evidence that accessory corpora lutea have the same secretory function as the normal glands (Guraya, 1985); however, their presence suggests a conserved pattern of cellular change.

The overarching point of these byproduct scenarios is that if a female produced only the eggs necessary for ovulation, e.g. if the human female produced only 400 eggs, she may not be able to produce as many offspring because (1) some of the eggs would be defective. Humans cull five million oogonia and oocytes prior to birth. (2) There would be no extra follicles to produce hormones during the pre-pubertal period, when waves of developing and degenerating follicles may be necessary to prime the hypothalamic–pituitary–ovarian axis. And (3) there would be no accessory follicles available to produce the hormones necessary to maintain fertile ovulatory cycles. The mammalian pattern of one-time oogenesis and follicle loss is cumbersome, and may not be an improvement over the continuous oogenesis practiced by fish, amphibians, and reptiles, but the system works. In fact, it works well enough to resist changes that would accommodate a longer lifespan with continued reproduction.

Conclusions

When it comes to explaining the evolution of menopause and post-reproductive life, there is more than one adaptationist perspective. Except for the mother and grandmother hypotheses, most of the scenarios described above have not been empirically tested. Most of the scenarios require the appearance of a long lifespan prior to or concurrent with the selection of post-reproductive life. This chapter did not address possible reasons for the evolution of a longer human life. Instead, longevity was accepted as a given in order to focus on ovarian physiology – something that is too often omitted from evolutionary arguments about menopause.

Many researchers argue that human post-reproductive life is so uniquely long that menopause and post-reproductive life should be explained as the direct result of natural selection; however, the processes that result in menopause – limited, one-time oogenesis and the loss of oocytes through atresia – are not human characteristics, but typical of all mammals. The genes regulating apoptosis (programmed cell death) involved in the mechanism of atresia have been well conserved across mammalian species (Rolaki *et al.*, 2005).

Cohen (2004) points out that any hypothesis about the evolution of post-reproductive life must address the physiological processes that are set in motion far earlier in the lifespan than at the point at which reproduction or menstruation ceases. A focus on ovarian physiology does just that. The human (or whale) lifespan lengthened. Why did the ovary not accommodate the change with continued oogenesis or a slower rate of atresia?

One could argue that menopause and post-reproductive life result from a *lack* of selective pressure to prolong reproduction, even within the context of an extended lifespan. It may be that because the process of atresia is comprised of many sub-traits (Guraya, 1985; Hsueh *et al.*, 1994; Rolaki *et al.*, 2005; Tapanainen & Vasikivuo, 2002), one small change may lead to negative consequences for other sub-traits. There are so many factors involved in the process of atresia, that the alteration of any one factor (e.g. estrogen production, progesterone production, gonadotropin receptors and responsiveness, IGF binding protein expression, cathepsin-D activity, expression of gap junction protein connexin-43, granulosa cell attachment to the basement membrane) may unbalance the system. Rather than positive selection, menopause may be the result of negative selection (Johnson, 2007), which acts to maintain the status quo of follicular atresia even as the lifespan of a species lengthens. The increase in lifespan set the stage for a human menopause and long post-reproductive life because the pattern of oogenesis and atresia did not change.

In summary, natural selection can be applied to explain menopause and post-reproductive life among human females through two categories of adaptationist scenarios. Some argue that menopause and post-reproductive life have been selected for directly. Alternatively, it can be argued that menopause and post-reproductive life are the neutral byproducts of a highly conserved mammalian pattern of oogenesis and atresia, coupled with a lengthened lifespan. In humans, short-finned whales, killer whales, and other mammals, the lengthened lifespan uncovered the exhaustion of female eggs. Menopause and post-reproductive life were the result.

Acknowledgements

I wish to thank Nick Mascie-Taylor, Lyliane Rosetta, and Rie Goto for giving me a deadline to get these thoughts on paper. Much of the research was done while I was a summer Guest Instructor at Ohio University and I thank Norman Gevitz for that opportunity. Further encouragement has come from Tom Clarkson and John Relethford. I especially thank Steve King and Norman Johnson for their reading of this manuscript, and I thank Nikolai Klibansky for his enthusiastic help regarding atresia in fish. I take credit for all mistakes.

References

Adams, C. E. (1966). Ovarian control of early embryonic development within the uterus. In: *Reproduction in the Female Mammal*, ed. G. E. Lamming and E. C. Amoroso, New York, NY: Plenum Press, pp. 532–548.

Alexander, R. D. (1974). The evolution of social behavior. *Annual Review of Ecology and Systematics*, 5, 325–383.

Atsalis, S. & Margulis, S. W. (2006). Sexual and hormonal cycles in geriatric *Gorilla gorilla gorilla. International Journal of Primatology*, 27, 1663–1687.

Baker, T. G. (1986). Gametogenesis. In: *Comparative Primate Biology, Vol 3: Reproduction and Development*, ed. W. R. Dukelow and J. Erwin, New York, NY: Alan R. Liss, Inc, pp. 195–213.

Beyene, Y. (1986). Cultural significance and physiological manifestations of menopause: a biocultural analysis. *Culture, Medicine and Psychiatry*, 10, 47–71.

Bribiescas, R. (1996). Testosterone levels among Ache hunter/gatherer men: a functional interpretation of population variation among adult males. *Human Nature*, 7, 163–188.

Bribiescas, R. G. (2006). *Men: Evolutionary and Life History*. Harvard University Press.

Bromley, P. J., Ravier, C. & Witthames, P. R. (2000). The influence of feeding regime on sexual maturation, fecundity and atresia in first-time spawning turbot. *Journal of Fish Biology*, 56, 264–278.

Byskov, A. G. (1979). Atresia. In: *Ovarian Follicular Development and Function*, ed. M. A. Rees and W. A. Sadler, New York, NY: Raven, pp. 41–57.

Byskov, A. G. & Hoyer, P. E. (1988). Embryology of mammalian gonads. In: *The Physiology of Reproduction*, ed. E. Knobil and J. Neill, New York, NY: Raven Press, pp. 265–302.

Cohen, A. A. (2004). Female post-reproductive lifespan: a general mammalian trait. *Biological Reviews*, 79, 733–750.

Comfort, A. (1979). *The Biology of Senescence*, 3rd edn., New York, NY: Elsevier.

Crisp, T. M. (1992). Organization of the ovarian follicle and events in its biology: oogenesis, ovulation or atresia. *Mutation Research*, 296, 89–106.

Croze, H., Hillman, A. K. K. & Lang, E. M. (1981). Elephants and their habitats: how do they tolerate each other? In: *Dynamics of Large Mammal Populations*, ed. C. W. Fowler. New York, NY: John Wiley, pp. 297–316.

Dawkins, R. (1976). *The Selfish Gene*. New York, NY: Oxford University Press.

de Bruin, J. P., Bovenhuis, H., van Noord, P. A. H. *et al.* (2001). The role of genetic factors in age at natural menopause. *Human Reproduction*, 16, 2014–2018.

DeLille Henderson, K., Bernstein, L., Henderson, B., Kolonel, L. & Pike, M. C. (2008). Predictors of the timing of natural menopause in the multiethnic cohort study. *American Journal of Epidemiology*, 167, 1287–1294.

Deslypere, J. P. & Vermeulen, A. (1984). Leydig cell function in normal men: effect of age, lifestyle, residence, diet, and activity. *Journal of Clinical and Endocrinology and Metabolism*, 59, 955–962.

Eaton, S. B & Eaton, S. B. (1999). Breast cancer in evolutionary context. In: *Evolutionary Medicine*, ed. W. R. Trevathan, E. O. Smith and J. J. McKenna, New York, NY: Oxford University Press, pp. 429–442.

Ellison, P. T., Bribiescas, R. G., Bentley, G. R. *et al.* (2002). Population variation in age-related decline in male salivary testosterone. *Human Reproduction*, 17, 3251–3253.

Fisher, R. A. (1930). *The Genetical Theory of Natural Selection*. Oxford: Calarendon Press.

Foote, A. D. (2008). Mortality rate acceleration and post-reproductive lifespan in matrilineal whale species. *Biology Letters*, 4, 189–191.

Forbes, L. S. (1997). The evolutionary biology of spontaneous abortion in humans. *Trends in Ecology and Evolution*, 12, 446–50.

Freeman, S. B., Yang, Q., Allran, K., Taft, L. F. & Sherman, S. L. (2000). Women with a reduced ovarian complement may have an increased risk for a child with Down syndrome. *American Journal of Human Genetics*, 66, 1680–1683.

Gaulin, S. J. C. (1980). Sexual dimorphism in the human post-reproductive life-span: Possible causes. *Journal of Human Evolution*. 9, 227–232.

Ghiselin, M. T. (1987). Evolutionary aspects of marine invertebrate reproduction. In: *Reproduction of Marine Invertebrates*, ed. A. C. Giese, J. S. Pearse and V. B. Pearse, Palo Alto, CA: Blackwell Scientific, pp. 609–665.

Gosden, R. G. & Faddy, M. J. (1994). Ovarian aging, follicular depletion, and steroidogenesis. *Experimental Gerontology*, 29, 265–274.

Gougeon, A., Ecochard, R. & Thalabard, J. C. (1994). Age-related changes of the population of human ovarian follicles: increase in the disappearance rate of non-growing and early-growing follicles in aging women. *Biology of Reproduction*, 50, 653–663.

Greenwald, G. S. & Terranova, P. F. (1988). Follicular selection and its control. In: *The Physiology of Reproduction*, ed. E. Knobil and J. Neill, New York, NY: Raven Press, 387–445.

Guraya, S. S. (1985). *Biology of Ovarian Follicles in Mammals*. New York, NY: Springer–Verlag.

Guraya, S. S. (1986). *The Cell and Molecular Biology of Fish Oogenesis*. Monographs in Developmental Biology, Vol 18. New York, NY: Karger.

Guraya, S. S. (1989). *Ovarian Follicles in Reptiles and Birds*. New York, NY: Springer–Verlag.

Guraya, S. S. (1998). *Cellular and Molecular Biology of Gonadal Development and Maturation in Mammals: Fundamentals and Biomedical Implications*. New York, NY: Springer–Verlag.

Habibi, H. R. & Andreu-Vieyra, C. V. (2007). Hormonal regulation of follicular atresia in teleost fish. In: *The Fish Oocyte: From Basic Studies to Biotechnological Applications*, ed. P. J. Babin, J. Cerda and E. Lubzens, Dordrecht: Springer, pp. 235–253.

Hall, R. (2004). An energetics-based approach to understanding the menstrual cycle and menopause. *Human Nature*, 15, 83–99.

Hamilton, W. D. (1966). The moulding of senescence by natural selection. *Journal of Theoretical Biology*, 12, 12–45.

Hansen, K. R., Knowlton, N. S., Thyer, A. C. *et al.* (2008). A new model of reproductive aging: the decline in ovarian non-growing follicle number from birth to menopause. *Human Reproduction*, 23, 699–708.

Hardardottir, K., Kjesbu, O. S. & Marteinsdottir, G. (2003). Atresia in Icelandic cod (*Gadus morhua L.*) prior to and during spawning. *Fisen og. Havet*, 12, 51–55.

Hawkes, K. (2007). The evolutionary legacy of post-menopausal longevity & a grandmother hypothesis. *Menopause*, 14, 1074.

Hawkes, K., O'Connell, J. F. & Blurton Jones, N. G. (1997). Hadza women's time allocation, offspring provisioning and the evolution of post-menopausal lifespans. *Current Anthropology*, 38, 551–578.

Hill, K. & Hurtado, A. M. (1991). The evolution of premature reproductive senescence and menopause in human females: an evaluation of the 'grandmother hypothesis'. *Human Nature*, 2, 313–350.

Hogarth, P. T. (1976). *Viviparity*, London: Edward Arnold.

Holman, D. J., Wood, J. W. & Campbell, K. L. (2000). Age-dependent decline of female fecundity is caused by early fetal loss. In: *Female Reproductive Ageing*, Studies in Profertility Series, ed. E. R. Velde, F. Broekmans and P. Pearson, Camforth: Parthenon, pp. 123–136.

Hrdy, S. B. (1999). *Mother Nature: a History of Mothers, Infants, and Natural Selection*. New York, NY: Pantheon.

Hsueh, A. J. W., Billig H. & Tsafriri A. (1994). Ovarian follicle atresia: a hormonally controlled apoptotic process. *Endocrine Reviews*, 15, 707–724.

Hunter, R. H. F. (2003). *Physiology of the Graafian Follicle and Ovulation*. New York, NY: Cambridge University Press.

Jamison, C. S., Cornell, L. L., Jamison, P. L. & Nakazato, H. (2002). Are all grandmothers equal? A review and a preliminary test of the "grandmother hypothesis" in Tokugawa Japan. *American Journal of Physical Anthropology*, 119, 67–76.

Johnson, N. A. (2007). *Darwinian Detectives: Revealing the Natural History of Genes and Genomes*. New York, NY: Oxford University Press.

Johnson, J., Canning, J., Kaneko, T., Pru, J. K. & Tilly, J. L. (2004). Germline stem cells and follicular renewal in the postnatal mammalian ovary. *Nature*, 428, 145–150.

Jones, K. P., Walker, L. C., Anderson, D. *et al.* (2007). Depletion of ovarian follicles with age in chimpanzees: similarities to humans. *Biology of Reproduction*, 77, 247–251.

Kaplan, H., Hill, K., Lancaster, J. & Hurtado, A. M. (2000). A theory of human life history evolution: diet, intelligence, and longevity. *Evolutionary Anthropology*, 9, 156–185.

Kelsey, J. L., Gammon, M. D. & John, E. M. (1993). Reproductive factors and breast cancer. *Epidemiologic Reviews*, 15, 36–47.

Kline, J. & Levin, B. (1992). Trisomy and age at menopause: predicted associations given a link with rate of oocyte atresia. *Paediatric and Perinatal Epidemiology*, 6, 225–239.

Kline, J., Kinney, A., Levin, B. & Warburton, D. (2000). Trisomic pregnancy and earlier age at menopause. *American Journal of Human Genetics*, 67, 395–404.

Kohn, R. R. (1978). *Principles of Mammalian Aging*. Englewood Cliffs, NJ: Prentice–Hall.

Korenman, S. G., Morley, J. E., Morradian, A. D. *et al.* (1990). Secondary hypogonadism in older men: its relation to impotence. *Journal of Clinical Endocrinology and Metabolism*, 71, 963–969.

Kuhle, B. X. (2007). An evolutionary perspective on the origin and ontogeny of menopause. *Maturitas*, 57, 329–337.

Lahdenperä, M., Lummaa, V., Helle, S., Tremblay, M. & Russell, A. F. (2004). Fitness benefits of prolonged post-reproductive lifespan in women. *Nature*, 428, 178–181.

Lancaster, J. B. & Lancaster, C. S. (1983). The parental investment: the hominid adaptation. In: *How Humans Adapt: A Biocultural Odyssey*, ed. D. J. Ortner, Washington, DC: Smithsonian Institution Press.

Langdon, J. H. (2005). *The Human Strategy: An Evolutionary Perspective on Human Anatomy*. New York, NY: Oxford University Press.

Leidy, L. E., Godfrey, L. R. & Sutherland, M. R. (1998). Is follicular atresia biphasic? *Fertility and Sterility*, 70, 851–859.

Madrigal, L. & Meléndez-Obando, M. (2008). Grandmothers' longevity negatively affects daughters' fertility. *American Journal of Physical Anthropology*, 136, 223–229.

Margulis, S. W., Atsalis, S., Bellem, A. & Wielebnowski, N. (2007). Assessment of reproductive behavior and hormonal cycles in geriatric western lowland gorillas. *Zoo Biology*, 26, 117–139.

Martin, C. R. (1985). *Endocrine Physiology*. New York, NY: Oxford University Press.

Martin, R. (2008). Meiotic errors in human oogenesis and spermatogenesis. *Reproductive BioMedicine Online*, 16, 523–531.

Marsh, H. & Kasuya, T. (1986). Evidence for reproductive senescence in female cetaceans. *Report of the International Whaling Commission*, 8, 57–74.

Merry, B. J. & Holehan, A. M. (1994). Aging of the male reproductive system. In: *Physiological Basis of Aging and Geriatrics*, ed. P. S. Timiras, 2nd edn, Ann Arbor, MI: CRC Press, pp. 171–178.

Mossman, H. W. & Duke, K. L. (1973). *Comparative Morphology of the Mammalian Ovary*. Madison, WI: The University of Wisconsin Press.

Murua, H., Kraus, G., Saborido-Rey, F. *et al.* (2003). Procedures to estimate fecundity of marine fish species in relation to their reproductive strategy. *Journal of Northwest Atlantic Fishing Science*, 33, 33–54.

Murua, H. & Saborido-Rey, F. (2003). Female reproductive strategies of marine fish species of the North Atlantic. *Journal of Northwest Atlantic Fishing Science*, 33, 23–31.

Nishida, T., Corp, N., Hamai, M. *et al.* (2003). Demography, female life history, and reproductive profiles among the chimpanzees of Mahale. *American Journal of Primatology*, 59, 99–121.

Novak, E. R. (1970). Ovulation after fifty. *Obstetric Gynecology*, 36, 903–910.

O'Connell, J. F., Hawkes, K. & Blurton Jones, N. G. (1999). Grandmothering and the evolution of *Homo erectus*. *Journal of Human Evolution*, 36, 461–485.

Olesiuk, P. F., Bigg, M. A. & Ellis, G. M. (1990). Life history and population dynamics of resident killer whales (*Orcinus orca*) in the coastal waters of British Columbia and Washington State. *Report to the International Whaling Commission* 12, 209–243.

O'Rourke, M. T. & Ellison, P. T. (1993). Menopause and ovarian senescence in human females. *American Journal of Physical Anthropology*, Suppl. 16, 154.

Packer, C., Tatar, M. & Collins, A. (1998). Reproductive cessation in female mammals. *Nature*, 392, 807–811.

Parker, K. L. & Schimmer, B. P. (2006). Embryology and genetics of the mammalian gonads and ducts. In: *Knobil and Neill's Physiology of Reproduction*, ed. J. D. Neill, 3rd edn, Boston, MA: Elsevier, pp. 313–336.

Pavard, S., Metcalf, C. J. E. & Heyer, E. (2008). Senescence of reproduction may explain adaptive menopause in humans: a test of the "mother" hypothesis. *American Journal of Physical Anthropology*, 136, 194–203.

Pavelka, M. S. M. & Fedigan, L. M. (1999). Reproductive termination in female Japanese monkeys: a comparative life history perspective. *American Journal of Physical Anthropology*, 109, 455–464.

Peccei, J. S. (1995). A hypothesis for the origin and evolution of menopause. *Maturitas*, 21, 83–89.

Peccei, J. S. (1999). First estimates of heritability in the age of menopause. *Current Anthropology*, 40, 553–558.

Peccei, J. S. (2001). Menopause: adaptation or epiphenomenon? *Evolutionary Anthropology*, 10, 43–57.

Peters, H. & McNatty, K. P. (1980). *The Ovary: A Correlation of Structure and Function in Mammals*. New York, NY: Granada.

Pollard, I. (1994). *A Guide to Reproduction: Social Issues and Human Concerns*. Cambridge: Cambridge University Press.

Price, G. R. (1970). Selection and covariance. *Nature*, 227, 520–521.

Rajkovic, A., Pangas, S. A. & Matzuk, M. M. (2006). Follicular development: mouse, sheep and human models. In: *Knobil and Neill's Physiology of Reproduction*, ed. J. D. Neill, 3rd edn, Boston, MA: Elsevier, pp. 383–423.

Reynolds, R. F. & Obermeyer, C. M. (2005). Age at natural menopause in Spain and the United States: Results from the DAMES project. *American Journal of Human Biology*, 17, 331–340.

Richardson, S. J, Senikas, V. & Nelson, J. F. (1987). Follicular depletion during the menopausal transition: evidence for accelerated loss and ultimate exhaustion. *Journal of Clinical Endocrinology and Metabolism*, 65, 1231–1237.

Ricklefs, R. E. (2008). The evolution of senescence from a comparative perspective. *Functional Ecology*, 22, 379–392.

Ricklefs, R. E., Scheuerlein, A. & Cohen, A. (2003). Age-related patterns of fertility in captive populations of birds and mammals. *Experimental Gerontology*, 38, 741–745.

Rolaki, A., Drakakis, P., Millingos, S., Loutradis, D. & Makrigiannakis, A. (2005). Novel trends in follicular development, atresia and corpus luteum regression: a role for apoptosis. *Reproductive BioMedicine Online*, 11, 93–103.

Rothchild, I. (2003). The yolkless egg and the evolution of eutherian viviparity. *Biology of Reproduction*, 68, 337–357.

Ryan, R. J. (1981). Follicular atresia: some speculations of biochemical markers and mechanisms. *Dynamics of Ovarian Function*, ed. N. B. Schwartz and M. Hunzicker-Dunn, New York, NY: Raven Press, pp. 1–11.

Sadleir, R. (1973). *The Reproduction of Vertebrates*. New York, NY: Academic Press.

Sear, R., Mace, R. & McGregor, I. A. (2000). Maternal grandmothers improve nutritional status and survival of children in rural Gambia. *Proceedings of the Royal Society B*, 267, 1641–1647.

Sear, R., Steele, F., McGregor, A. A. & Mace, R. (2002). The effects of kin on child mortality in rural Gambia. *Demography*, 39, 43–63.

Shanley, D. P., Sear, R., Mace, R. & Kirkwood, T. B. L. (2007). Testing evolutionary theories of menopause. *Proceedings of the Royal Society B*, 274, 2943–2949.

Sievert, L. L. (2001). Aging and reproductive senescence. In: *Reproductive Ecology and Human Evolution*, ed. P. Ellison, Hawthorne, NY: Aldine de Gruyter, pp. 267–292.

Sievert, L. L. (2006). *Menopause: A Biocultural Approach.* Rutgers University Press.

Sievert, L. L. & Hautaniemi, S. I. (2003). Age at menopause in Puebla, Mexico. *Human Biology*, 75, 205–226.

Snieder, H., MacGregor, A. J. & Spector, T. D. (1998). Genes control the cessation of a woman's reproductive life: a twin study of hysterectomy and age at menopause. *Journal of Clinical Endocrinology and Metabolism*, 83, 1875–1880.

Strassman, B. I. (1996). The evolution of endometrial cycles and menstruation. *Quarterly Review of Biology*, 71, 181–220.

Tapanainen, J. S. & Vasikivuo, T. (2002). Apoptosis in the human ovary. *Reproductive Medicine Online*, 6, 24–35.

Thomas, F., Renaud, F., Benefice, E., De Meeus, T. & Guegan, J-F. (2001). International variability of ages at menarche and menopause: patterns and main determinants. *Human Biology*, 73, 271–290.

Thorsen, A., Marshall, C. T. & Kjesbu, O.S. (2006). Comparison of various potential fecundity models for north-east Arctic cod *Gadus morhua, L.* using oocyte diameter as a standardizing factor. *Journal of Fish Biology*, 69, 1709–1730.

Treloar, S. A., Do, K.-A. & Martin, N. G. (1998). Genetic influences on the age at menopause. *Lancet*, 352, 1084–1085.

Uchida, A. R., Bribiescas, R. G., Kanamori, M. *et al.* (2002). Salivary testosterone levels in healthy 90-year-old Japanese males: implications for endocrine senescence. Paper presented at International Society for Human Ethology, Montreal.

Vihko, R. & Apter, D. (1981). Endocrine maturation in the course of female puberty. In: *Hormones and Breast Cancer*, Banbury Report #8, ed. M. C. Pike, P. K. Siiteri and C. W. Welsch, Cold Spring Harbor, NY: Cold Spring Harbor Laboratory, pp.57–69.

Voland, E. & Beise, J. (2002). Opposite effects of maternal and paternal grandmothers on infant survival in historical Krummhörn. *Behavioral Ecology and Sociobiology*, 52, 435–443.

Voland, E. & Beise, J. (2005). "The husband's mother is the devil in house" Data on the impact of the mother-in-law on stillbirth mortality in historical Kummhorn (1750–1874) and some thoughts on the evolution of postgenerative female life. In: *Grandmotherhood: The Evolutionary Significance of the Second Half of Female Life*, ed. E. Voland, A. Chasiotis, W. Schiefenhovel, New Brunswick, New Jersey:Rutgers University Press, pp. 239–55.

Voland, E., Chasiotis, A. & Schiefenhovel, W. (2005). *Grandmotherhood: The Evolutionary Significance of the Second Half of Female Life*. New Brunswick, NJ:Rutgers University Press.

Wasser, S. K. & Place, N. J. (2001). Reproductive filtering and the social environment. In: *Reproductive Ecology and Human Evolution*, ed. P. T. Ellison, Hawthorne, NY: Aldine de Gruyter, pp.137–157.

World Health Organization (WHO). (1996). *Research on the Menopause in the 1990s*. WHO Technical Report Series No. 866, Geneva: WHO.

Williams, G. C. (1957). Pleiotropy, natural selection, and the evolution of senescence. *Evolution*, 11, 398–411.

Wood, J. W. (1994). *Dynamics of Human Reproduction: Biology, Biometry, Demography*. New York, NY: Aldine de Gruyter.

Wood, J. W., Holman, D. J. & O'Connor, K. A. (2001). Did menopause evolve by antagonistic pleiotropy? In: *Homo unsere Herkunft und Zukunft*. Proceedings 4. Kongress der Gesellschaft für Anthropologie (GfA). Göttingen: Cuvillier Verlag, pp. 483–490.

9 *Analysing the characteristics of the menstrual cycle in field situations in humans: some methodological aspects*

JEAN-CHRISTOPHE THALABARD, LAURENCE JOUBIN, LYLIANE ROSETTA AND C. G. NICHOLAS MASCIE-TAYLOR

Introduction

Reproductive function represents a rather unique endocrine system since its "goal" is the perpetuation of the species, not the individual. Hence its regulation in humans, which integrates both exogenous and endogenous factors, and represents a complex system with large within- and between-women variability in both the menstrual cycle length and the ovulatory cycle.

In human clinical practice various tools are routinely used to assess female reproductive status including keeping a menstrual diary, charting body temperature, or follicular ultrasound monitoring (Ecochard *et al.*, 2001), but these methods are not appropriate for field situations, especially with less literate or illiterate subjects. Although measuring progesterone and oestradiol levels in serum is the gold standard, obtaining blood samples can be difficult in studies involving serial sampling, especially in developing countries, owing to ethical, logistical and cultural constraints. So the focus here is on the use of changes in hormonal profiles, collected by non-invasive methods, to examine the likelihood of ovulation occurring. In field situations this methodology can be applied to both literate and illiterate women.

The World Health Organisation (WHO) Collaborating Centre developed non-invasive immunoassay methods for measuring human steroid hormones from saliva and urine in the 1980s (Sufi and Donaldson, 1988; Sufi *et al.*, 1985). This technique uses enzyme immunoassay to detect urinary pregnanediol-3-alpha-glucuronide (PdG) and oestrone-3-glucuronide (E_1G), the urinary metabolites of progesterone and estradiol, respectively, throughout the menstrual cycle.

A luteal rise in progesterone is characteristic of normal ovulation, but there is no definitive threshold of progesterone metabolites below which ovulation can be said not to have occurred (Ellison, 1988) and both absolute and relative threshold methods have been used to determine luteal activity evidence.

Reproduction and Adaptation, eds. C. G. Nicholas Mascie-Taylor and Lyliane Rosetta.
Published by Cambridge University Press.

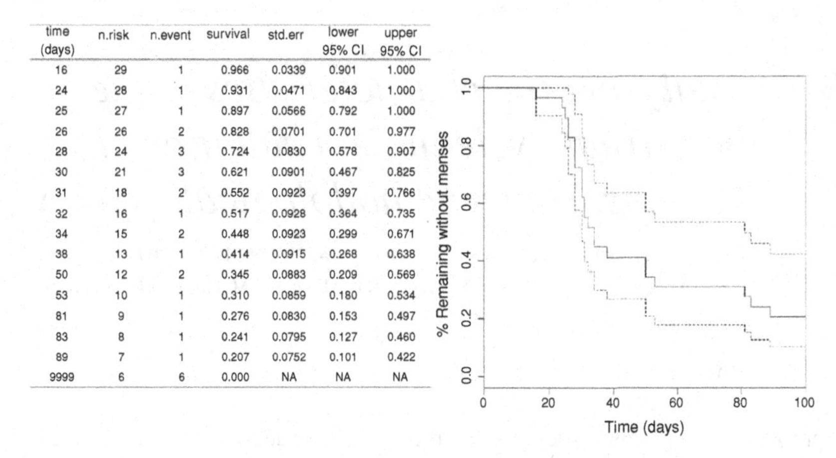

time (days)	n.risk	n.event	survival	std.err	lower 95% CI	upper 95% CI
16	29	1	0.966	0.0339	0.901	1.000
24	28	1	0.931	0.0471	0.843	1.000
25	27	1	0.897	0.0566	0.792	1.000
26	26	2	0.828	0.0701	0.701	0.977
28	24	3	0.724	0.0830	0.578	0.907
30	21	3	0.621	0.0901	0.467	0.825
31	18	2	0.552	0.0923	0.397	0.766
32	16	1	0.517	0.0928	0.364	0.735
34	15	2	0.448	0.0923	0.299	0.671
38	13	1	0.414	0.0915	0.268	0.638
50	12	2	0.345	0.0883	0.209	0.569
53	10	1	0.310	0.0859	0.180	0.534
81	9	1	0.276	0.0830	0.153	0.497
83	8	1	0.241	0.0795	0.127	0.460
89	7	1	0.207	0.0752	0.101	0.422
9999	6	6	0.000	NA	NA	NA

Figure 9.1. Summary of menstrual cycle durations, using the Kaplan–Meier non-parametric representation. The median value (solid line (survival line), with the 95% confidence intervals (dotted lines)) corresponding to the 50% ordinate is a valid unbiased non-parametric summary of menstrual cycle length using the entire sample.

The length of the menstrual cycle is defined as the time elapsed between the first day of two consecutive bleeding periods. However, in some circumstances, data collection for a particular subject can end before the return of menses and the length of the menstrual cycle can no longer be calculated. The only available information is that the length of the corresponding cycle is at least longer than that elapsed since the last menses. It corresponds to a right-censored variable. A summary of the menstrual cycle duration can take into account censored data by using percentiles and the Kaplan–Meier method (Figure 9.1).

This chapter reviews the various methods which can discern whether ovulation has occurred and estimates the lengths of both the follicular and luteal phases. Additional methods, easily implemented using freely available statistical packages, are developed.

Subjects and methods

A group of French recreational runners (Rosetta *et al.*, 2001) provided urine samples every other day throughout one menstrual cycle, starting on day 2, following WHO procedures (Sufi and Donaldson, 1988). Amenorrhoeic women were asked to provide urine samples every other day for 28 days commencing on a day set by the investigators, but within the same time period as the other women. Four pH strips impregnated with the first urine flow after waking were collected, and after allowing the strips to dry in the open air, the women were asked to store each set in a plastic bag labelled with their identification number on which they recorded the date. At the end of the cycle they were provided with a large secure envelope to send their samples to the laboratory where they were stored at −20°C until assayed.

For the French runners data were collected every other day, starting with the second day of the cycle and continuing until the next menses. However, generally in field studies two difficulties are likely to be encountered: unequally spaced collections – for instance, on days 2 and 3, or days 2 and 5, instead of days 2 and 4, or missing data, i.e. days 2 and 6. In the present analyses, if a runner had more than two missing consecutive data points, all the data for that cycle were not used in the subsequent analysis. In case of unequally spaced collections, a small automatic linear interpolating procedure was written to produce equally spaced data from the raw data.

In the laboratory the quantity of pregnanediol-3-alpha-glucuronide and estrone-3-glucuronide was determined using a standard enzyme immunoassay technique designed by the World Health Organisation for use in developing countries as mentioned earlier. The technique initially used saliva (Sufi *et al.*, 1985) and subsequently urine (Sufi and Donaldson, 1988). The use of impregnated paper instead of liquid urine has been shown to give very similar results for both PdG and E_1G, although the paper impregnated with urine gave slightly higher values (Wasalathanthri *et al.*, 2003).

Ethical approval was obtained from the ethics committee of the Hôpital Cochin-Port-Royal (CPPRB), Paris, in accordance with the French bioethics regulations. All subjects provided written informed consent.

Ovulation detection methods

Most of the proposed algorithms in the literature for detecting ovulation are based on daily urinary collection (for reviews, see Kesner *et al.*, 1992; McConnell *et al.*, 2002; Santoro *et al.*, 2003) limiting their potential use in field studies, when daily collection faces logistic difficulties. Indeed, the average level of progesterone metabolites during the first 5/6 days of the menstrual cycle is commonly used as an estimate of the noise level, when ovulation-associated progesterone secretion into the bloodstream is physiologically impossible. This noise level is then used to calculate a threshold above which ovulation is predicted. Hence, when only every-other-day collections were available, the number of time points included in the averaging process was adapted to correspond to the same initial 5/6 day period.

Simple empirical threshold methods

(a) Progesterone urinary metabolite-based methods

The occurrence of ovulation based on progesterone metabolite (PdG) can be detected using either an absolute threshold like PdG values above either 3 ng/mg creatinine, when adjusting on urinary creatinine level, or 3 ng/ml assuming a normal 1g/l urinary creatinine level for 3 consecutive days, which we transformed to two consecutive records for every-other-day collections, or a relative threshold. For the latter, starting with a follicular reference value defined as the average of the PdG values corresponding to day 1 to day 5 of the menstrual cycle (APdG5) with its standard deviation SD (or the first three consecutive values), Brown (1977) proposed a threshold T_B = APdG5 + 3*SD for two consecutive days to signify ovulation.

Kassam *et al.* (1996) proposed different methods, depending on whether sampling was twice-per-cycle, weekly, twice-weekly or every other day. In the context of the human studies described here in which data were collected every other day, average values of the corresponding concentrations in consecutive overlapping 5 day-periods were calculated as well as the minimum PdG value (APdGMin) of these 5 day-period means. The proposed threshold (T_K) for ovulation is 3*APdGMin. Waller (1998, Waller 1 in Table 9.1) slightly modified this algorithm to $T_W = Tk + 1 + T_k^{1/2}$ and imposed at least three out of the five daily PdG values to validate ovulation. In all other situations, the cycle is declared anovulatory.

(b) Methods based on E_1G/PdG ratio

As the LH peak, which precedes the ovulation, is triggered by a species-specific sustained oestradiol level, followed by an oestradiol decrease at the time of ovulation, while progesterone is released into the bloodstream, a concomitant decrease of the ratio of the corresponding main urinary metabolites has been proposed as a good ovulation predictor. In addition, as both the E_1G and PdG metabolites are measured in the same sample, the ratio is independent of the urinary concentration.

Baird *et al.* (1991) introduced the notion of overlapping sequences along the time-axis of five consecutive daily urinary ratio values (denoted as EP1, EP2, EP3, EP4, and EP5), starting at day 0 up to the last day of the cycle. From them, a selected sequence was identified if the maximum of the EP4 and EP5 values was inferior or equal to EP1*DC, where DC is a descent criterion corresponding to the expected decrease of the E_1G/PdG ratio after ovulation occurred. The

Table 9.1 *Day of ovulation based on 10 different methods for the same menstrual cycle for 29 French runners – signifies that the algorithm failed to detect a change point.*

Identifier	Baird	Cusum	Brown	Kassam	Waller 1	Waller 2	Blackwell95	Blackwell99	Piecewise	NLR	MBP
10	–	12	–	–	–	–	25	13	18	18	18
12	–	10	18	16	16	10	17	17	15	15	–
13	–	–	–	–	–	–	13	15	3	–	4
14	–	10	10	10	8	14	9	9	15	15	16
15	20	–	–	–	–	20	15	17	5	–	6
16	–	18	22	16	14	14	17	19	16	16	14
17	18	24	24	20	18	18	19	23	21	21	18
18	–	20	18	18	16	16	17	19	16	16	16
19	8	10	–	–	–	8	15	15	9	9	8
20	16	14	16	16	14	16	19	19	18	18	–
50	–	12	12	12	10	12	15	15	13	13	14
52	–	18	18	18	18	14	17	19	16	17	18
54	–	–	–	8	8	–	15	15	29	–	28
55	–	24	24	24	22	20	23	23	22	21	22
57	–	–	–	–	–	28	13	13	2	–	–
58	20	20	26	6	24	20	15	15	3	–	4
60	–	20	20	20	18	16	19	19	17	17	18
61	–	–	–	–	–	2	25	27	9	–	8
62	18	20	20	20	18	18	19	19	18	18	–
63	14	12	8	16	14	14	15	17	12	12	16
65	–	20	20	20	18	18	19	21	16	16	18
67	–	16	16	14	14	10	15	17	13	12	–
68	–	–	–	–	–	–	–	–	3	–	–
69	–	12	20	20	18	16	15	23	16	17	16
70	–	–	20	20	18	16	9	13	16	–	–
71	28	32	32	28	26	28	27	29	26	25	–
73	–	18	16	16	14	16	17	19	16	16	–
75	–	24	26	24	22	20	23	25	22	22	24
76	–	24	24	20	18	16	21	21	19	19	20

DC has been set to 0.4 in Baird's algorithm. The day of luteal phase transition is then simply EP2. In case of multiple selected sequences, a specific decision rule (DR) compares the means (EP0 + EP2)/2 for all the selected sequences. Waller (1998) modified the previous algorithm by using (a) the ratio E1/(PdG + 1) to handle low PdG values and (b) a descent criterion DC = 0.6; (c) a DR reduced to the maximum of the respective means of the EP0 and the EP1 values (Waller 2).

Standard statistical methods for detection of a breakpoint

Predicting the time of luteal transition when the cycle is still ongoing, as is commonly used for artificial insemination (on line detection) exposes the risk of false detection; whereas an a posteriori change point detection (off line), as is the case in epidemiological studies, uses the entire signal including additional clues, namely the next menses or some other physiological signs, limiting the previous risk. The raw urinary PdG data concentrations defined as X_t (where $t = 1, 2, 3 \ldots k$ samples) are usually log-transformed into $Y_t = \log(X_t)$ ($t = 1, 2, 3 \ldots k$), in order to take into account the heteroscedasticity of the variance. In the absence of ovulation, the urinary PdG log-transformed signal is assumed to represent a white noise centred around a constant mean value.

(a) Cusum

CUSUM (**CU**mulative **SUM**) is a statistical method used to detect a significant underlying change in a time series. This method has been successfully applied to ovulation detection by Royston (Royston, 1982) and Cekan *et al.* (1986). A baseline value BV = mean (Y_3, Y_4, Y_5, Y_6, Y_7, Y_8), where Y_j refers to day j of the cycle ($j = 3 \ldots 8$), with its corresponding baseline SD is calculated, allowing the derivation of a reference level, Ref = BV + SD. The CUSUM algorithm is then applied to the residual Yt – Ref, using as a stopping rule 2*SD. The method is directly implemented in R statistical free software package (R Development Core Team, 2004) using the *qcc* procedure.

In the present analysis, BV was restricted to the available observed Y values within the period day 3–day 8. The day 3–day 8 period was selected, like in the reference article, as it corresponds from the physiological standpoint to a non-ovulatory period, without progesterone secretion, giving baseline levels of progesterone urinary metabolites, as previously mentioned.

(b) Blackwell method (Blackwell95 & Blackwell99)

The method based on Trigg's method was initially proposed by Cembrowski *et al.* (1975) and implemented in the context of ovulation detection by Blackwell *et al.* (1998), and Blackwell and Brown (1992). An exponential smoothed average is run on the initial signal $ESA(t) = \alpha.Y(t) + (1 - \alpha).ESA(t - 1)$, where α is a smoothing constant ($0 \leq \alpha \leq 1$), related to the number of points N included in a moving average (MA) by $\alpha = 2/(N + 1)$.

It allows for calculating a sequentially forecast error $FE = Y(t) - ESA(t - 1)$, which represents the difference between the observed data and the forecast value. Its corresponding forecast error is exponentially smoothed using the same parameter α giving the smoothed forecast error $SFE(t) = \alpha.FE(t) + (1 - \alpha).SFE(t - 1)$. Similarly, a smoothed absolute deviation $MAD(t) = \alpha. | SFE(t) | + (1 - \alpha).MAD(t - 1)$ is calculated.

The procedure is initiated using FE(0) = SFE(0) = 0, MA(0)= SD, $ESA(0) = gmean\ (D1 - D6)$, where *gmean* is the geometric mean. A tracking signal $TS(t) = SFE(t)/MAD(t)$ is then calculated and graphed. Its fluctuations under an absence of shift have been tabulated. Ovulation corresponds to the time when the signal deviates from the expected profile.

The method is referred to as Blackwell95 and Blackwell99 according to the choice of the type I error (0.05 and 0.01, respectively) for detection of an outlier.

(c) Piecewise regression (Piecewise method, non-linear regression (NLR), multiple breakpoints (MBP))

The basic idea of the method is to assume that in the absence of ovulation, the cumulated (log-transformed) progesterone urinary metabolite values exhibit a linear dependence on time. If ovulation occurred, a slope change should be observed. The underlying model with four unknown parameters is the following

$$X(t) = b_0 + b_1{}^*t \quad \text{if} \quad t < T \quad \text{or}$$
$$X(t) = b_0 + b_1{}^*t + b_2{}^*(t - T) \quad \text{if } t = T$$

As the number of possible T values is limited, it is always possible to perform a Piecewise linear regression, for all possible T values and select the model with the minimum residual sum of squares (Piecewise method). An alternative is to use any non-linear regression package (e.g. *optim*, *nlm* in R) using as minimisation criteria the residual sum of squares between the observed values and the theoretical values (NLR method).

However, when graphing the PdG values in relation to the day of the cycle, in some individual recordings, a secondary shift can be observed, in relation to the late luteal phase, and a slowing down of progesterone secretion in the absence of pregnancy. This phenomenon could interfere with the performance of a Piecewise regression method based on only one slope change. In the absence of pregnancy, suppressing backwards the last points could be used, with the risk of masking a short luteal phase. However, several algorithms have been proposed to detect multiple breakpoints and are already implemented: we used the function *breakpoints* in the R package {*strucchange*}, as well as the *confint* function. The selection of the optimal number of breakpoints relies on the residual sum of squares or the Bayesian information criteria (BIC). We called the corresponding method adapted for luteal phase onset detection, the MBP method.

All calculations were carried out using the R statistical free software (R Development Core Team, 2004). Some specific procedures were adapted. The following packages *qcc*, *strucchange*, *lmtest* were used. The corresponding text files are available at http://www.math-info.univ-paris5.fr/~thalabard/.

Data were analysed using all these methods, adapted to every-other-day collection. The results were compared using pairwise correlation analysis. Finally, in order to examine the changing patterns across methods, repeated measures analysis of variance was run using nine out of the ten methods, keeping the methods concerning PdG only, thus excluding Baird.

Results

A total of 93 runners were asked to participate in the study. Forty-five volunteered and only 29 completed the study, with usable data. Table 9.1 presents a summary of the days of ovulation as detected by the 10 different methods in one menstrual cycle for each of these 29 French runners.

Simple threshold method analysis of the PdG profiles are illustrated in Figure 9.2. Methods relying on the E1/PdG ratio are illustrated in Figures 9.3 and 9.4. As already mentioned, they appear highly sensitive to the E2 level, with a risk of false negative (subject 52) and false positive (subject 15) ovulation detection.

The Blackwell's method with the two detection levels applied to the PdG values and relying on the Trigg's signal is illustrated in Figure 9.5. Figures 9.6 and 9.7 illustrate various methods for analysing the structural changes in time-evolution of the cumulated PdG signal. In Figure 9.6, two approaches for a two-part Piecewise linear regression were carried out, using the R-standard functions *lm* and *optim* respectively, whereas Figure 9.7 corresponds to more specific statistical tools, like *cusum*, *qcc*, *ewma* from the {*qcc*}package in R.

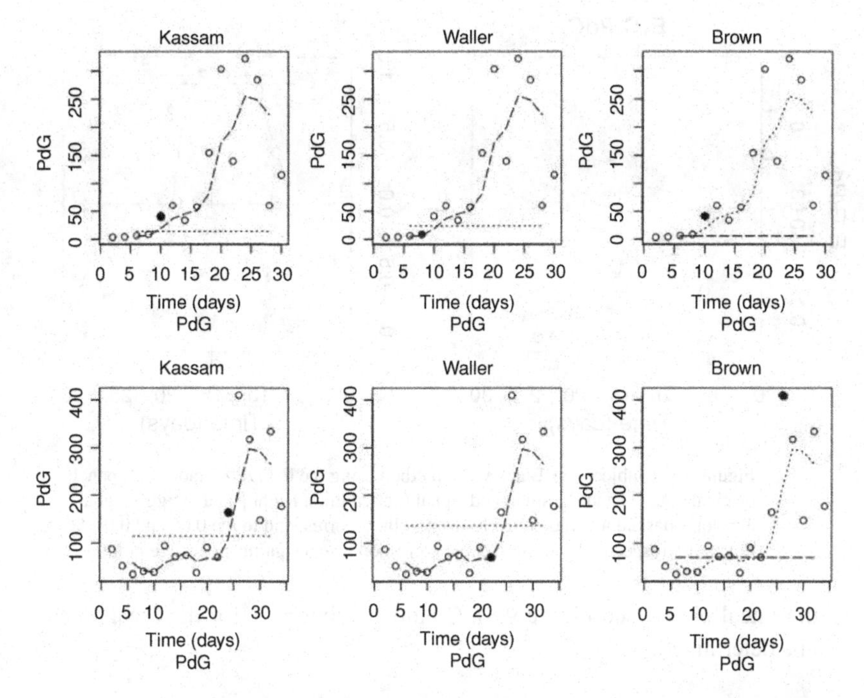

Figure 9.2. Simple threshold methods. Illustration of their respective outputs on two menstrual cycle urinary collections in two different women. Axis: day 0 is the first day of the cycle, i.e. the first day of menses. The empty circles are the measured PdG. The filled circles correspond to the day the luteal phase onset was detected. Horizontal lines: left and middle panels refer to the absolute threshold; right panel refers to the relative noise level.

Iterative testing of possible changing points, the best model being selected on the residuals sum of square (RSS) or the Akaike Information Criterion (AIC), does not allow derivation of a confidence interval for the detected day of luteal phase onset T_{ov}.. Non-linear regression with T_{ov}. as an unknown parameter along with the three parameters corresponding to the two linear parts of the PdG signal can provide an estimate of the standard error of T_{ov}., when the variance–covariance matrix is not singular, although it is known to be often too optimistic.

The PdG time-profile along a menstrual cycle often exhibits more than one change, leading to different potential algorithms and was not suitable in detecting a single point change in time. However the use of the *breakpoints* function from the R- package {*strucchange*}provided a very simple and efficient way of determining the breakpoint, reflecting the upward PdG surge at the beginning

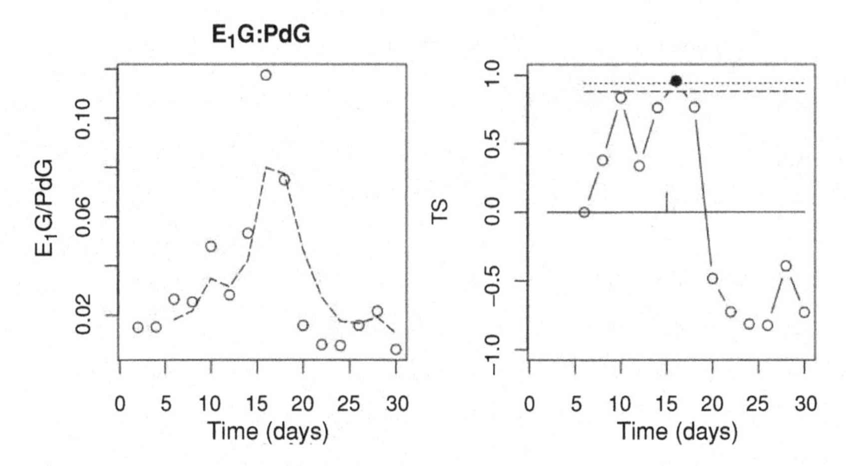

Figure 9.3. (Subject 65) Blackwell's method using the E_1G /PdG ratio. Left panel: original data (○) and the smoothed signal (dotted line). Right panel: Trigg's signal. The long-dashed and the dotted horizontal lines correspond to the 0.05 and 0.01 threshold, respectively. A filled circle corresponds to a significant change point.

of the luteal phase (see Figure 9.8). Confidence intervals for the breakpoints can be determined.

Correlation analysis

The results of the correlation analyses measuring the association between day of ovulation by method are shown in Table 9.2, but as so many of the results using the Baird method were missing, no analyses were attempted using this method. The pairwise method showed positive correlations with all the other methods, significantly so for all but the two Waller methods. The highest correlations were with the *nlr* and *Piecewise* methods.

Paired t-tests of the pairwise method with the other methods revealed that the day of ovulation was significantly later, using both Blackwell methods (pairwise – Blackwell95 = −2.393 days, $p = 0.045$ and pairwise – Blackwell99 = −3.393 days, $p = 0.008$) and the Brown method (pairwise – Brown = −3.048 days, $p = 0.018$).

Repeated measures analysis of variance across methods

The results showed overall that there were significant differences in day of ovulation within-subjects ($F_{(7,126)} = 9.808$, $p < 0.001$) but the pairwise method differed only from Blackwell, 1998 (mean difference of pairwise = −2.421,

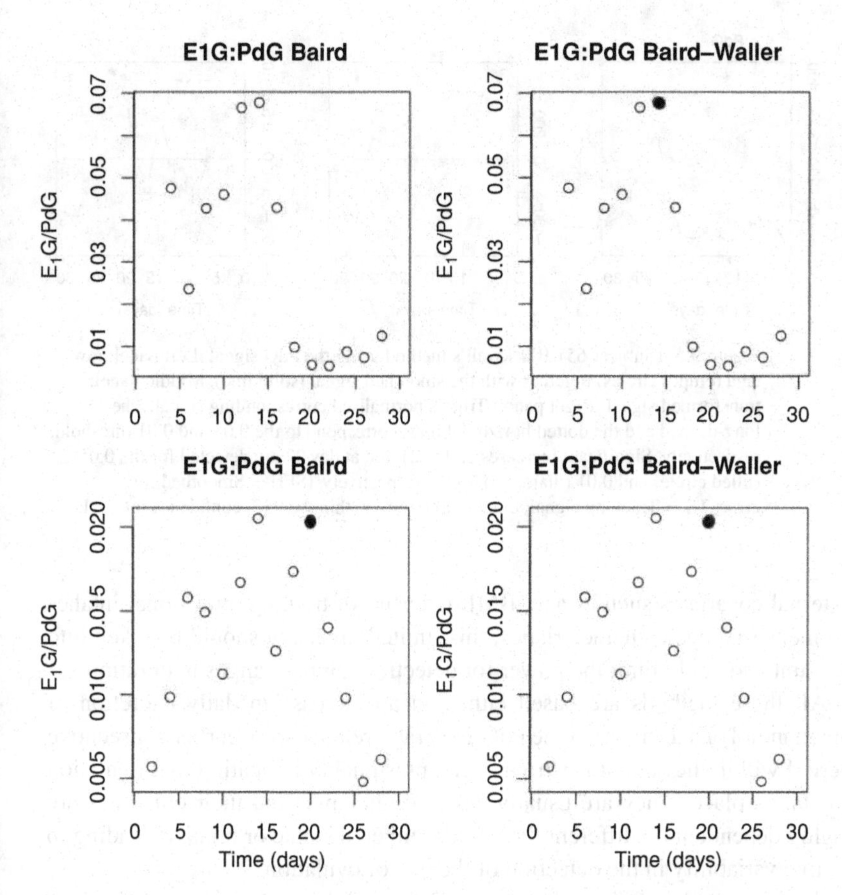

Figure 9.4. E_1G/PdG ratio. Baird's algorithm (left panel) and its modified version by Waller (right panel). Empty circles correspond to the raw data. Filled circles correspond to the detected change. Compare the lower panel with the corresponding PdG profile in Figure 9.6.

$p = 0.017$). As expected there was significant heterogeneity between subjects ($F_{(1,18)} = 400.382$, $p \ll 0.001$).

Discussion and conclusions

Our results show that the determination of the time of ovulation in humans appears to be highly dependent on the algorithms used, which can greatly impact on the estimation of the ovulatory rate and hence measurements of both the follicular and luteal phase durations. Hence, when studying the effects of

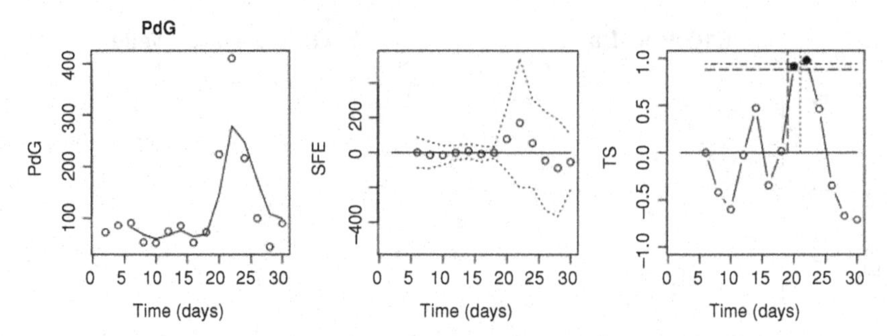

Figure 9.5. (Subject 65) Blackwell's method using the PdG signal. Left panel: raw data (empty circles) together with the smoothed signal (solid line). Middle panel: transformed signal. Right panel: Trigg's normalised corresponding signal. The long-dashed and the dotted horizontal lines correspond to the 0.05 and 0.01 threshold, respectively. Significant increases at day 21 and at day 22 are detected for the 0.05 (filled circle) and 0.01 (diamond) levels, respectively. (SFE = Smoothed forecast error; TS = Tracking signal; empty circles are within the 95% confidence interval.)

external covariates such as age, BMI, intensity of training, nutritional intakes on menstrual cycle characteristics, this limited precision should be taken into account and could limit the power for detecting minor changes in duration.

All these methods are based either on an increase in daily excretion of pregnandiol or a decrease of the PdG/E_1G ratio relative to an earlier consecutive period within the same menstrual cycle, presumably indicating that ovulation has taken place. They are usually easy to implement but their outcomes are highly dependent on different thresholds, either absolute or relative, leading to a large variability in the detection of the day of ovulation.

Some methods appear to be more satisfactory than others, i.e. more robust in detecting an upsurge in pregnanediol-3-alpha-glucuronide even in the presence of a less regular signal. The changing point detection methods appear to have more solid theoretical and statistical grounds and wide applications in different fields. Although they may appear more complex, they can be easily implemented now, using freely available software.

Field studies with urinary regular collection inevitably face the possibility of missing collection and/or time-lags in the schedule. Provided these protocol deviations are sporadic, simple procedures for objectively correcting the data set can be used. In addition the Piecewise regression method on log-transformed data seems to overcome some of these difficulties, except, of course, when missing data are around the ovulation time.

Collecting samples every other day instead of every day introduces a decrease in the precision of ovulation detection, if it occurs, but should not change the performances of the methods, provided the algorithms are adapted to this

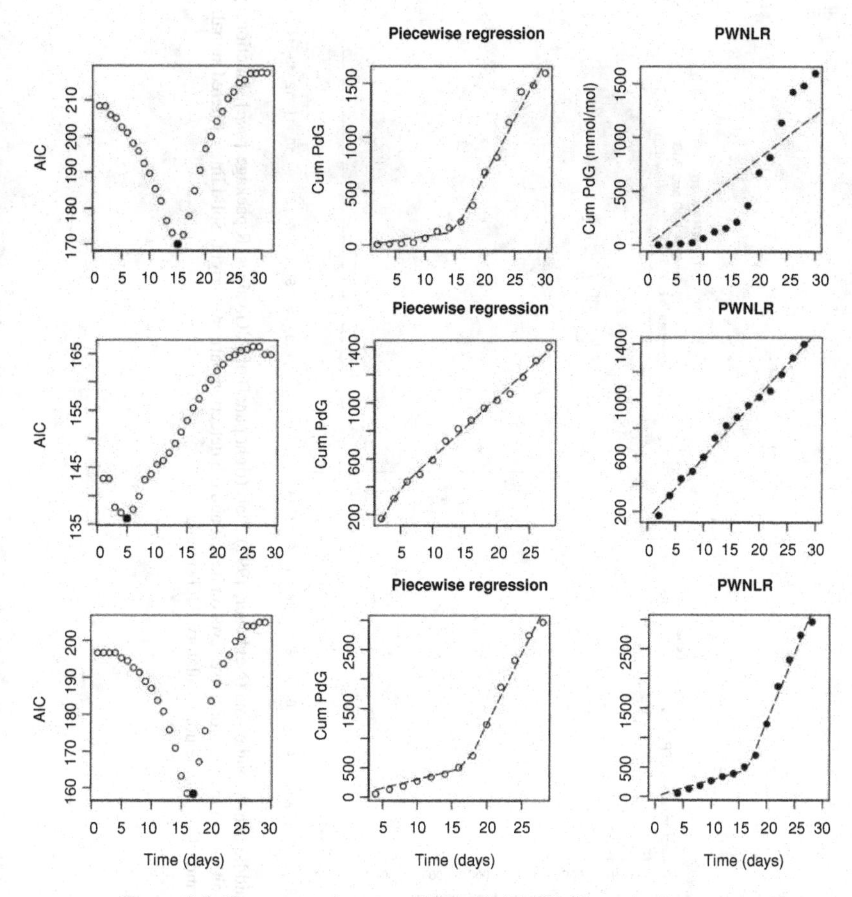

Figure 9.6. (From top to bottom: subject 14, 15, 52) Piecewise linear regression on the cumulated PdG values (Cum PdG). Right panel: use of a Piecewise non-linear regression (PWNLR) fit. Middle panel: iterative simple linear regression with best fit selected on the Akaike Information Criterion (AIC) (left panel). Only an upward slope change corresponds to a possible luteal phase onset.

sampling schedule. In contrast to the clinical situation, where an exact timing drives the medical attitude, this lack of precision appears acceptable in field studies. However, this imprecision should be taken into account when displaying summary results and performing comparisons.

The performances of these non-invasive methods for ovulation detection have often been criticised when compared with repeated blood samplings, which are validated in clinical settings but so far barely adapted to field studies, considering not only the acceptance by the woman but also all the logistics for conditioning, storing and transporting the collected samples under safe procedures.

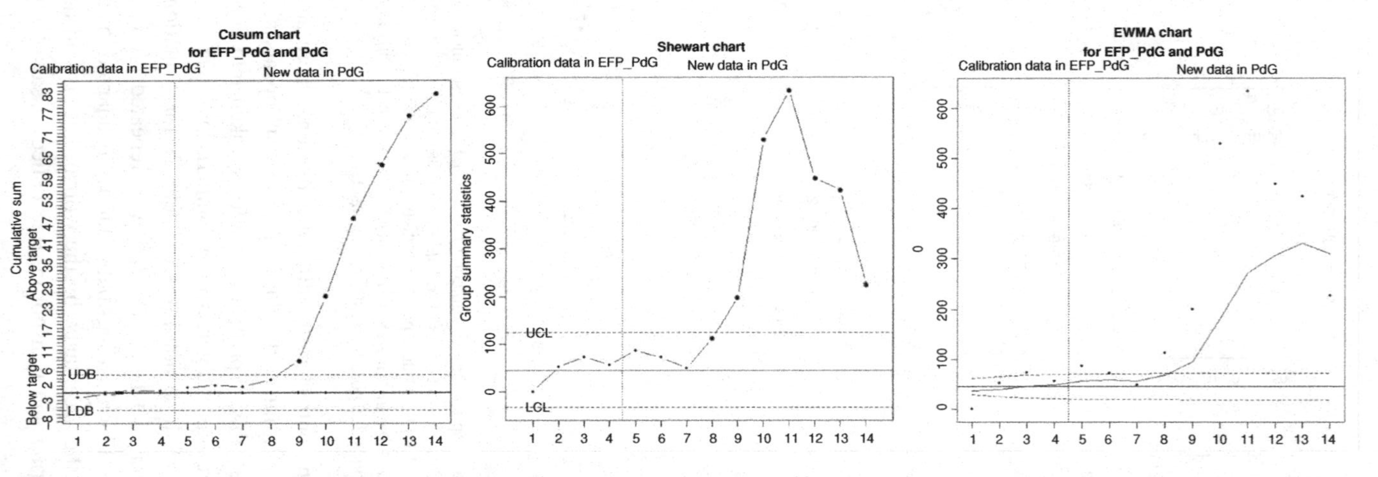

Figure 9.7. Cusum chart (left panel), Shewart chart (middle panel), and ewma (Neubauer, 1997) chart (right panel) of PdG values. R package {*qcc*}, functions *cusum*, *qcc*, *ewma*. The vertical line differentiates the initial training part for noise level on its left and the detection part on its right. Solid lines: filtered signal. Dotted lines: upper and lower confidence limits (UCL and LCL) for the non-ovulatory PdG signal.

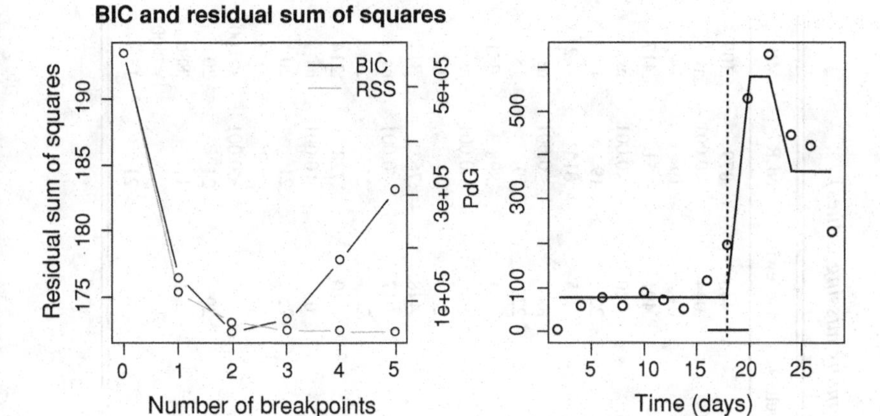

Figure 9.8. PdG urinary concentrations. Use of the R package *strucchange* for multiple breakpoints detection in a time series. BIC: Bayesian information criteria; RSS: residual sum of squares.The first upward change point is selected as the luteal phase onset (vertical dotted line).

However, the respective performances of serum, saliva, and urine collection in determining ovulation have already been extensively compared: Metcalf, Evans, and Mackenzie (Metcalf *et al.*, 1984) took urine, plasma, and saliva samples collected over a 24-hour period from 20 women during both the follicular and the luteal phases. The 24-hour excretion of pregnanediol-3-alpha-glucuronide was compared with (a) the concentration of progesterone in plasma, (b) the concentration of progesterone in saliva, (c) the concentration of pregnanediol-3-alpha-glucuronide in small urine samples, (d) the rate of excretion of pregnanediol-3-alpha-glucuronide, and (e) the ratio of pregnanediol-3-alpha-glucuronide to creatinine in small urine samples.

Although each analyte increased substantially during the luteal phase, the median increase, i.e. the ratio of luteal to follicular phase values, was 14.8, 3.2, 10.6, 11.9, and 11.1 for (a) to (e) respectively, while the median increase in the 24-hour pregnanediol-3-alpha-glucuronide output was 9.2. When the other analytes were used instead of the 24-hour excretion of pregnanediol-3-alpha-glucuronide to assess ovulation, the incidence of misclassification (follicular samples classified as luteal and vice versa) was 0, 12.8, 5.9, 2.0, and 1.0% for (a) to (e) respectively. The most satisfactory method was concentration in plasma and the least satisfactory was concentration of progesterone in saliva.

Urinary collection represents a satisfactory compromise in longitudinal studies over a long period of time, with limited and rather well-accepted constraints.

It is worth mentioning that these algorithms can easily incorporate additional species-specific physiologic knowledge, such as for instance a minimal

Table 9.2 *Correlation(r) matrix between nine methods (Baird excluded because of too many missing values).*

		Brown	Kassam	Waller 1	Waller 2	Blackwell95	Blackwell99	Piecewise	NLR	MBP
Brown	r	1	.557	.880	.752	.673	.704	.443	.879	.109
	p		.009	<0.001	<0.001	.001	<0.001	.044	<0.001	ns*
	N	21	21	21	21	21	21	21	19	14
Kassam	r		1	.630	.567	.684	.805	.491	.851	.417
	p			.002	.007	<0.001	<0.001	.020	<0.001	ns
	N		22	22	21	22	22	22	19	15
Waller 1	r			1	.783	.631	.688	.015	.815	−.213
	p				<0.001	.002	<0.001	ns	<0.001	ns
	N			22	21	22	22	22	19	15
Waller 2	r				1	.106	.100	.154	.898	.351
	p					Ns	Ns	ns	<0.001	ns
	N				25	25	25	25	20	17
Blackwell95	r					1	.758	.448	.762	.314
	p						<0.001	.017	<0.001	ns
	N					28	28	28	21	20
Blackwell99	r						1	.410	.728	.214
	p							.030	<0.001	ns
	N						28	28	21	20
Piecewise	r							1	.992	.972
	p								<0.001	<0.001
	N							29	21	20
NLR	r								1	.886
	p									<0.001
	N								21	15

* p ns.

period for follicular duration or, in animal studies, some external clues, such as in monkey ano-genital skin-colour change and coitus behaviour, in order to restrain the possible times of ovulation.

Appendix

Data were entered using a single Excel®-like spreadsheet and subsequently saved into a csv or text format, with a pre-defined semi-colon separator. Each subject was identified using a unique identifier "Iden":52, 17 . . .

An example of the data file is shown below.

A set of R-procedures has been developed allowing for 1) reading the data set; 2) extracting the data corresponding to one particular individual recording; 3) checking for missing or null data, with the possibility of either dismissing the data or correcting for missing values and/or unequal spacing between two samples, before running the various detection methods.

Iden	Days	PdG	E_1G
52	2	0	0.55
52	4	52.6	2.5
52	6	73.32	1.73
52	8	56.84	2.43
52	10	86.48	3.98
52	12	72.52	4.83
52	14	49.12	3.33
52	16	112.56	4.83
52	20	528.66	2.95
52	22	634.66	3.38
52	24	448.3	3.93
52	26	423.1	3.1
52	28	227.44	2.83
17	2	150.72	1.65
17	4	91.44	2.55
17	6	59	2.03
17	8	35	2.13
17	10	13.48	1.08
17	26	592.66	10.83
17	28	545.9	5.4
15	2	171.56	0.98
15	4	145.32	1.43
15	6	120.24	1.9
15	24	117.4	1.15
15	26	120.35	0.58
15	28	97.6	0.6
16	2	58	0.35
16	4	38.6	0.35

```
Syntax for using program R
# JCT September 2007- April 2009
# Main procedure part1.r
#.............................................................
# Procedures are supposed stored in a "Procedures" directory, the name has
to be adapted Procedures = "/media/JCT31012009/Travaux/LRosetta/
Procedures/"
#
# Loading utility functions, corresponding to the different algorithms
source(paste(Procedures,"Part0.r",sep="",collapse=NULL))
source(paste(Procedures,"Part2.r",sep="",collapse=NULL))
source(paste(Procedures,"Part3.r",sep="",collapse=NULL))
source(paste(Procedures,"Part4.r",sep="",collapse=NULL))
source(paste(Procedures,"Part5.r",sep="",collapse=NULL))
source(paste(Procedures,"Part7.r",sep="",collapse=NULL))
source(paste(Procedures,"Part8.r",sep="",collapse=NULL))
source(paste(Procedures,"Part9.r",sep="",collapse=NULL))
#
#.............................................................


# Data input (adapt the directory and filename)
# Input file stored in csv format from Excel
# First line: names of the variables
# Next lines: individual concentrations, separated by ";"
# One line for each measurement time
# WARNING : should be adapted to the location of the data set
#
# Name of the directory where the data sets are stored. Give the complete path
Location ="/media/JCT31012009/Travaux/LRosetta/Donnees/DonneesLR/
Data_Joggueuses/"
# File Name
Donnees_Totales = "Datalyr_30082007.csv"
#
fichier = paste(Location,Donnees_Totales,sep="",collapse=NULL)
# File data input. Here each record is composed of data separated by
semi-columns donnees =read.table(fichier,header=TRUE,sep=";") # adapt ";"
if another separation mark
#
# List of all the different individuals in the file
Subject_List = unique(donnees$Iden)
```

```
#
#........................................................

# Selection of the number of points for averaging according to sample
collection frequency
#
# N = 6    # Urinary samples collected every day
# N = 3    # Urinary samples collected every other day
N = 3
directory="/media/JCT31012009/Travaux/LRosetta/"
#
# Choice of one or several particular individuals in the subjects list
#The values (52, 15, 17,..) are the identifiers of the selected subjects
#
Recording_ID = c(52,15,17,65,75)

Individual = matrix(apply(matrix(Recording_ID,ncol=1),1,Question,
Subject_List),ncol=1)
#
# End of data preparation
# ==============================================


# Call for the various methods. Can be single and/or in any order
# Call for Blackwell method on PdG values for the records in the Individual list
Hormone = "PdG"
Blackwell = data.frame(t(apply(Individual,1,part2,donnees,directory,
Hormone,N)))
names(Blackwell) = c("Iden","Hormone","Blackwell95","Blackwell99")
# Results display
Blackwell
#........................................................
# Call for the group of the Simple Threshold methods (Kassam, Waller, Brown)
Hormone = "PdG"
KWB_Threshold = data.frame(t(apply(Individual,1,part3,donnees,directory,
Hormone,N)))
names(KWB_Threshold) = c("Iden","Hormone","Kassam","Waller","Brown")
# Partial results display
KWB_Threshold
#........................................................
# Call for Blackwell method applied to E1/PdG ratio
```

```
Hormone = "E1G:PdG"
BlackwellE1PdG = data.frame(t(apply(Individual,1,part4,donnees,directory,
Hormone,N)))
names(BlackwellE1PdG) = c("Iden","Hormone","Blackwell95","Blackwell99")
# Partial results display
BlackwellE1PdG
#.................................................................
# Call for Baird & Waller methods applied to E1G:PdG data
BairdWaller = data.frame(t(apply(Individual,1,part5,donnees,directory,
"E1G:PdG",N)))
names(BairdWaller) = c("Iden","Hormone","Baird","BairdWaller")
# Partial results display
BairdWaller
#.................................................................
# Call for Multiple Break Points Detection
library(strucchange)
MultipleBP = data.frame(t(apply(Individual,1,MultipleBreakPoints,donnees,
directory,"PdG")))
names(MultipleBP) = c("File_Iden","Subject","Hormone","ChangePoint",
"Inf_Lim","Sup_Lim")
# Partial results display
MultipleBP
#.................................................................
# Call for Piecewise linear regression on the cumulated PdG values
#
PiecewiseLR = data.frame(t(apply(Individual,1,Piecewise,donnees,directory,
"PdG")))
names(PiecewiseLR) = c("File_Iden","Subject","Hormone","PWLR_
ChangePoint","Slope1","Slope2")
# Partial results display
PiecewiseLR
#.................................................................
# Call for a NLRegression model
NLRegression = data.frame(t(apply(Individual,1,Detector,donnees,directory,
"PdG",0)))names(NLRegression) = c("Iden","Sujet","Hormone","Origin",
"Slope1","Slope2","T","Sigma_T")
# Partial results display
NLRegression
#
# Call for the cusum method
```

```
Cusum.res = apply(matrix(Individual,ncol=1),1,part9,donnees,directory,
"PdG",3)

Cusum.res = data.frame(t(Cusum.res))

Cusum.res = data.frame(Cusum.res[,1],Subject_List[Individual],
Cusum.res[,2:4])


# Summary of the all the results from the previous methods corresponding to
the subjects listed in Individual
Recap_Total=data.frame(KWB_Threshold, Blackwell[,2:4],
BlackwellE1PdG[,2:4],BairdWaller[,2:4], MultipleBP[,2:5])
Recap_Total


# Part0. JCT July 2007- April 2009
# Some utility functions
# Data preparation
#Function for replacing missing data using linear interpolation
interpol = function(t,temps,x)
{
temps = c(temps,999)
x = c(x,1)
u_inf = max(which(temps <= t))
u_sup = min(which(temps > t))
z = x[u_inf]+ (x[u_sup]-x[u_inf])*(t-temps[u_inf])/(temps[u_sup]-temps[u_inf])
return(z)
}
# . . . . . . . . . . . . . . . . . . . . . . . . . . . . . . . . . .
# Small procedure to avoid null values
# It simply replaces a null value with the mean of the two adjacent values.
When either the first
# or the last values of the vector is null, the vector is reduced by the
corresponding value
#
missing_data = function(temps,x)
{
x = ifelse(x <= 0.0, NA, x)
if(is.na(x[1])) {x = x[-1]; temps=temps[-1]}
if(is.na(x[length(x)])) {x = x[1:(length(x)-1)]; temps =
```

```
temps[1:(length(temps)-1)]}
y1 = c(x[-1],x[length(x)])
y2 = c(x[1],x[-length(x)])
y = ifelse(is.na(x), (y1+y2)*0.5,x)
z = cbind(temps,y)
return(z)
}
#. . . . . . . . . . . . . . . . . . . . . . . . . . . . . . . . .
CleaningLady = function(time,x)
# Procedure which suppresses missing data and create a new data set, with
equally spaced datapoints over the same period of time
#
{
plot(time,x)
time = time[!is.na(x)]
x = x[!is.na(x)]
sequence = as.matrix(seq(time[1],time[length(time)],length=length(time)),
ncol=1)
y = apply(sequence,1,interpol,time,x)
points(sequence,y,col="red")
time = sequence
z = data.frame(time,y)
return(z)
}
#
++++++++++++++++++++++++++++++++++++++++++++++++++++++++++
+++++
# Analysis of urinary hormonal profile
# Routines for the Blackwell's Method
# ============================
# Exponential smoothing
# Conceptually analogous to a Moving Average MA(N) on the same data set
# alpha is related to the MA(N) by the relation alpha= 2/(N+1)
#
esa = function(x,y,alpha)
{
z = alpha*x +(1-alpha)*y
return(z)
}
# Difference between the forecast and the actual value
#
```

```
fe = function(x,y)
{
z = x - y
return(z)
}
# Exponentially smoothed forecast error = measure of the departure from the
established values
#
sfe = function(x,y,alpha)
{
z = alpha*x +(1-alpha)*y
return(z)
}
# Mean absolute deviation
# Exponentially smoothed absolute forecast error
# Index of the degree of random variation of the data
#
mad = function(x,y,alpha)
# mad= Mean Absolute Deviation MAD
# x = Absolute value of the latest error
# y = previous MAD
{
z = alpha*abs(x)+(1-alpha)*y
return(z)
}
# Tracking signal
# TS(n) = SFE(n)/MAD(n)
#
ts = function(x,y)
# x = smoothed forecast error SFE
# y = MAD
{
z = x/y
return(z)
}
# Standard deviation at time n
#
ds= function(x,alpha)
{
z = sqrt(pi/2)*sqrt((2-alpha)/2)*x
return(z)
```

```
}
# +++++++++++++++++++++++++++++++++++++++++++++++++++++++++++
# Functions used by the Baird's algorithm
critere1 = function(x)
{
flag= ifelse(x[1] == max(x),1,0)
return(flag)
}
#
critere2 = function(x,threshold,size)
{
taille = length(x)
z = max(x[(taille-1*is.numeric(size > 3)):taille])
flag=ifelse(z <= threshold*x[1],1,0)
return(flag)
}
# +++++++++++++++++++++++++++++++++++++++++++++++++++++++++++
# Function used by the Piecewise regression method
#
critere3 = function(x,y)
{
z = (x-y)*as.numeric(x > y)
return(z)
}
# +++++++++++++++++++++++++++++++++++++++++++++++++++++++++++
# Retrieve the index of a row corresponding to a specific recording within a
table Question = function(x,tableau)
{
z = which(tableau == x)
return(z)
}
#


# Part2.r July 2007- April 2009
# JCT
# Blackwell's algorithm for a single hormone
# Analysis of the data set corresponding to a specific Individual
# N should be adapted to the urinary collection frequency
# ====================================
part2 = function(Individual,donnees,directory,Hormone,N)
```

```
# Input variables
# Individual: file number of the selected subject
# donnees: name of the data set file
# directory: location of the data set file
# Hormone: name of the urinary metabolite of concern with quotes "PdG" or
"E1G"
# N          : number of selected points for averaging (3 or 6)
{
# Data extraction for both the selected individual and the
# selected hormone metabolite
#
data_set = donnees[donnees$Iden == Subject_List[Individual],]
Titre = paste(Hormone,"F",data_set$Iden[1],sep="_",collapse=NULL)
data.out = paste(directory,"Sorties/Blackwell",Titre,sep="",collapse=NULL)
data.out = paste(data.out,"jpeg",sep=".")
Variable = colnames(data_set,do.NULL=FALSE)==Hormone
#
# Correction for missing data
data_rec1 = missing_data(data_set$Jours,data_set[,Variable])
# Data set ready. Analysis according to Blackwell method
#
temps = data_rec1[,1]
u=data_rec1[,2]
nmax = length(u)
t_max = max(temps,30)
#t_max = max(temps,40)
Seuil3 = c(0.88,0.691)
Seuil6 = c(0.94,0.804)
Size = c(3,6)
Seuil = data.frame(Size,Seuil3,Seuil6)
# Initialisation
#
alpha = 2/(N+1)
ESA = exp(mean(log(u[1:N])))
FE = 0
SFE = 0
TS = 0
lambda_sigma = sd(log(u[1:N]))
sigma = sqrt((ESA^2)*exp((log(10)*(lambda_sigma^2)/0.4343)-1))
MAD = sqrt(2/pi)*sigma
SD = sqrt(pi/2)*sqrt((2-alpha)/2)*MAD
```

```
#
for(n in (N+1):nmax)
{
taille = length(ESA)
FE = c(FE, fe(u[n],ESA[taille]))
SFE = c(SFE,sfe(FE[taille+1],SFE[taille],alpha))
MAD =c(MAD, mad(FE[taille+1],MAD[taille],alpha))

ESA =c(ESA,esa(u[n],ESA[taille],alpha))

TS = c(TS, ts(SFE[taille+1],MAD[taille+1]))
SD =c(SD,ds(MAD[taille+1],alpha))
}
# Graphing the results
op <- par(mfrow = c(1, 3), pty = "s") # Graph window partitioned into 3
horizontal subplots


plot(temps,u,main=Titre,xlim=c(1,t_max),xlab="Time (days)",ylab=Hormone)
temps_mod = temps[N:nmax]
lines(temps_mod,ESA,col="red")
Limsup = SFE+1.96*SD
Liminf = SFE-1.96*SD
Y_Scale = max(c(Limsup[!is.na(Limsup)],abs(Liminf[!is.na(Liminf)])))
#
plot(temps_mod,SFE,xlim=c(1,t_max),ylim =c(-Y_Scale,Y_Scale),xlab=
"Time (days)")
lines(temps,rep(0,length(temps)),col="red")
lines(temps_mod,Limsup,col="purple")
lines(temps_mod,Liminf,col="purple")
plot(temps_mod,TS,col="blue",type="b",xlim=c(1,t_max),ylim=c(-1,+1),xlab=
"Time (days)",pch=21)
lines(temps,rep(0,length(temps)),col="red")
conflevel95=rep(Seuil[Seuil$Size == N,2],length(TS))
conflevel99 =rep(Seuil[Seuil$Size == N,3],length(TS))
lines(temps_mod,conflevel95,col="red")
lines(temps_mod,conflevel99,col="purple")
temps95 = ifelse(is.finite(min(temps_mod[TS[!is.na(TS)] >= Seuil[Seuil$Size
== N,2]])),min(temps_mod[TS[!is.na(TS)] >= Seuil[Seuil$Size == N,2]]),999)
temps99 = ifelse(is.finite(min(temps_mod[TS[!is.na(TS)] >= Seuil[Seuil$Size
== N,3]])),min(temps_mod[TS[!is.na(TS)] >= Seuil[Seuil$Size == N,3]]),999)
```

```
if(temps95 < 999)
{
points(temps95,TS[(temps_mod==temps95)&!is.na(TS)],col="red",pch=19)
Temps_Ov95 = temps95-1
lines(c(Temps_Ov95,Temps_Ov95),c(0,Y_Scale),lty="dashed",col="pink")
}
if(temps99 < 999)
{
points(temps99,TS[(temps_mod==temps99)&!is.na(TS)],col="blue",pch=19)
Temps_Ov99 = temps99-1
lines(c(Temps_Ov99,Temps_Ov99),c(0,Y_Scale),lty="dotted",col="purple")
}
par(op)
savePlot(data.out,"jpeg") # Graph output in jpeg format
#
# Detected times returned
Blackwell95 =ifelse(temps95 < 999,Temps_Ov95,999)
Blackwell99 =ifelse(temps99 < 999,Temps_Ov99,999)
z = matrix(c(Individual,Hormone,Blackwell95,Blackwell99),ncol=1)
return(z)
}


# September 2007. JCT. Revised april 2008
#
# Threshold detection algorithms
# Kassam, Waller (modified Kassam), Brown
# Part3.r
#
=============================================
===
#
part3 = function(Individual,donnees,directory,Hormone,N)
# Input variables
# Individual: file number of the selected subject
# donnees: name of the data set file
# directory: location of the data set file
# Hormone: name of the urinary metabolite of concern with quotes "PdG" or
"E1G"
# N          : number of selected points for averaging (3 or 6)
{
```

```
#
# Data extraction
data_set = donnees[donnees$Iden == Subject_List[Individual],]
Titre = paste(Hormone,"F",data_set$Iden[1],sep="_",collapse=NULL)
data.out = paste(directory,"Sorties/ThresholdKWB",Titre,sep="",collapse=
NULL)
data.out = paste(data.out,"jpeg",sep=".")
Variable = colnames(data_set,do.NULL=FALSE)==Hormone
data_set[,Variable]
# Correction for missing data
#
data_rec1= missing_data(data_set$Jours,data_set[,Variable])
# The graph window is split into 1 x 3 subwindows
op <- par(mfrow = c(1, 3), pty = "s")
# Algorithm 2: relative threshold (Kassam)
#
temps = data_rec1[,1]
u=data_rec1[,2]
t_max=max(temps)
#
y = u
signal = NULL
indice = length(y)-N
for(i in 1:indice)
{
signal = c(signal,mean(y[i:(i+N-1)]))
}
baseline = min(signal)
day = 999
plage= length(y)-(N-2)
for(i in N:plage)
{
flag=ifelse(min(y[i:(i+ N-2)]) >= 3*baseline,1,0)
if(flag==1) {day = i; break}
}
plot(temps,y,xlab="Time (days)",ylab="PdG",xlim=c(1,t_max))
title(main="Kassam",sub=Titre)
lines(temps[N:(length(temps)-1)],signal,col="red")
lines(temps[N:(length(temps)-1)],rep(3*baseline,length(signal)),col="blue")
points(temps[day],y[day],col="darkblue",pch=19)
PdG_Kassam = ifelse(day < 999, temps[day], 999)
```

```
#
# Algorithm 3: Waller (Modified Kassam)
#
signal = NULL
indice = length(y)-N
for(i in 1:indice)
{
signal = c(signal,mean(y[i:(i+N-1)]))
}
baseline = min(signal)
threshold = baseline + 1 + sqrt(baseline)

day = 999
plage= length(y)-N
for(i in N:plage)
{
aux1 = sort(y[i:(i+N)])[2]
aux2 = max(y[i:(i+N)])
flag=ifelse((aux1 >= 3*threshold)&(aux2>= 3*threshold),1,0)
if(flag==1) {day = i; break}
}
plot(temps, y,xlab="Time (days)",ylab="PdG",xlim=c(1,t_max))
title(main="Waller",sub=Titre)
lines(temps[N:(length(temps)-1)],signal,col="darkblue")
lines(temps[N:(length(temps)-1)],rep(3*threshold,length(signal)),col="pink")
points(temps[day],y[day],col="darkblue",pch=19)

PdG_Waller = ifelse(day < 999, temps[day], 999)
#
# Algorithm 4: Brown Threshold/duration method
# ===================================
signal = mean(y[1:N])
SD = sd(y[1:N])
threshold=3*SD
indice = length(y)-N
for(i in 2:indice)
{
signal = c(signal,mean(y[i:(i+N-1)]))
SD = c(SD,sd(y[i:(i+N-1)]))
}
day = 999
```

```
plage= length(y)-N+1
for(i in N:plage)
{
aux1 = y[i:(i+N-3)]
aux2 = signal[i-2]
flag=ifelse(min(aux1-rep(aux2,N)) >= threshold,1,0)
if(flag==1) {day = i; break}
threshold=3*mean(SD[1:(i-2)])
}
plot(temps,y,main="Brown",xlab="Time (days)",ylab="PdG",xlim=c(1,t_max))
title(main="Brown",sub=Titre)
lines(temps[N:(length(temps)-1)],signal,col="darkblue")
lines(temps[N:(length(temps)-1)],rep(threshold,length(signal)),col="pink")
points(temps[day],y[day],col="darkblue",pch=19)
PdG_Brown = ifelse(day < 999, temps[day],999)
#
# End of the partitioned window
par(op)
savePlot(data.out,"jpeg")
z = c(Individual,Hormone,PdG_Kassam,PdG_Waller,PdG_Brown)
return(z)
# End of part3.r
}


# Part4.fr
# October 2007, JCT
==================================================
====
# Study of the E1G/PdG ratio. Blackwell. Threshold 0.95 and 0.99
#
part4 = function(Individual, donnees,directory,Hormone,N)
# Input variables
# Individual: file number of the selected subject
# donnees: name of the data set file
# directory: location of the data set file
# Hormone: name of the urinary metabolite of concern with quotes "E1GPdG"
# N          : number of selected points for averaging (3 or 6)
{
# Data extraction
#
```

```
data_set = donnees[donnees$Iden == Subject_List[Individual],]
Titre = paste(Hormone,"F",data_set$Iden[1],sep="_",collapse=NULL)
data.out = paste(directory,"Sorties/Blackwell_",Titre,sep="",collapse=NULL)
data.out = paste(data.out,"jpeg",sep=".")
Variable = colnames(data_set,do.NULL=FALSE)== "PdG"
data_set[,Variable]
data_rec1= missing_data(data_set$Jours,data_set[,Variable])
Variable = colnames(data_set,do.NULL=FALSE)== "E1G"
data_set[,Variable]
# Data checking and correction
#

data_rec2= missing_data(data_set$Jours,data_set[,Variable])
temps = data_rec2[data_rec2[,1] %in% data_rec1[,1],1]
u=data_rec2[data_rec2[,1] %in% data_rec1[,1],2]/data_rec1[,2]
# Initialisation
#
nmax = length(u)
t_max = max(temps,30)
Seuil3 = c(0.88,0.691)
Seuil6 = c(0.94,0.804)
Size = c(3,6)
Seuil = data.frame(Size,Seuil3,Seuil6)
alpha = 2/(N+1)
ESA = exp(mean(log(u[1:N])))
FE = 0
SFE = 0
TS = 0
lambda_sigma = sd(log(u[1:N]))
sigma = sqrt((ESA^2)*exp((log(10)*(lambda_sigma^2)/0.4343)-1))
MAD = sqrt(2/pi)*sigma
SD = sqrt(pi/2)*sqrt((2-alpha)/2)*MAD
#
# Filtering the signal

for(n in (N+1):nmax)
{
taille = length(ESA)
FE = c(FE, fe(u[n],ESA[taille]))
SFE = c(SFE,sfe(FE[taille+1],SFE[taille],alpha))
MAD =c(MAD, mad(FE[taille+1],MAD[taille],alpha))
```

```
ESA =c(ESA,esa(u[n],ESA[taille],alpha))
TS = c(TS, ts(SFE[taille+1],MAD[taille+1]))
SD =c(SD,ds(MAD[taille+1],alpha))
}
#
# Graphing the results

op <- par(mfrow = c(1, 2), pty = "s")
plot(temps,u,main=Titre,xlim=c(1,t_max),xlab="Time (days)",ylab=Hormone)
temps_mod = temps[N:nmax]
lines(temps_mod,ESA,col="red")
Limsup = SFE+1.96*SD
Liminf = SFE-1.96*SD
Y_Scale = max(c(Limsup[!is.na(Limsup)],abs(Liminf[!is.na(Liminf)])))
#plot(temps_mod,SFE,xlim=c(1,t_max),ylim =c(-Y_Scale,Y_Scale),xlab=
"Time (days)")
#lines(temps,rep(0,length(temps)),col="red")
#lines(temps_mod,Limsup,col="purple")
#lines(temps_mod,Liminf,col="purple")
plot(temps_mod,TS,col="blue",type="b",xlim=c(1,t_max),ylim=c(-1,+1),xlab=
"Time (days)",pch=21)
lines(temps,rep(0,length(temps)),col="red")
conflevel95=rep(Seuil[Seuil$Size == N,2],length(TS))
conflevel99 =rep(Seuil[Seuil$Size == N,3],length(TS))
lines(temps_mod,conflevel95,col="red")
lines(temps_mod,conflevel99,col="purple")
#
# Comparing the suject's filtered values with the 95% and 99% threshold
values
# Set to 999 si no value above the threshold
#

temps95 = ifelse(is.finite(min(temps_mod[TS[!is.na(TS)] >= Seuil[Seuil$Size
== N,2]])),min(temps_mod[TS[!is.na(TS)] >= Seuil[Seuil$Size == N,2]]),999)
temps99 = ifelse(is.finite(min(temps_mod[TS[!is.na(TS)] >= Seuil[Seuil$Size
== N,3]])),min(temps_mod[TS[!is.na(TS)] >= Seuil[Seuil$Size == N,3]]),999)
if(temps95 < 999)
{
points(temps95,TS[(temps_mod==temps95)&!is.na(TS)],col="red",pch=19)
Temps_Ov95 = temps95-1
lines(c(Temps_Ov95,Temps_Ov95),c(0,Y_Scale),lty="dashed",col="pink")
```

```
}
if(temps99 < 999)
{
points(temps99,TS[(temps_mod==temps99)&!is.na(TS)],col="blue",pch=19)
Temps_Ov99 = temps99-1
lines(c(Temps_Ov99,Temps_Ov99),c(0,Y_Scale),lty="dotted",col="purple")
}
par(op)
#
savePlot(data.out,"jpeg") # Saving the plot into ªjpeg" format
#
Blackwell95 =ifelse(temps95 < 999,Temps_Ov95,999)
Blackwell99 =ifelse(temps99 < 999,Temps_Ov99,999)
#
# Returning the two time points, together with the file id and the corresponding
# identification number
z = matrix(c(Individual,Subject_List[Individual],Hormone,Blackwell95,
Blackwell99),ncol=1)
return(z)
}
#
=====================================================
===
# JCT August 2007
# Package part5.r
# DD Baird's method and Waller corrected Baird's method
# ====================
part5 = function(Individual,donnees,directory,Hormone,N)
# Input variables
# Individual: file number of the selected subject
# donnees: name of the data set file
# directory: location of the data set file
# Hormone: name of the urinary metabolite of concern with quotes "E1GPdG"
# N          : number of selected points for averaging (3 or 6)
{
# Data extraction
Subject_List = unique(donnees$Iden)
data_set = donnees[donnees$Iden == Subject_List[Individual],]
Titre = paste(Hormone,paste("F",data_set$Iden[1],sep=""),
sep="_",collapse=NULL)
data.out = paste(directory,"Sorties/BairdWaller",Titre,sep="",collapse=NULL)
```

```
data.out = paste(data.out,"jpeg",sep=".")
Variable = colnames(data_set,do.NULL=FALSE)== "PdG"
data_rec1= missing_data(data_set$Jours,data_set[,Variable])
#
Variable = colnames(data_set,do.NULL=FALSE)== "E1G"
data_rec2= missing_data(data_set$Jours,data_set[,Variable])
#
# Complete values with both PdG and E1G values
# Calculation of the E1G/ PdG ratios
#
temps = data_rec2[data_rec2[,1] %in% data_rec1[,1],1]
y = data_rec2[data_rec2[,1] %in% data_rec1[,1],2]/data_rec1[,2]
#
seuil_baird = 0.40
seuil_waller= 0.60
t_max = max(temps,30)
Longueur_y=length(y)
Day1=NULL
Indice_shift       = NULL
Value_shift        = NULL
Baird              = NULL
Waller             = NULL
#...............................

for(i in 1:(Longueur_y-N))
{
k=i+N
x = y[i:k]
z = critere1(x)*critere2(x,seuil_baird,N)
          if(z == 1)
          {
          Day1 = c(Day1,i)
# points(temps[i],y[i],col="red",pch=19)
          Indice_shift = c(Indice_shift,(i+1))
Value_shift = c(Value_shift,x[2])
#         i = i+ N
          }
}
Value_Comp = (y[Day1-1]+y[Day1+1])/2
if(length(Value_Comp) == 1)
{
```

```
        Baird = 1
        Waller = 1
Recap=data.frame(data_set$Iden[1],temps[Day1],Indice_shift,Value_shift,
Value_Comp,Baird,Waller)
}

if(length(Value_Comp) > 1)
{
for(i in 1:length(Value_Comp))
        {
        Baird = c(Baird,ifelse(Value_Comp[i] > 2*max(Value_Comp[-i]),1,0))
        Waller = c(Waller,ifelse(Value_Comp[i] > max(Value_Comp[-i]),1,0))
        }
Recap=data.frame(data_set$Iden[1],temps[Day1],Indice_shift,Value_shift,
Value_Comp,Baird,Waller)
}
Recap
BValue = ifelse(max(Baird) > 0,y[Day1[Baird == 1]],-1)
BTime = ifelse(max(Baird) > 0,temps[Day1[Baird == 1]],999)
WValue = ifelse(max(Waller) > 0,y[Day1[Waller == 1]],-1)
WTime = ifelse(max(Waller) > 0,temps[Day1[Waller == 1]],999)

# Graph of the detected change point
op= par(mfrow=c(1,2),pty = "s")
Titre5 = paste(Titre,"Baird",sep=" - ")
plot(temps,y,xlim=c(1,t_max),main= Titre5, xlab="Time (days)",ylab=
Hormone)
points(BTime,BValue,col="blue",pch=19)
Titre5 = paste(Titre,"Baird- Waller",sep=" - ")
plot(temps,y,xlim=c(1,t_max),main= Titre5, xlab="Time (days)",ylab=
Hormone)
points(WTime,WValue,col="red",pch=19)
par(op)
#
savePlot(data.out,"jpeg") # Saving the graph into a jpeg formatted file
#
# Detected points are returned
Baird_Time = BTime
Waller_Time = WTime
z = matrix(c(Individual,Subject_List[Individual],Hormone,Baird_Time,
Waller_Time),ncol=1)
```

```
return(z)
}


# Part6.r    JCT September 2007. Revised 04/2009
#
 ================================================
==
# Necessary packages:
# library(lmtest)
# library(strucchange)
# Rob Hyndman & Muhamed Abkam
#.................................
MultipleBreakPoints = function(Individual,donnees,directory,Hormone)
# Input variables
# Individual: file number of the selected subject
# donnees: name of the data set file
# directory: location of the data set file
# Hormone: name of the urinary metabolite of concern with quotes "PdG"
# N          : number of selected points for averaging (3 or 6)
{
# Data extraction
Subject_List = unique(donnees$Iden)
data_set =donnees[donnees$Iden == Subject_List[Individual],]
#
Titre = paste("Strucchange","_F",Subject_List[Individual],sep="",collapse=
NULL)
Dir_Sortie = paste(directory,"/Sorties/lmtest/",sep="",collapse=NULL)
Out = paste(Dir_Sortie,Titre,".jpeg",sep="",collapse=NULL)
data.out = paste(directory,"/Sorties/",Titre,".csv",sep="",collapse=NULL)
#
lmtest.res = rep(999,length(Subject_List))
lmtest.inf = lmtest.res
lmtest.sup = lmtest.res
#
Variable = colnames(data_set,do.NULL=FALSE)== Hormone
data_set[,Variable]
# Data preparation
Extrait= missing_data(data_set$Jours,data_set[,Variable])
#
Var_sup = cumsum(Extrait[,2])
Extrait = data.frame(Extrait,Var_sup)
```

```
#
# Graphing the results (2 horizontal graphs)
#
op <- par(mfrow = c(1, 2), pty = "s")
  y = Extrait[,2]
  bp <- breakpoints(y ~ 1,h = 3)
  plot(bp)
  fm.seg <- lm(y ~ 0 + breakfactor(bp))
  IC95 = confint(bp)
  IC95$confint = 2*IC95$confint
  IC95$datatsp = c(1/max(Extrait[,1]),1,max(Extrait[,1]))
  IC95$nobs= max(Extrait[,1])
#
  plot(Extrait[,1],y,xlab="Time (days)",ylab="PdG",main=Titre)
  lines(Extrait[,1], fitted(fm.seg), col = 4)
  lines(IC95)
  lines(IC95$confint[1,c(1,3)],c(5,5),col="red")
  lines(rep(IC95$confint[1,2],2),c(-1,600),col="red",lty="dotted")
  savePlot(Out,type="jpeg") # Saving the plot in ªjpeg" format
par(op)
#
lmtest.res=IC95$confint[1,2]
lmtest.inf=IC95$confint[1,1]
lmtest.sup=IC95$confint[1,3]
result = c(Individual,Subject_List[Individual],Hormone,lmtest.res,lmtest.inf,
lmtest.sup)
return(result)
}


# Part7.r    JCT September 2007- April 2009
# ====================
## Piecewise_Regr.r
#
Piecewise = function(Individual,donnees,directory,Hormone)
#
# Input variables
# Individual: file number of the selected subject
# donnees: name of the data set file
# directory: location of the data set file
# Hormone: name of the urinary metabolite of concern with quotes "E1GPdG"
```

```
# N          : number of selected points for averaging (3 or 6)
{
# Data extraction
Subject_List = unique(donnees$Iden)
Result = NULL
data_set=donnees[donnees$Iden == Subject_List[Individual],]
Titre = paste("Piecewise Regression","_F",Subject_List[Individual],sep="",
collapse=NULL)
Variable = colnames(data_set,do.NULL=FALSE)==Hormone
#
# Data preparation
data_rec1= missing_data(data_set$Jours,data_set[,Variable])
#
# Individual cumulated hormone level over time
Var_sup = cumsum(data_rec1[,2])
Extrait = data.frame(data_rec1,Var_sup)

#
===========================================================
===
# Piecewise regression
# Selection using the AIC criterion
# Only Possible T values are investigated
#. . . . . . . . . . . . . . . . . . . . . . . . . . . . . . . . .
# Select the directory you want your outputs redirected to
Dir_Sortie = paste(directory,"/Sorties/Piecewise/",sep="",collapse=NULL)
Out = paste(Dir_Sortie,"PR_F",Subject_List[Individual],sep="",collapse=
NULL)
data.out = paste(Out,".jpeg",sep="",collapse=NULL)
#
AIC_Choice = NULL
for(i in 1:(max(Extrait$temps)+1))
{
T_Ov = i
xsup = critere3(Extrait[,1],T_Ov)
Resultat.lm = lm(Var_sup~ Extrait[,1] + xsup -1 ,data=Extrait)
AIC_Choice= c(AIC_Choice,AIC(Resultat.lm))
}
AIC_Choice    #List of the variou AICs corresponding to each possible T value
T_Ov = seq(1,(max(Extrait$temps)+1))[AIC_Choice == min(AIC_Choice)]
T_Ov
# Graphing the results
```

```
#
op <- par(mfrow = c(1, 2), pty = "s")
plot(seq(1,(max(Extrait$temps)+1),1),AIC_Choice,xlab="Time (days)",
ylab="AIC")
points(T_Ov,min(AIC_Choice),pch=19,col="red")
plot(Extrait$temps,Extrait$Var_sup,xlab="Time (days)",ylab="Cum PdG",
main=Titre)
xsup = critere3(Extrait$temps,T_Ov)
Resultat.lm = lm(Var_sup~ temps + xsup -1 ,data=Extrait)
lines(Extrait$temps,fitted(Resultat.lm),col="red")
#
savePlot(Out,type="jpeg") # Saving the plot in "jpeg" format
par(op)
#
# Return the changing point value and the 2 slopes
z = c(Individual,Subject_List[Individual],Hormone,T_Ov,coefficients
(Resultat.lm)[1],coefficients(Resultat.lm)[2])
return(z)
}


# Part8.r    JCT September 2007

# =================

# Detection of a change in the cumulated profile of urinary Pg metabolites

#

# Subroutines needed: rupture and critere

#

rupture <- function(t,parameter)

# Input

# t: time elapsed since beginning of the current menstrual cycle

# parameter: 3 parameters define the 2 lines. To avoid negative values,

# the adjusted parameters are the log(parameters). They are secondarly
```

```
# back converted to the functional parameters using exponentiation

#

# Only one rupture is expected

# . . . . . . . . . . . . . . .

{

parameter <- exp(parameter)

a <- parameter[1]

b <- parameter[2]

c <- parameter[3]

T1 <- parameter[4]

z1 <- ifelse(t < T1,a+b*t,0)

z2 <- ifelse(t >= T1,a+b*T1+c*(t-T1),0)

z <- z1 + z2

return(z)

}


critere4 <- function(param,temps,observes)

{

# criterium to minimize

#

z <- rupture(temps,param)

total <- sum((z-observes)*(z-observes))
```

```
return(total)

}


Detector = function(Individual,donnees,directory,Hormone,Flag)

{

t <- seq(1,35,by=1)

#Individual=3

#Hormone ="PdG"

#Flag=0


Subject_List = unique(donnees$Iden)


Titre = paste("PWNLR_F",Subject_List[Individual],sep="",collapse=NULL)

sortie = paste(directory,"Sorties/PWNLR/",Titre,sep="",collapse=NULL)

data.extrait = donnees[donnees$Iden == Subject_List[Individual],]


Variable = colnames(data.extrait,do.NULL=FALSE)== Hormone

Days <- data.extrait$Days

Observe_raw <- data.extrait[,Variable]

#

Extrait = missing_data(Days,Observe_raw)

temps = Extrait[,1]

Observe = Extrait[,2]
```

```
Observe = cumsum(Observe)

#

parametre <- c(100, 54.82, 300, 15)

parametre<-log(parametre)

#resultat <-
optim(parametre,critere4,gr=NULL,method="BFGS",control=
list(trace=Flag,maxit=10000,alpha=1.0,beta=0.5,gamma=2.0,reltol=
1e-10),hessian=TRUE,observes=Observe,temps=temps)

resultat <- optim(parametre,critere4,gr=NULL,method="Nelder-
Mead",control=list(trace=Flag,maxit=10000,alpha=1.0,beta=0.5,gamma=
2.0,reltol=1e-10),hessian=TRUE,observes=Observe,temps=temps)

op <- par(mfrow = c(1, 3), pty = "s")

plot(temps,Observe,main=Titre,xlab="Time(days)",ylab="Cum PdG
(mmol/mol)",pch=19,xlim=c(0,30))

z <- rupture(t,resultat$par)

lines(t,z,lty="longdash",col="black")

par(op)

savePlot(sortie,type="jpeg")

#

T = exp(resultat$par)

sigma_T = ifelse(det(resultat$hessian) != 0, T*sqrt(solve(resultat$
hessian)[4,4]),-1)

z= c(Individual,Subject_List[Individual],Hormone,T,sigma_T)

return(z)
```

```
}

# Part9.r July 2007

# JCT
Cusum method and related algorithms
# Analysis of the data set corresponding to a specific Individual

#

# N should be adapted to the urinary collection frequency

#

# ===========================================

#

library(qcc)

#

part9 = function(Individual,donnees,directory,Hormone,N)

{

#

# Data extraction(9(

#directory ="/media/JCT31012009/Travaux/LRosetta/"

#Hormone = "PdG"

#Individual = 18

data_set = donnees[donnees$Iden == Subject_List[Individual],]

Titre = paste(Hormone,"F",data_set$Iden[1],sep="_",collapse=NULL)

data.out = paste(directory,"Sorties/Cusum/",Titre,sep="",collapse=NULL)
```

```
data.out = paste(data.out,"jpeg",sep=".")

Variable = colnames(data_set,do.NULL=FALSE)==Hormone

data_set[,Variable]

# Correction for missing data

#

data_rec1= missing_data(data_set$Days,data_set[,Variable])

temps = data_rec1[,1]

u=data_rec1[,2]


#

# Utilisation du package {qcc}

# ======================

# 1) Function qcc


#

op = par(mfrow = c(3, 1))

A =
qcc.options(bg.margin="white",bg.figure="white","violating.runs"=
list(pch=22,col="black"),"beyond.limits"=list(pch=25,col="black"))

EFP_PdG = u[1:4]

PdG = u[5:length(u)]

obj = qcc(EFP_PdG,type="xbar.one",newdata=PdG,data.name="PdG",title=
"Shewart chart",chart.all=TRUE,add.stats=FALSE)

#op <- par(mfrow = c(2, 1), pty = "s")
```

```
obj.cusum= cusum(obj,title="Cusum chart",add.stats=F,xlab="Time",ylab="")

Detect.cusum=temps[obj.cusum$violations$beyond.limits]

PdG_Cusum= min(Detect.cusum)

# 2) Function ewma

#

obj.ewma= ewma(obj,title="Ewma chart",add.stats=F,xlab="Time",ylab="")

Detect.ewma=temps[obj.ewma$violations$beyond.limits]

PdG_Ewma= min(Detect.ewma)

par(op)

savePlot(data.out,"jpeg")


#


z = matrix(c(Individual,Hormone,PdG_Cusum,PdG_Ewma),ncol=1)

return(z)

}
```

References

Baird, D. D., Weinberg, C. R., Wilcox, A. J., McConnaughey, D. R. & Musey, P. I. (1991). Using the ratio of urinary oestrogen and progesterone metabolites to estimate day of ovulation. *Statistics in Medicine*, 10, 255–66.

Blackwell, L. F. & Brown, J. B. (1992). Application of time-series analysis for the recognition of increase in urinary estrogens as markers for the beginning of the potentially fertile period. *Steroids*, 57, 554–62.

Blackwell, L. F., Brown, J. B. & Cooke, D. (1998). Definition of the potentially fertile period from urinary steroid excretion rates. Part II. A threshold value for pregnanediol-3-alpha-glucuronide as a marker for the end of the potentially fertile period in the human menstrual cycle. *Steroids*, 63, 5–13.

Brown, J. B. (1977). Timing of ovulation. *Medical Journal of Australia*, 2, 780–3.

Cekan, S. Z., Beksac, M. S., Wang, E. *et al.* (1986). The prediction and/or detection of ovulation by means of urinary steroid assays. *Contraception*, 33, 327–45.

Cembrowski, G., Westgard, J. O., Eggert, A. A. & Torem, E. D. (1975). Trend detection in control data: Optimization and interpretation of trigg's technique for trend analysis. *Clinical Chemistry*, 21, 1396–405.

Ecochard, R., Boehringer, H., Rabilloud, M. & Marret, H. (2001). Chronological aspects of ultrasonic, hormonal, and other indirect indices of ovulation. *British Journal of Obstetrics and Gynaecology*, 108, 822–9.

Ellison, P. T. (1988). Human salivary steroids: methodological considerations and applications in physical anthropology. *Yearbook of Physical Anthropology*, 31, 115–42.

Kassam, A., Overstreet, J. W., Snow-Harter, C. *et al.* (1996). Identification of anovulation and transient luteal function using a urinary pregnanediol-3-glucuronide ratio algorithm. *Environmental Health Perspectives*, 104, 408–13.

Kesner, J. S., Wright, D. M., Schrader, S. M., Chin, N. W. & Krieg, J. (1992). Methods of monitoring menstrual function in field studies: efficacy of methods. *Reproductive Toxicology*, 6, 385–400.

McConnell, H. J., O'Connor, K. A., Brindle, E. & Williams, N. Y. (2002) Validity of methods for analyzing urinary steroid data to detect ovulation in athletes. *Medicine & Science in Sports & Exercise*, 34, 1836–44.

Metcalf, M. G., Evans, J. J. & Mackenzie, J. A. (1984). Indices of ovulation: comparison of plasma and salivary levels of progesterone with urinary pregnanediol. *Journal of Endocrinology*, 100, 75–80.

Neubauer, A. S. (1997). The EWMA control chart: properties and comparison with other quality-control procedures by computer simulation. *Clinical Chemistry*, 43, 594–601.

R Development Core Team (2004). R: A language and environment for statistical computing. R Foundation for Statistical Computing, Vienna, Austria. (ISBN 3-900051-00-3; <http://www.Rproject.org>.)

Rosetta, L., Conde da Silva Fraga, E. & Mascie-Taylor, C. G. N. (2001). Relationship between self-reported food and fluid intake and menstrual disturbance in female recreational runners. *Annals of Human Biology*, 28, 444–54.

Royston, J. P. (1982). Basal body temperature, ovulation and the risk of conception, with special reference to the lifetimes of sperm and egg. *Biometrics*, 38, 397–406.

Santoro, N., Crawford, S. L., Allsworth, J. E. *et al.* (2003). Assessing menstrual cycles with urinary hormone assays. *American Journal of Physiology – Endocrinology and Metabolism*, 284, E521–30.

Sufi, S. B. & Donaldson, A. (1988). Design of immunoassays for use in developing countries. *Progress in Clinical and Biological Research*, 285, 331–41.

Sufi, S. B., Donaldson, A., Gandy, S. C. *et al.* (1985). Multicenter evaluation of assays for oestradiol and progesterone in saliva. *Clinical Chemistry*, 31, 101–3.

Waller, K., Swan, S. H., Windham, G. C. *et al.* (1998). Use of urine biomarkers to evaluate menstrual function in healthy premenopausal women. *American Journal of Epidemiology*, 147, 1071–80.

Wasalathanthri, S., Tennekoon, K. H. & Sufi, S. B. (2003). Feasibility of using paper impregnated with urine instead of liquid urine for assessing ovarian activity. *Ceylon Medical Journal*, 48, 4–6.

10 *An insidious burden of disease: the pathological role of sexually transmitted diseases in fertility*

GEOFF P. GARNETT

Introduction

There is a perception that the demographic impact of infectious disease is largely determined by the mortality associated with pandemics or the death of children caused by infections early in life (Murray and Lopez, 1997), but infectious diseases can also reduce the birth rate through their impact on fertility (Weisenfeld and Cates, 2008). Widely spreading epidemics of fatal infections can cause sudden population declines. The ability of populations to recover from such shocks depends upon whether these are transitory epidemics or whether endemic spread leads to a continued heavy toll (Garnett and Lewis, 2007). The great mortalities like the Black Death and Spanish Influenza have been associated with directly transmitted infections which can spread widely. Infections such as measles, malaria and diarrhoeal disease, which cause many infant deaths in poorly nourished populations, have a high incidence in the young with common exposure caused by direct transmission or transmission via water or insect vectors. Apart from some intrauterine mortality associated with infections like tuberculosis and malaria (Mayaud, 2001), the infections generally associated with reductions in fertility are more restricted to a sexual route of transmission. That reductions in fertility are caused by sexually transmitted infections can be explained by their association with the genital and reproductive tract. The demographic impact of the reductions in fertility associated with these infections is less widely recognised than that associated with mortality because sterility and early pregnancy loss are less noticeable events. Nonetheless, reduced fertility can potentially greatly alter demography. At its simplest a model of population growth includes births and deaths, and growth can be reduced by more deaths or fewer births.

A crude generalisation would differentiate the epidemiological and evolutionary characteristics of those infections responsible for increased mortality and those infections reducing fertility. Widespread fatal pandemics and

Reproduction and Adaptation, eds. C. G. Nicholas Mascie-Taylor and Lyliane Rosetta.

continued child mortality are particularly associated with pathogens whose evolutionary strategy is to spread quickly before an individual's immunity can control the infection. In moving quickly to new hosts the pathogens do not pay a high evolutionary price for rapid replication and pathogenicity, which may be correlated with transmissibility (May and Anderson, 1983). Often the infections of childhood are such infections, which then lead to lasting acquired immunity if the child survives. This effective resistance to repeat infections has allowed the development of immunisation, which in turn has facilitated the development of vaccines that have greatly reduced infant and child deaths globally (Borysiewicz, 2010). In contrast, the infections responsible for reductions in fertility are much less likely to generate acquired immunity. Their sexual route of transmission severely restricts potential contacts for onward transmission, necessitating a long duration of infection if such infections are to persist in the host population, and such duration is only possible if the infections can avoid being cleared by immunity for some time (Garnett, 2008). This leads in some cases to low fatality rates as is the case for gonorrhoea, chlamydia and genital herpes and the human papilloma virus types leading to genital warts, or an extremely long period before fatal consequences as is the case for HIV, human papilloma virus types leading to cervical cancer (Clifford *et al.*, 2003), or syphilis, which before the availability of penicillin caused 30% mortality over a period of decades (Garnett *et al.*, 1997)

Despite the clear link between sexually transmitted infections and infertility, the scale of the problem and its importance in demography is very difficult to estimate. In part infertility is difficult to measure and in part the rate of infection with sexually transmitted infections (STIs) and the rate of complications associated with those infections are poorly assessed globally. Here, the currently available evidence and how it is incorporated into a demographic framework of the proximate determinants of fertility is reviewed. A description of the different STIs and how they influence fertility is followed by calculations based on the prevalence of infection and the incidence of sequela which are presented to illustrate the factors influencing the importance of STIs in demography.

The proximate determinants of fertility and the role of sterility

A theoretical framework has been developed in demography to explain the total fertility rate (TFR). The TFR is not strictly a rate, but is the number of children born to the average women during her reproductive life. It is reduced from a theoretical maximum by a number of factors that directly influence the expectations of conception and childbirth, called the proximate determinants. For the period between menarche and sterility a women could theoretically

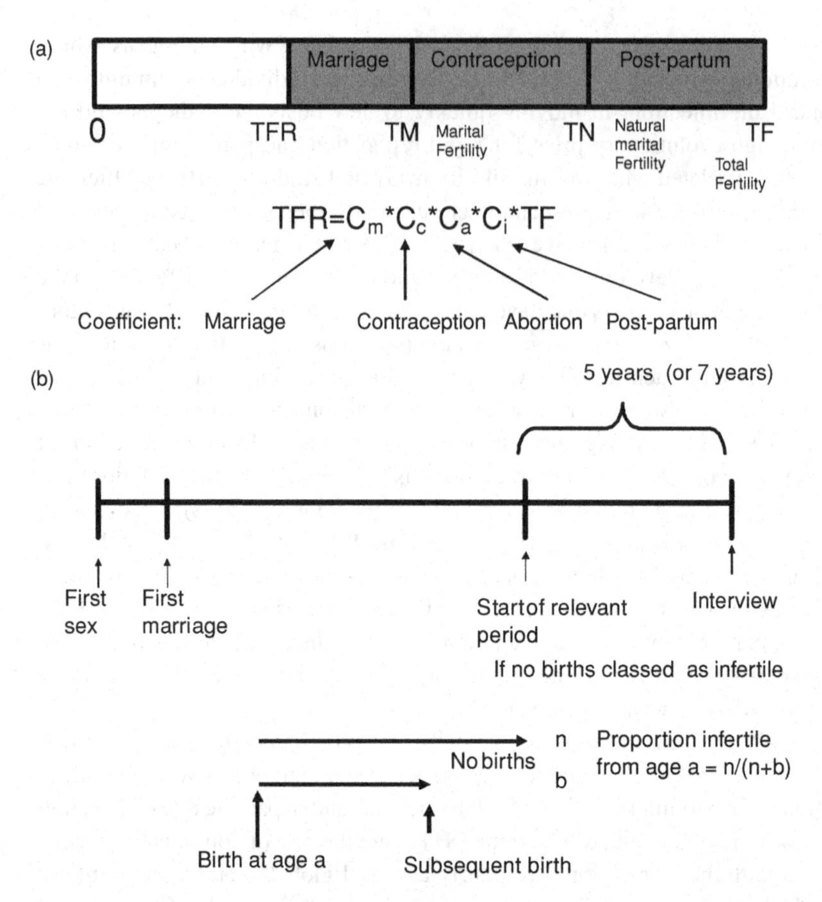

Figure 10.1a) and b). Schematics illustrating proximate determinant models and the method of deriving estimates of infertility from surveys: (a) The relationship between total fertility, natural fertility and the actual total fertility rate within populations, which is influenced by coefficients representing the major proximate determinants; (b) the timing of births in relation to the timing of surveys and the determination of infertility by age.

have one child a year, but this requires intercourse, conception and gestation to be uninterrupted year after year. Davies and Blake (1956) identified a set of determinants which reduced the number of births. From these Bongaarts (1978) developed a simple model with proximate determinants that could be measured in a population and used to explain the fertility observed. This model included four proximate determinants: the proportion married, post-partum infecundability, abortion and contraception. In the model calculations the potential of total fertility is reduced to the total fertility rate by a set of coefficients representing the fertility that remains after the suppression by the factor (Figure 10.1a;

Bongaarts and Potter, 1983). Shortly after the initial description of the model, pathological sterility was added (Bongaarts *et al.*, 1984) and the model was also allowed to be broken down into age groups (Bongaarts and Stover, 1986).

The proximate determinants framework allows us to consider the mechanisms via which sexually transmitted infections could influence fertility, the most obvious of which is via sterility. Sterility though is a difficult concept to define and measure in a population. Sterility is the physiological incapacity to produce a live birth, which includes the inability to conceive or to carry a pregnancy through to term. It is actually a property of a couple, but is normally considered as referring to a woman because she could have the option of changing partner (Bongaarts and Potter, 1983). Primary sterility occurs when the ability to produce a child never develops, whereas secondary sterility involves becoming infecund after first being able to have a child (Boerma and Mgalla, 2001). Distinguishing between primary sterility (never developing the ability to have a child) versus secondary sterility (becoming infecund) is nearly impossible before the birth of a first child. Sterility after a first child is obviously secondary sterility, so whether a woman has previously had a child is used to distinguish secondary from primary sterility. All women become infecund eventually through menopause, but some women have a pathological sterility which reduces the age at which sterility occurs and reduces their reproductive lifespan.

A number of methods exist to estimate the extent of infertility and sterility across populations which create different estimates that under- or overestimate its extent (Boerma and Mgalla, 2001). Medical diagnosis provides a lower bound as it relies on reporting and the facilities to provide a diagnosis being available. Medically, infertility is often diagnosed after a year of unprotected intercourse without conception, but the World Health Organisation used a definition of two years exposure without conception (World Health Organisation, 1975). Expanding beyond a diagnosis of infertility to include as infertile all women seeking care for infertility will also provide an underestimate because many infertile women will not consider it a problem, and if they do many will not seek care. In surveys it is possible to ask women if they believe they are infertile, but this could lead to underestimating or overestimating depending upon perceptions of what is normal fertility. In demographic studies the pattern of fertility recorded in surveys has been a way of estimating infertility (Larsen and Menken, 1989; 1991). This allows a reasonable population level estimate when contraception is rare, but once modern contraception is available the observed patterns of fertility are very different from the natural fertility that would be possible in the absence of pathological sterility. If contraceptive use is widespread two distinct assumptions are possible when using survey methods.

First, to assume contraceptive users are fertile: this will generate a lower bound on sterility as some contraceptive users will be sterile. Second, to ignore contraceptive users: this will generate an upper bound because most contraceptive users will not have children and will not be sterile (Boerma and Mgalla, 2001).

The method of using surveys to record the birth history of women within a population allows the systematic comparison of infertility across populations. Retrospectively analyses of patterns of childbirth allow sterility to be identified, but a period of some time without children before the time of interview is required to be confident that a woman is sterile (Larsen and Menken, 1989; 1991). Periods of 5 and 7 years have been used in studies across populations based on data from World Fertility Surveys (WFS) and Demographic and Health Surveys (DHS). Women who have not had a child before this period of 5 (or 7) years after the age 'a' are classed as infertile from age a. Dividing the number infertile beyond age a by all women allows the proportion of infertile women to be calculated (Figure 10.1b). This is not the same as sterility but an age-specific adjustment to translate the estimate of infertility into an estimate of sterility if possible (Larsen and Menken, 1989). As already stated such estimates are more readily applicable where contraceptives are rarely used.

All of the measures of infertility and sterility described require a history of childbirth to allow the calculation of a level of sterility. An alternative approach would be to look at the causes of sterility, and estimating the risk of sterility associated with each cause and the extent to which the causes are present in a population. Such an approach has been used for STIs as a cause of infertility (Brunham *et al.*, 1992; White *et al.*, 2001). To estimate the risk of sterility in the individual, prospective cohort or retrospective case control studies can be used (Wiesenfeld and Cates, 2008). At a population level regression analyses have allowed the relationship between STIs and sterility to be quantified (Arya *et al.*, 1980). In both cases the measurement of sterility is still required. Thus, while using mathematical models as a framework to estimate the influence of sexually transmitted infections in a population, it should be remembered that the parameters used in these models have to be estimated at some point from observational studies.

Influence of sexually transmitted diseases on fertility

The major sexually transmitted diseases chlamydia, gonorrhoea and syphilis are caused by the bacteria *Chlamydia trachomatis*, *Neisseria gonorrhoea* and *Treponema palidum* respectively. The protozoan infection *Trichonomias vaginalis* also causes local inflammation. Herpes simplex virus types 1 and 2 are the cause of genital herpes. Although type 2 is better adapted to sexual transmission, type 1 – which is more often associated with cold sores – is often found as a cause of genital herpes. Human papilloma viruses are a necessary but not sufficient cause of cervical cancer, with the majority of cervical cancer cases associated with HPV type 16, but a number of other HPV types are also able to cause cervical cancer. Not all HPV types are associated with cancer, with some types, especially HPV 6 and HPV 11, causing genital warts. Human immunodeficiency virus (HIV) is the cause of acquired immune deficiency syndrome (AIDS), which had an incubation period of, on average, 10 years and a remarkably high fatality rate. Effective combined antiviral therapy became available in 1996 and has greatly reduced the incidence of AIDS, with recent expansion of antiretroviral (ARV) coverage to many of those in need in developing countries (UNAIDS, 2009).

The sexually transmitted infections can change fertility in four general ways: 1) by reducing the chances of conception; 2) by causing intrauterine mortality; 3) by altering patterns of post-partum infecundability through altering patterns of breast feeding, either through neonatal deaths or through attempting to prevent transmission to the infant; 4) by changing patterns of sexual behaviour and levels of unprotected sexual intercourse.

The bacterial infections gonorrhoea and chlamydia lead to inflammation. In women this is often a cause of pelvic inflammatory disease and both symptomatic and asymptomatic infection can cause scarring of the fallopian tubes (Wiesenfeld and Cates, 2008). Bilateral tubal occlusion, where both fallopian tubes are blocked, is a pathological reason for sterility, whereas any scarring could reduce the probabilities of conception. The evidence that chlamydia and gonorrhoea cause infertility is clear. Case control studies comparing the presence of antibodies to gonorrhoea and chlamydia which indicate a past infection find a significantly greater fraction of those infertile have been infected than those not (Wiesenfeld and Cates, 2008). In addition the damage caused by the infections can be identified. Whether the risks of damage are associated with an infection are greater the longer the infection lasts or are similar across infections of a given severity, is unclear. This is important if treating cases detected through screening is used to reduce the risks of infertility, since without controlling the spread of the infection such an approach would only work if risks of damage and infertility accumulate as the infection persists. Recent studies also suggest that the inflammation associated with bacterial STIs can cause reduced fertility in men (Fernandez *et al.*, 2007).

The influence of tubal occlusion, if infecundability is complete, is a reduced age at which sterility occurs, which, if the STIs are acquired at a young age, could be many years, or decades, before sterility would normally occur. This dramatic shift in the reproductive lifespan could be an important proximate determinant of fertility if it is widespread in a population (Brunham *et al.*, 1992). However, the role of tubal occlusion needs to be considered in the light of the other proximate determinants and the extent to which STIs are restricted to certain individuals within the population (Garnett *et al.*, 1992).

1) Intrauterine mortality

How well recorded fetal loss is depends upon how early in the pregnancy the loss takes place. Within 6 weeks of conception the event is more likely to be experienced as a delay in conception. Later in pregnancy the time pregnant along with any period of abstinence or infecundability associated with lactation could be a significant part of a woman's reproductive lifespan. More general infectious disease can lead to fetal loss with malaria and tuberculosis being non-sexually transmitted infections which can cause spontaneous abortion (Mayaud, 2001). These infections, along with sexually transmitted syphilis and HIV, have in common a high incidence in young adults who are in the reproductive ages, and symptoms involving systemic disease. The reduced pregnancy rates associated with HIV became apparent when testing the validity of sentinel surveillance in women attending antenatal clinics where, in population-based surveys, women who were HIV infected were found to be less often pregnant (Carpenter *et al.*, 1997; Zaba and Gregson, 1998). A similar pattern was found in the Rakai, Uganda, for those who had syphilis and the effect was greater when both infections, syphilis and HIV, were combined (Gray *et al.*, 1998). The influence of HIV when the woman is severely ill is likely to be through reduced sexual intercourse, but in the asymptomatic incubation period intrauterine mortality is the presumed cause. The widespread roll out of antiretroviral treatment should have increased fertility of those who are treated, with anecdotal evidence suggesting that family planning need has increased as ARV roll out has taken place.

2) Neonatal death and patterns of breastfeeding

Syphilis and HIV, along with herpes simplex virus -1 (HSV-1) and HSV-2, can be transmitted from the mother to the neonate, leading to rapid infant mortality. The risk of vertical transmission of HSV per episode is greater if the HSV infection is acquired during pregnancy, and this is particularly true in the case of genital HSV-1, where the mother is unlikely to have previously acquired

HSV antibodies (Brown *et al.*, 2003). However, because more neonates are exposed to mothers who have an existing HSV-2 infection, then despite the higher risk associated with a new infection more neonatal herpes cases are associated with prevalent infections. Neonatal herpes leads to death in 57% of untreated disseminated infections, which represent around a third of neonatal infections (Whitley *et al.*, 1991). Screening pregnant women for herpes is not really an option since it is unclear whether having an existing infection is better than being at risk of acquiring a new infection (Barnabas *et al.*, 2002). The disparate rates of neonatal herpes between populations are not understood. Neonatal acquisition of syphilis and HIV can be prevented through screening and treatment and only still occurs because of a failure in care (Shafii *et al.*, 2008). Nonetheless in many developing countries vertically transmitted HIV and syphilis are frequent (UNAIDS, 2009). Survival of infants with neonatal syphilis and HIV depends upon the severity of the disease. In the case of HIV there appears to be a bimodal survival distribution with a fraction of children dying rapidly, which is associated with acquisition of the infection in utero (Newell *et al.*, 2004).

Normally neonatal and infant mortality will be a devastating loss to parents and community, with the greatest loss of potential life years associated with death. Demographically though, the infant does not join the population for long and the main effect is that the period of the woman's reproductive life is lost. This will include any period of post-partum abstinence or lactational infecundability that occurs following the birth of the dying child. Viewed in another way the period of post partum abstinence and breast feeding will likely be shortened and subsequent conception and childbirth accelerated (Bongaarts and Potter, 1983). Thus, while decreasing population growth through mortality, neonatal death can increase the fertility rate.

In addition to the death of a child altering the potential influence of breast feeding and associated lactational amenorrhoea on fertility, interventions to prevent vertical transmission of HIV infection through breast feeding could also alter its influence. The scale of this effect will depend upon both the prevalence of HIV and the extent to which mothers know of their infection. Further the direction of the effect will depend on the strategy employed to prevent vertical transmission. Where supplies of formula feed and clean water are reliable and exposure to diarrhoeal disease is limited it is recommended that breast feeding be avoided to prevent HIV infection of the child (UNAIDS, 2009). This will reduce the period of breast feeding and infecundability. Alternatively, mixed feeding exposes the infant to a greater risk of HIV acquisition than only breast feeding (Newell, 1999). In places where there are problems with supplies of formula feed and more particularly the access to clean water then consistent breast feeding can be a preferred option. This would extend the period of lactational amenorrhoea and infecundability, thereby increasing the interval between births.

3) Changes in patterns of unprotected sexual intercourse

Lactational infecundability is a biological factor altering a woman's ability to conceive, but HIV-positive women who know of their status could also choose not to risk a pregnancy, increasing infertility. This may also apply when a woman is uninfected but their male partner infected where unprotected sex is avoided to reduce the risk of HIV transmission. In developed countries there is the possibility of in vitro fertilisation to avoid transmission to the women. The use of antiretroviral treatment to reduce the HIV viral load and thereby reduce the likelihood of HIV transmission is a potential alternative which could allow unprotected sex and conception,especially in resource-limited settings where in vitro fertilisation is not really an option (Garnett and Gazzard, 2008).

Evidence for the rate of sterility associated with gonorrhoea and chlamydia

While the causal link between gonorrhoea and chlamydia is not in doubt, the level of risk of infertility associated with an episode of chlamydia and gonorrhoea is uncertain for a number of reasons. In a systematic review of the influence of chlamydia on fertility the authors found no studies they consider to give good estimates of the risks of infertility (Wallace *et al.*, 2008). The retrospective case control method, which identifies an exposure for cases of disease, fails to measure the risk associated with an exposure. To explain this further, the odds ratio (OR) is the ratio of exposed to unexposed with disease compared to the ratio of exposed to unexposed in a control population without disease. Putting this in an equation the odds ratio (OR) is given by: $OR = (Ed/Nd)/(Ec/Nc)$, where Ed and Nd are those exposed and not exposed with disease and Ec and Nc those exposed and not exposed without disease. The relative risk (RR) on the other hand, is the proportion developing disease in those exposed compared to the proportion developing disease in those unexposed: $RR = (Ed/(Ed + Ec))/(Nd/(Nd + Nc))$, but the comparison needs to be made with a denominator population not selected because of their disease state. Unfortunately, the lack of prospective studies of those exposed means that the increased risk associated with chlamydia is not well quantified (Wiesenfeld and Cates, 2008).

The majority of prospective studies linking disease to infertility are based on frank disease cases. Diagnostic tests have changed. Antibody testing through Enzyme Linked Immunosorbent Assays (ELISA) detected a history of infection rather than the current infection which Nucleic Acid Amplification Tests (NAATs) now detect. Unfortunately, antibody tests were very insensitive, leading many past infections to go undetected (Kuypers *et al.*, 2008). This would lead to misclassification of exposed as unexposed, reducing our ability to link infertility to the past infection. The increased sensitivity of NAATs is for current presence of the bacterial antigens, which requires studies to follow up those infected. Those not infected at the time of screening may have previous or subsequent infections, again leading to misclassification of the exposure. In addition, the nature of the infection and how severe it is for a NAATs positive is unclear, so currently detected asymptomatic cases of chlamydia and gonorrhoea may not be equivalent to the disease cases observed in earlier studies. This problem is likely less with gonorrhoea, which is more often symptomatic, than it is with chlamydia. Arguments for the control of gonorrhoea and chlamydia through screening rely on the prevention of infertility and ectopic pregnancy (Low *et al.*, 2009). However, those arguments are poorly founded while we have no good estimates of the risk of infertility associated with the cases detected.

The scale of the problem

The magnitude of the influence of sexually transmitted infections on fertility is directly dependent upon the level of sexually transmitted infections within the population. Prevalence is the proportion of the population with the infection or disease at a particular moment, measured per person or for a denominator population, whereas incidence is the rate at which infection or disease is acquired, and is measured per denominator population per unit of time. The level of STIs in the population, in turn, is determined by a set of biological and behavioural proximate determinants responsible for the spread of infection through the population (Lewis *et al.*, 2007). Particularly important for infections that can be treated and cured is the pattern of access to treatment. In developed countries gonorrhoea, chlamydia and syphilis are more likely treated when symptomatic than in developing countries (Aral *et al.*, 2008). Further, screening for asymptomatic infections often takes place. This control explains why gonorrhoea and syphilis are often more prevalent in developing countries. Chlamydia does not seem to differ as much in its prevalence, in part this may be because it is more likely to remain asymptomatic, but also the influence of acquired immunity seems to limit its spread (Brunham *et al.*, 2005).

Differences in the prevalence of herpes and HIV are less easily explained in terms of the treatment available, but clearly the prevalence of HSV-2 is greater in developing countries, particularly in Africa (Looker *et al.*, 2008), and HIV has a remarkably high prevalence in Southern and Eastern African Countries where it is widespread amongst heterosexuals (UNAIDS, 2009). In many countries HIV is focussed in those with behaviours exposing them to infection, such as injecting drug users, men having sex with men or unprotected commercial sex (UNAIDS, 2009). This focus will limit their impact on fertility, whereas where HIV is hyperendemic with prevalences of over 10% in heterosexual adults, the fertility impact could be great. Why genital herpes and HIV are so much more prevalent is some parts of Africa is not certain.

The presence of other sexually transmitted infections such as gonorrohea, chlamydia and syphilis which increase both susceptibility and transmissibility for HIV is part of the explanation of the initial outbreaks in cities (Røttingen *et al.*, 2001). However, control of STIs has largely failed to stop the spread of HIV, which in many places has moved beyond those with STIs (Institute of Medicine, 2008). Genital herpes, which is widespread, may also play a role, but again this is being questioned following trials to control HIV through herpes treatment or suppression, and one also needs to explain the high prevalence of the genital herpes. There is debate over the relative role of those with a high risk of acquiring and transmitting infection, such as sex workers and their clients, versus the role of overlapping, or concurrent partnerships, both of which may be more of a norm in some societies (Morris *et al.*, 2009). Whatever the cause of difference in STI prevalence, understanding the role of STIs in fertility and demography is more straightforward through looking at levels of sterility, or patterns of STI prevalence, than in trying to infer a role from patterns of sexual behaviour.

Estimating levels of sterility from reported birth histories

Larsen and Raggers (2001) conducted a detailed analysis of birth histories from World Fertility Surveys and Demographic and Health surveys, which show marked differences in the age specific levels of sterility. A subset of these estimates is illustrated in Figure 10.3a with some e.g. Togo in 1988 showing low levels at younger ages with a rapid rises in the 40s. Others, e.g. Cameroon in 1978, show high levels of sterility at younger ages. It is believed that prevalence of bacterial STIs was high at that time (Arya *et al.*, 1980). Comparing this pattern with the classic historical pattern (Figure 10.2b) we can see how different it is, and the likely scale of influence of bacterial STIs.

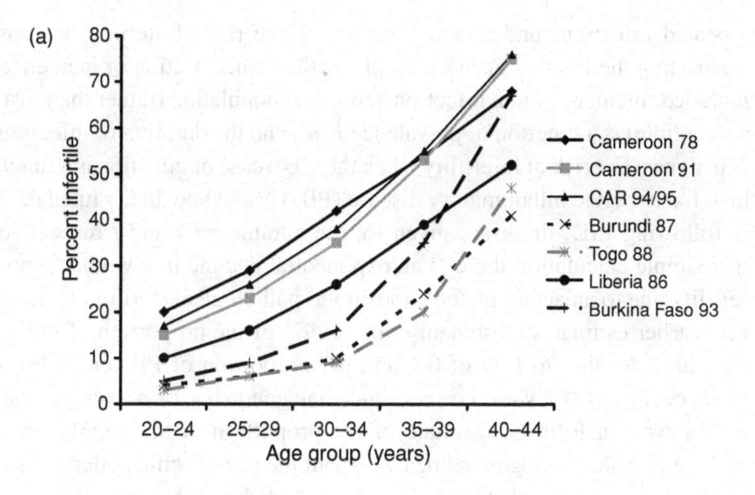

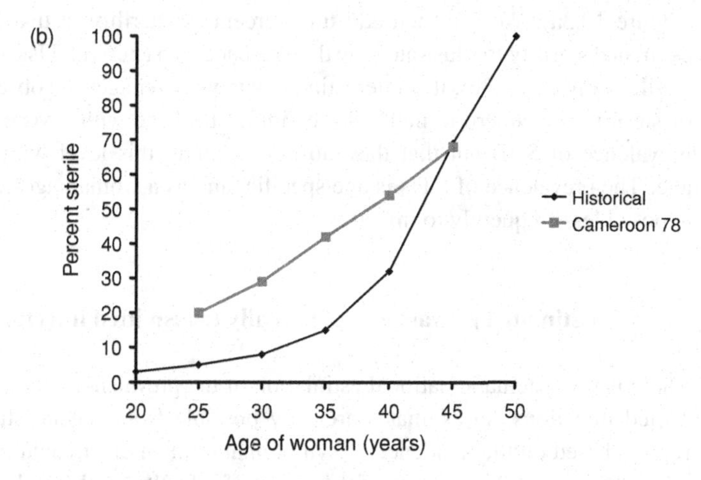

Figure 10.2. Patterns of infertility as a function of age estimated from population based surveys by Larsen and Raggers (2001) using a subsequent infertility estimator. (a) Age specific infertility in 5 year age classes for a selection of countries in given years. (b) A comparison of the historical pattern with that estimated for Cameroon in 1988 by Larsen and Raggers (2001).

A simple model to estimate the age-specific level of pathological secondary sterility $s(a)$ can be constructed from prevalence estimates if we assume that the STI is at equilibrium.

$$S(a) = k\left(1 - e^{-\int_{\tau}^{a}\left[\frac{P(a')}{D}.i.f\right]da'}\right)$$

Here k is the proportion of the population at risk of acquiring STIs. Since some are more likely than others to acquire STIs, due to their risk behaviours, repeated infections and repeat exposure to the risk of sterility are possible. Restricting the fraction at risk to k allows this concentration of incidence to be included. Incidence of an infection across the population (rather than just those susceptible) is a function of prevalence $P(a')$ and the duration of infectiousness D per year. The risk of infertility when there is a case of infection is a function of how likely pelvic inflammatory disease (PID) is and how likely tubal occlusion is following PID; these are given by the parameters i and f respectively. In this simple calculation there is an exponential decline in a woman's potential fertility due to an accumulated exposure to pathological sterility. If we assume that earlier estimates (Brunham *et al.*, 1992) of the proportion of gonorrhoea infections leading to PID of 0.6 and the proportion of PID cases leading to tubal occlusion 0.2 were correct, and that gonorrhoea on average lasts half a year we can form an estimate of the proportion of the population at risk and the prevalence of gonorrhoea work out the age-specific pattern of sterility, in this case assume that prevalence of gonorrhoea is constant across ages (Figure 10.3a). We can then add this percentage sterility to that historically estimated sterility to illustrate how the presence of bacterial STIs can influence sterility (Figure 10.3b). It is interesting that at younger ages the observed levels of sterility in Cameroon in 1978 are similar to those with a very high (20% prevalence of STI) but that they do not maintain this level with increasing age. The prevalence of STIs is age specific, and as a woman ages, she will be exposed less frequently to infection.

Estimated prevalences of sexually transmitted infections

There is no systematic national estimation of the prevalence of sexually transmitted infections, so estimates are only possible from special studies. Case reports based on the incidence of symptomatic infections treated in clinics do not provide a good indication of the extent of infection as they rely on patients seeking care and have no clear denominator. In addition, they will not include asymptomatic infections, which can only be identified through contact tracing or from screening. A better indication of prevalence comes from surveys of the general population, but these are few. Recent nucleic acid amplification tests have increased the sensitivity of methods to detect asymptomatic gonorrhoea and chlamydia so the comparison of surveys over time needs to take account of the method of test used. In a systematic search of the PubMed data base in October 2007 using the terms gonorrhoea (and gonorrhea its American spelling), chlamydia and syphilis along with epidemiology and prevalence studies from African populations from 1995 were identified. These are shown in Table 10.1. There is a wide range of prevalence some of which could have a

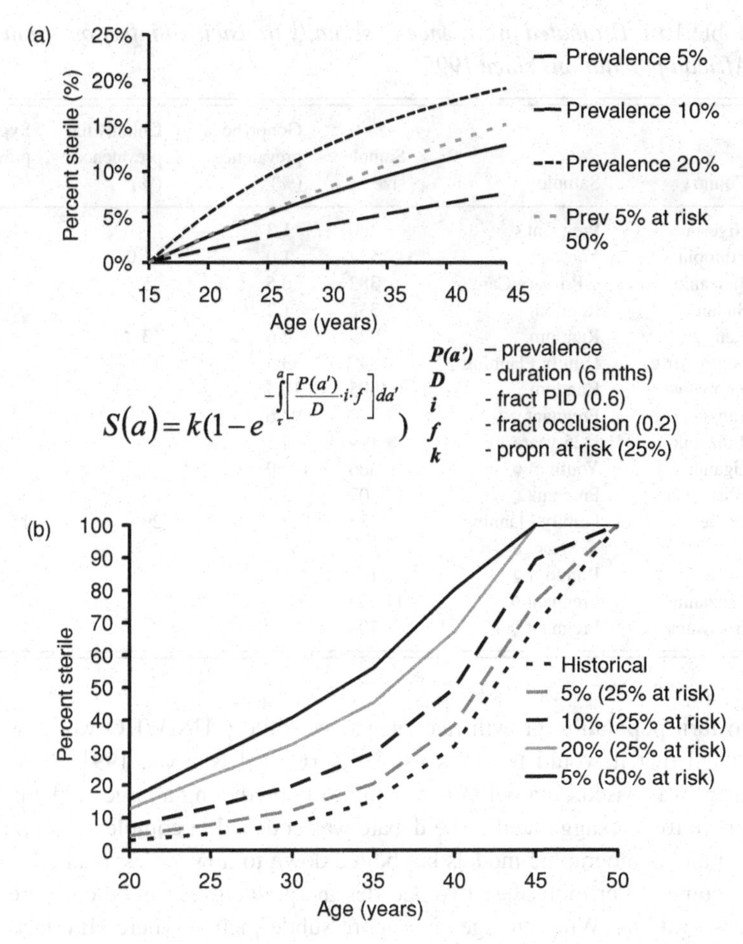

$$S(a) = k\left(1 - e^{-\int_{\tau}^{a}\left[\frac{P(a')}{D} \cdot i \cdot f\right]da'}\right)$$

Figure 10.3. The percentage sterile as a function of age, derived from a simple model of secondary pathological sterility using the prevalence of gonorrhoea. (a) Modelled pathological sterility for different prevalences of infection with 25% and 50% of the population at risk. (b) underlying sterility in the absence of pathological sterility from the historical pattern and with pathological sterility for different prevalences of infection.

substantial impact on fertility, but none of the scale seen before the widespread use of antibiotics and in earlier studies from central and west Africa.

The impact of HIV

Early during the spread of HIV there was debate about the potential demographic impact of the virus with some claiming the epidemic had the potential

Table 10.1 *Estimated prevalence of sexually transmitted infections from African populations since 1995*

Country	Sample	Sample size	Gonorrhoea prevalence (%)	Chlamydia prevalence (%)	Syphilis prevalence (%)
Nigeria	Pregnant ♀	230	1.3		1.7
Ethiopia	Youth ♂ ♀	522	1.0	1.0	
Tanzania	♀ Primary Care	382	0.5	5	4
Sudan	Random	338	1.2		0.9
Kenya	Random	1929	2.6	3.2	
South Africa	Family Planning ♀	249	3	12	
Zimbabwe	Random	1005	18.4		
Tanzania	Pregnant ♀	777	3.6		4
Tanzania	<25 years	199	2.5		
Uganda	Youth ♂ ♀	306	9.0	4.5	4.0
Botswana	Pregnant ♀	703	3	8	
Cameroon	Family Planning Married ♀	783	7	29	15
Zambia	Random ♀	2107			6.5
Tanzania	Pregnant ♀	17,323			7.3
Botswana	Pregnant ♀	703			5%

to turn population growth rates negative, while a UN/WHO working group stated that it would be 'at worst 30%' (United Nations, 1991). This statement was based on a belief that parameters predicting a large epidemic 'may prove to be exaggerated'. The debate was couched in complex epidemiological and demographic models but boiled down to how widespread HIV would become. After more than two decades the predictions can be compared with observations. What emerges is a more subtle picture where HIV has spread to prevalences that could turn population growth rates negative, but only in certain countries, and then, only in the parts of the population with the highest initial fertility levels (Sewankambo *et al.*, 1994; Gregson *et al.*, 2007). This means that HIV has had a large demographic impact, in part due to its impact on fertility, but that it has not turned population growth rates negative. This is illustrated in a study in rural Zimbabwe where births and deaths in an open cohort in 12 communities stratified into towns, commercial estates and villages were followed (Gregson *et al.*, 2007). The influence of HIV in the population was estimated by comparing the crude birth and death rates for those with and without HIV infection (Figure 10.4).

In a similar manner to that used in calculating the effect of bacterial STIs we can derive an estimate of the influence of HIV and syphilis on fertility via its effects on exposure to conception and intrauterine mortality. If HIV infection

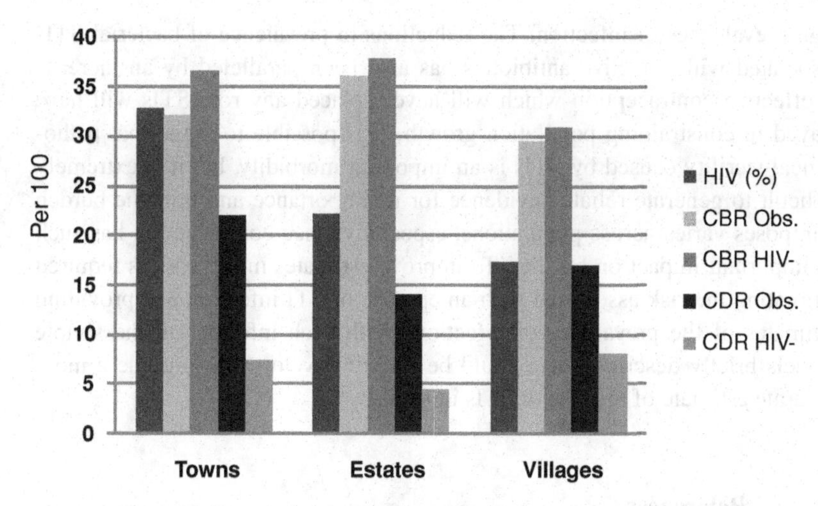

Figure 10.4. The influence of HIV on crude birth and death rates in Manicaland, rural Zimbabwe 1998–2005. The prevalence of HIV averaged over 3 survey rounds is presented for towns, commercial estates. The crude birth rate CBR is compared for all those observed and for the HIV negatives in the population and the crude death rate (CDR) compared for the population as a whole and HIV negatives (Gregson *et al.*, 2007).

reduces fertility on average by 25% then for each 10% increase in the prevalence of HIV we expect a 2.5% decrease in fertility, which is roughly in line with that observed in villages and estates in the Manicaland (Zimbabwe) study, but half that observed in towns (Gregson *et al.*, 2007). However, in making this comparison not only must we consider the influence of chance we also have to consider that the contraceptive usage may differ between those infected with HIV and those uninfected differentially, especially if the women are aware of their infection. Furthermore the influence of HIV is likely to have changed dramatically with the introduction of antiretroviral treatment.

Conclusions

Clinically sexually transmitted infections are an important cause of sterility and infertility, reducing both the reproductive lifespan of women and the fecundity of women. At young ages the loss of fertility can have important social and psychological consequences. Intrauterine and neonatal mortality can have devastating consequences and tubal occlusion is a cause of ectopic pregnancies as well as sterility. Beyond the individual, STIs, which include HIV, can have a substantial impact on fertility. It has been argued that they could have a controlling influence on population growth, but this would only have been at very high prevalences of infection. The reductions in prevalence of bacterial STIs associated with effective antibiotics has also been paralleled by an increase in effective contraception which will have replaced any role STIs will have played in constraining population growth. It is possible to argue that pathological sterility caused by STIs is an important morbidity, but it is extremely difficult to generate reliable evidence for its importance and how the burden it imposes varies across populations, especially since contraception has such an important impact on fertility. To improve estimates more work is required estimating the risk associated with an episode of STI infection and providing estimates of the prevalence of infection. With such information the simple models briefly described here could be further developed to provide a more accurate estimate of the role of STIs in fertility.

References

Aral, S.O., Hogben, M. & Wasserheit, J.N. (2008). STD related health-care seeking and health service delivery. In *Sexually Transmitted Diseases*, ed. K.K. Holmes, F.P. Sparling, W.E. Stamm, P. Piot *et al.* New York: McGraw-Hill, pp. 1803–20.

Arya, O.P., Taber, S.R. & Nsanze, H. (1980). Gonorrhea and female infertility in rural Uganda. *American Journal Obstetrics and Gynecology*, 138, 929–32.

Barnabas, R.V., Carabin, H. & Garnett, G.P. (2002). The potential role of suppressive therapy for sex partners in the prevention of neonatal herpes: a health economic analysis. *Sexually Transmitted Infections*, 78, 425–9.

Boerma, J.T. & Mgalla, Z. (2001). Introduction. In *Women and Infertility in sub-Saharan Africa: A Multi-disciplinary Perspective*, ed. J.T. Boerma & Z. Mgalla. Amsterdam: KIT Publishers, pp. 13–23.

Bongaarts, J. (1978). A framework for analysing the proximate determinants of fertility. *Population and Development Review*, 4, 105–32.

Bongaarts, J., Frank, O. & Lesthaeghe, R. (1984). The proximate determinants of fertility in sub-Saharan Africa. *Population and Development Review*, 10, 511–37.

Bongaarts, J. & Potter, R.G. (1983). *Biology and Behaviour: An Analysis of the Proximate Determinants*. New York: Academic Press.

Bongaarts, J. & Stover, J. (1986). The population council target-setting model: a user manual. The Population Council New York.

Borysiewicz, L.K. (2010). Prevention is better than cure. *Lancet*, 375, 513–23.

Brown, Z.A., Wald, A., Morrow, R.A., Selke, S. *et al.* (2003). Effect of serological status and caesarean delivery on transmission rates of herpes simplex virus from mother to infant. *Journal American Medical Association*, 289, 203–9.

Brunham, R.C., Garnett, G.P., Swinton, J. & Anderson, R.M. (1992). Gonococcal infection and human fertility in sub-Saharan Africa. *Proceedings of the Royal Society B*, 246, 173–7.

Brunham, R.C., Pourbohloul, B., Mak, S., White, R. & Rekart, M.L. (2005). The unexpected impact of a Chlamydia trachomatis infection control program on susceptibility to reinfection. *Journal of Infectious Diseases*, 192, 1836–44.

Carpenter, L.M., Nakiyingi, J.S., Ruberantwari, A., Malama, S.S. *et al.* (1997). Estimates of the impact of HIV infection on fertility in a rural Ugandan population cohort. *Health Transition Reviews*, 7: Suppl 2, 113–26.

Clifford, G.M., Smith, J.S., Plummer, M., Munoz, N. & Franceschi, S. (2003). Human papillomavirus types in invasive cervical cancer worldwide: a meta-analysis. *British Journal Cancer*, 88, 63–73.

Davies, K. & Blake, J. (1956). Social structure and fertility: an analytic framework. *Economic Development and Cultural Change*, 4, 211–35.

Fernandez, J.L., Ramoz, B., Santiso, R., Agarwal, A. *et al.* (2007). Frequency of sperm cells with fragmented DNA in males infected with Chlamydia trachomatis and mycoplasma Sp determined with sperm chromatin dispersion (SCD) test. Abstract 1O-12 American Society for Reproductive Medicine 63rd Annual Meeting Washington DC October 13–17, 2007.

Garnett, G.P. (2008). Transmission dynamics of sexually transmitted Diseases. In *Sexually Transmitted Diseases*. ed. K.K. Holmes, F.P. Sparling, W.E. Stamm, P. Piot *et al.* New York: McGraw-Hill, pp. 27–39.

Garnett, G.P., Aral, S., Hoyle, D.V., Cates, Jr. W. & Anderson, R.M. (1997). The natural history of syphilis: its implications for the transmission dynamics and control of infection. *Sexual Transmitted Diseases*, 24, 185–200.

Garnett, G.P. & Gazzard, B. (2008). Risk of HIV transmission in discordant couples. *Lancet*, 372, 270–1.

Garnett, G.P. & Lewis, J.J.C. (2007). The impact of population growth on the epidemiology and evolution of infectious diseases. In *HIV, Resurgent Infections and Population Change in Africa*, ed. M. Caraël and J.R. Glynn. New York: Springer Series: International Studies in Population, Volume 6.

Garnett, G.P., Swinton, J., Brunham, R.C. & Anderson, R.M. (1992). Gonococcal infection, infertility and population growth: II. The influence of behavioural heterogeneity. *IMA Journal of Mathematics Applied in Medicine and Biology*, 9, 127–44.

Gray, R.H.,Wawer, M.J., Serwadda, D., Sewankamdo, N. *et al.* (1998). Population-based study of fertility in women with HIV infection in Uganda. *Lancet*, 351, 98–103.

Gregson, S., Nyamukapa, C., Lopman, B., Mushati, P., *et al.* (2007). Critique of early models of the demographic impact of HIV/AIDS in sub-Saharan Africa based on contemporary empirical data from Zimbabwe. *Proceedings National Academy Sciences*, 104, 14586–91.

Institute of Medicine. (2008). *Methodological Challenges in Biomedical HIV Prevention Trials*. Washington, DC: National Academies Press.

Kuypers, J., Gaydos, C.A. & Peeling, R.W. (2008). Principals of laboratory diagnosis of STIs. In *Sexually Transmitted Diseases*, ed. K.K. Holmes, F.P. Sparling, W.E. Stamm, P. Piot *et al.* New York: McGraw-Hill, pp. 937–58.

Larsen, U. & Menken, J. (1989). Measuring sterility from incomplete birth histories. *Demography*, 26, 185–201.

Larsen, U. & Menken, J. (1991). Individual level sterility: a new method of estimation with application to sub-Saharan Africa. *Demography*, 28, 229–47.
Larsen, L. & Raggers, H. (2001). Levels and trends in infertility in sub-Saharan Africa. In *Women and Infertility in sub-Saharan Africa: A Multi-disciplinary Perspective*. Amsterdam: KIT Publishers, pp. 25–69.
Lewis, J.J.C., Donnelly, C.A., Mare, P., Mupembireyi, Z. *et al.* (2007). Evaluating the proximate determinants framework for HIV infection in rural Zimbabwe. *Sexually Transmitted Infections*, 83, i61–9.
Looker, K.J., Garnett, G.P. & Schmid, G.P. (2008). An estimate of the global prevalence and incidence of herpes simplex virus type 2 infection. *Bulletin World Health Organization*, 86, 805–12.
Low, N., Bender, N., Nartey, L., Shang, A. & Stephenson, J.M. (2009). Effectiveness of chlamydia screening: systematic review. *International Journal of Epidemiology*, 38, 435–48.
May, R.M. & Anderson, R.M. (1983). Epidemiology and genetics in the coevolution of parasites and hosts. *Proceedings Royal Society Series* B, 219, 281–313.
Mayaud, P. (2001). The role of reproductive tract infections. *Women and Infertility in sub-Saharan Africa: A Multi-disciplinary Perspective*. Amsterdam: KIT Publishers, pp. 71–107.
Morris, M., Kurth, A.E., Hamilton, D.T., Moody, J. & Wakefield, S. (2009). Concurrent partnerships and HIV prevalence disparities by race: linking science and public health practice. *American Journal of Public Health*, 99, 1023–31.
Murray, C.J. & Lopez, A.D. (1997). Mortality by cause for eight regions of the world: global burden of disease study. *Lancet*, 349, 1269–76.
Newell, M.L. (1999). Infant feeding and HIV-1 transmission. *Lancet*, 354, 442–3.
Newell, M.L., Coovadia, H., Cortina-Borja, M., Rollins, N. *et al.* Ghent International AIDS Society (IAS) Working Group on HIV Infection in Women and Children (2004). Mortality of infected and uninfected infants born to HIV-infected mothers in Africa: a pooled analysis. *Lancet*, 364, 1236–43.
Røttingen, J.A., Cameron, D.W. & Garnett, G.P. (2001). A systematic review of the epidemiological interactions between classical STDs and HIV. *Sexually Transmitted Diseases*, 28, 579–97.
Sewankambo, N.K., Wawer, M.J., Gray, R.H., Serwadda, D. *et al.* (1994). Demographic impact of HIV infection in rural Rakai district, Uganda: results of a population-based cohort study. *AIDS*, 8,1707–13.
Shafii, T., Radolf, J.D., Sanchez, P.J., Schultz, K.F. & Murphy, F.K. (2008). Congenital Syphilis. In *Sexually Transmitted Diseases*, ed. K.K. Holmes, F.P. Sparling, W.E. Stamm, P. Piot *et al.* New York: McGraw-Hill, pp. 1577–1612.
United Nations (1991). The AIDS Epidemic and its demographic consequences.
UNAIDS (Joint United Nations Programme on HIV/AIDS) (2009). AIDS Epidemic Update 2009. URL data.unaids.org/pub/Report/2009/JC1700_Epi_Update_2009_en.pdf
Wallace, L.A., Scoular, A., Hart, G., Reid, M. *et al.* (2008). What is the excess risk of infertility in women after genital chlamydia infection? A systematic review of the evidence. *Sexually Transmitted Infections*, 84, 171–5.

White, R.G., Zaba, B., Boerma, J.T. & Blacker, J. (2001). Modelling the dramatic decline of primary infertility in sub-Saharan Africa. In *Women and Infertility in sub-Saharan Africa: A Multi-disciplinary Perspective*, ed. J.T. Boerma & Z. Mgalla. Amsterdam: KIT Publishers, pp. 117–50.

Whitley, R., Arvin, A., Prober, C., Corey, L. *et al.* (1991). Predictors of morbidity and mortality in neonates with herpes simplex virus infections. The National Institute of Allergy and Infectious Diseases Collaborative Antiviral Study Group. *New England Journal of Medicine*, 324, 450–4.

Wiesenfeld, H.C. & Cates, Jr. W. (2008). Sexually transmitted diseases and infertility. In *Sexually Transmitted Diseases*. ed. K.K. Holmes, F.P. Sparling, W.E. Stamm, P. Piot *et al.* New York: McGraw-Hill, pp. 937–58, 1511–1527.

World Health Organisation (1975). The epidemiology of infertility. Report of a scientific working group. Technical Report Series no 582, Geneva 1975.

Zaba, B. & Gregson, S. (1998). Measuring the impact of HIV on fertility in Africa. *AIDS*, 12, S41–50.

11 *Family planning and unsafe abortion*

ANNA GLASIER

Family planning

Throughout history mankind has tried to limit family size (Glasier, 2002). Until the twentieth century this was achieved largely by abstinence, infrequent coitus, coitus interruptus and breast feeding. Although male condoms were described as long ago as 1350 BC and cervical caps were first produced in 1830, 'modern' methods of contraception have only been around for some one hundred years and hormonal contraception for only about fifty years.

Spurred on by fears about over-population, the Family Planning Movement can be said to have begun in the mid-nineteenth century when the Malthusian League argued the case for fertility control (AbouZahr, 1999). Motivated later by concerns for women whose lives were dominated by childbearing, the first family planning clinic was opened in Amsterdam in 1882. In the early twentieth century, clinics were opened in a number of developed countries by women whose names have become inexorably linked to family planning, such as Margaret Sanger in the USA and Marie Stopes in the UK (Leathard, 1980). Renewed concern about the 'population explosion' in the middle of the twentieth century led to the establishment of national family planning programmes, starting in India in 1952 and expanding rapidly over the next 40 years, so that by 1994 over 120 countries around the world had national family planning programmes (Cleland *et al.*, 2006). These programmes classically formed part of socio-economic development plans and were institutionalised in government ministries. Most had specific population targets which had to be met within a certain time-frame and budgetary allocations were linked to achieving the targets. As a result contraceptive use gradually increased in many countries and fertility rates began to fall. Between 1960 and 2000 the proportion of married women in the developing regions using contraception rose from less than 10% to around 60% with a parallel fall in the total fertility rate from six to three (Population Reference Bureau, 2004).

Reproduction and Adaptation, eds. C. G. Nicholas Mascie-Taylor and Lyliane Rosetta. Published by Cambridge University Press.

Inevitably the strategies used by some family planning programmes, particularly in Asia, were criticised as being coercive (famously, transistor radios were offered to poor men in India in exchange for having vasectomies). In addition the quality of family planning services in many countries was poor (Cleland *et al.*, 2006). At the fifth (and last) of a series of International Conferences an Population and Development (ICPD), which was held in Cairo in 1994, the agenda shifted away from national population policies and demography to considerations of the quality of life and the concept of reproductive health (United Nations, 1995). Almost all the governments represented at the Cairo conference signed up to a definition of reproductive health which formally recognised that all aspects of reproductive health were important for improving the quality of life of women, children and communities. The emphasis was on enabling women to achieve autonomy over their reproductive lives in the belief that this would reduce fertility rates and achieve the objectives of population policies.

Much of the success of the effect of rising contraceptive use on total fertility rates was thanks to the development of modern methods of contraception and particularly the oral contraceptive pill, 'invented' by Gregory Pincus and colleagues (Pincus, 1965). Alan Parkes recognised the enormous influence that the pill had had and suggested that had the work not been in such a sensitive area (sex) the invention of the pill would have been rewarded with a Nobel Prize (Parkes, 1985).

Today a wide range of highly effective methods of contraception is available, all of them cheap and safe and many of them offering other non-contraceptive health benefits such as an improvement in menstrual bleeding patterns. In the last five years the particular value of the so-called long-acting reversible methods of contraception (LARC – intrauterine contraceptives, implantable and injectable hormonal contraceptives), which are much less dependent on (or totally independent of) compliance for their effectiveness, has been recognised (National Institute for Health and Clinical Excellence, 2005). Falling teenage pregnancy rates in the USA in recent years have been attributed to increased use of the long-acting progestogen injectable contraceptive Depo Provera® (Santelli *et al.*, 2007) and in Scotland the government planned a campaign in 2009 to increase the uptake of all methods of LARC (Horton, 2008). In the twenty-first century currently available methods of contraception really should enable the elimination of almost all unintended pregnancies.

While family planning is considered to be a success story it is, to quote John Cleland in his recent article in the *Lancet*, 'an unfinished agenda' (Cleland *et al.*, 2006). In this article Cleland and colleagues point out that in half of the 75 larger low- and middle-income countries contraceptive practice remains low and fertility, unmet needs for contraception and population growth remain high. Most governments of poor countries, Cleland goes on to say, already have appropriate population and family planning policies but receive too little international encouragement and funding to implement them. The countries in the world with the highest total fertility rates (TFR) are almost all in sub-Saharan Africa and include Niger [TFR 8], Guinnea Bissau, Mali and Somalia [TFR 7 or more], Uganda, Angola, Burundi, Liberia and the Democratic Republic of Congo, all with total fertility rates of greater than 6.7 (Population Reference Bureau, 2005). In the ten countries with the highest TFR, the only country outside Africa is Afghanistan, where the average family size is 6.8 children. It is not a coincidence that these are also the poorest countries in the world. In many of these countries when contraception is used it tends to be the least effective methods. While in Great Britain 50% of couples (whose families are complete) are sterilised, 32% take the pill and 28% use condoms, only 2% of couples in sub-Saharan Africa are sterilised, 6% use the pill and only 2% use condoms (Population Reference Bureau, 2004). The cause of the slow pace of change in fertility rates in sub-Saharan Africa is complex. Conventional theory is that socio-economic development is a key driver to fertility decline, and the dramatic fall in total fertility rates in many South East Asian countries like Thailand is almost certainly a testament to the relationship. However, while during the late-twentieth century many countries experienced substantial economic growth the gross domestic product in sub-Saharan Africa declined (Bongaarts, 2008). It has also been suggested that the decline in life expectancy resulting from the AIDS pandemic may have contributed to the stalling of fertility decline in the worst-affected countries, most of which are in sub-Saharan Africa (Bongaarts, 2008).

In his introduction to the *Lancet* series on 'Sexual and Reproductive Health', the editor (Richard Horton) describes sexual and reproductive health as 'an issue that has been utterly marginalised from the global conversation about health and well-being during the past decade' (Horton, 2006). And he is correct. In 2000 the United Nations set eight goals for global development to be achieved by 2015 (United Nations, 2001). While three of the so-called millennium development goals (MDGs) were related to health (reducing child mortality, improving maternal health and combating HIV/AIDS, malaria and other diseases), they made no mention of sexual and reproductive health. While other MDGs also included eradicating extreme poverty and hunger and ensuring environmental stability, they made no reference to population growth and gave no recognition of its impact on poverty, hunger and environmental instability.

The world population has more than doubled, from less than three billion to more than six billion, in the second half of the twentieth century and is predicted to grow to between 8 and 10.5 billion by 2050. Ninety-nine per cent of growth in the twenty-first century will take place in developing countries and 90 per cent of it will be concentrated in the poorest countries. Despite these gloomy statistics and, arguably as a consequence of the change in emphasis promulgated at ICPD in Cairo, the focus on population growth was lost in the 1990s and the focus on family planning was lost with it.

Very recently the penny appears to have dropped and the effect of population growth on the environment has led to a resurgence of interest in the topic. Having been politically incorrect in the 1990s it is now okay to express concern about over-population. In 2006 the Stern review repeatedly acknowledged the effect of population growth on climate change (Stern, 2006). Also in 2006 the UN approved a new MDG, calling for universal access to reproductive health care by 2015 (United Nations, 2005) and in January 2007 the UK government published a report entitled Return of the Population Growth Factor: Its Impact Upon the Millennium Development Goals' (All Party Parliamentary Group on Population Development and Reproductive Health, 2007). In the report the threats of population growth to all the MDGs and their targets are emphasised. The report explicitly states that the ICPD plan of action, which recommended that governments should meet the family planning needs of the populations as soon as possible, 'seems a highly cost-effective way of addressing many of the world's problems'.

In the developed world, where access to the full range of contraceptive methods is the norm (free of charge in a few countries such as the United Kingdom), the main issue lies in getting people to use contraception effectively. While contraceptive prevalence is high – in the UK over 92% of couples wishing to avoid pregnancy say they are using contraception – methods which rely on consistent use and perfect compliance for their effectiveness predominate. In the UK in 2006, 32% of couples rely on the male condom and 28% on the oral contraceptive pill (Taylor *et al.*, 2006). While failure rates during the first year of use of these two methods if used perfectly is said to be 2% and 0.5% respectively, typical-use failure rates are 15% and 8% respectively (Trussell, 1998). Poor compliance with oral contraception is common. In one US study, 47% of women reported missing one or more pills each cycle and 22% reported missing two or more (Rosenberg & Waugh, 1999). In another, which used electronic diaries to record compliance, 63% of women in the first cycle of use and 74% in the second cycle missed one or more pills (Potter *et al.*, 1996). Inconsistent and incorrect use is even higher with methods of contraception like condoms, which rely on use with every act of intercourse. In addition discontinuation rates of all methods of contraception are high. In the USA around one-third of women who start the pill have stopped using it within one year; almost half of new users on condoms and Depo Provera®, and almost one in five women using intrauterine contraceptives or implants stop within a year (Trussell, 1998). Switching to a different method of contraception is common and often the new method chosen is a less effective one (Grady *et al.*, 2002).

Poor compliance and discontinuation of oral contraceptives together account for an estimated 700,000 unintended pregnancies in the USA each year (Frost *et al.*, 2008).

Unsafe abortion

Lack of access to contraception or inability or lack of motivation to use it consistently and correctly inevitably leads to unintended (unplanned) pregnancy and then, in many cases, to abortion. For centuries women with unwanted pregnancies have resorted to abortion as a means of fertility control. Recognition of the inevitability of unwanted pregnancy and therefore of abortion (which if illegal is often unsafe) led many countries in the developed world to legalise abortion in the second half of the twentieth century.

Of the 210 million pregnancies that occur annually worldwide, 80 million are estimated to be unplanned, 46 million are terminated, 19 million of them illegally (Sedgh *et al.*, 2007). In general where abortion is illegal, it is usually unsafe. There are only four countries in the world in which abortion is not permitted on any grounds; however, only in 52 of the 193 countries worldwide is it permitted on request and in only 63 is it available on social or economic grounds.

Unsafe abortion occurs almost exclusively in the developing world. It is estimated that there are some 68,000 deaths a year owing to unsafe abortion – more than 67,500 of them in developing countries (Sedgh *et al.*, 2007). In Africa, where an estimated 44.2 million abortions are undertaken each year, 30,000 women die from illegal abortion.

In developed countries abortion is extremely safe: it is safer in the USA to have an abortion than it is to be sterilised, have a vaginal delivery, a caesarean section or a hysterectomy. In their recent discussions the British Medical Association Medical Ethics Committee has publicly declared that in the UK it is safer to have a pregnancy terminated than it is to have a baby (British Medical Association, 2007). While in developed countries the mortality rate from abortion is less than one death per 100,000 abortions, in Africa it is 680. The Society of Obstetricians and Gynaecologists of Nigeria recently estimated that 20,000 Nigerian women die from unsafe abortions every year (Raufu, 2002). Half of these women are adolescents. Abortion complications are responsible for 72% of all deaths among teenagers below the age of 19. Despite very restrictive abortion laws, the abortion rate in Nigeria is 45/1000. In contrast in Sweden where abortion in the first trimester is available on request, the rate is 18.1/1000 (less than half). Of course, over 90% of women in Sweden use modern methods of contraception compared with only 6% in Nigeria (Population Reference Bureau, 2005). In a study of 144 women having an abortion in Nigeria in the early 1990s (half of them under 20 years of age) 9% died, 27% suffered sepsis, 13% haemorrhage, 9% per cent injury to the cervix, uterus or vagina, 4% injury to bowel and 1% vesico-vaginal fistula (Ikpeze, 2000).

As well as deaths, unsafe abortion accounts for substantial morbidity in many countries where abortion laws are restrictive. In the developing world as a whole, every year an estimated 5 million women (the entire population of Scotland) are admitted to hospital for treatment of complications from induced abortions. In Egypt 216,000 women are treated every year in hospital for the complications of unsafe abortion, that is more than the total number of women who have an abortion in the UK (Singh, 2006). In Brazil the number of women treated in hospital for complications is almost 300,000 every year (Singh, 2006).

The relationship between levels of contraceptive use and abortion in a region or country is complex. Intuitively one would anticipate that as use of contraception, and particularly use of effective contraception, rises there should be an associated decline in abortion rates. In reality the increased use of modern contraception in a country with high fertility rates tends to coincide with a widespread desire for smaller families – fewer children. Since universal contraceptive use is hard (perhaps impossible) to achieve, the desire to limit family size fuels an increase in induced abortion (Marston & Cleland, 2003). As fertility rates decrease to very low levels characteristic of industrialised societies, as age of first sex decreases and age of marriage increases so the duration of exposure to the risk of unintended pregnancy increases further. In a country like the UK where the average age of first sex is 16 years, most people have two children and the average age of the menopause is 51, most couples (women in reality) spend more than 30 years trying to avoid pregnancy. Once the use of highly effective contraception is high (over 80%), the potential demand for abortion *should* fall (Marston & Cleland, 2003). Unfortunately as most women, even in developed countries, are not protected by absolutely effective contraception at all times a demand for abortion always does and always will exist.

No country has ever achieved low fertility levels without access to abortion. Induced abortion is a simple and cheap procedure which can be competently undertaken by nurses (Warriner *et al.*, 2006). In countries where abortion has been legalised there has been a significant fall in maternal deaths (Sedgh *et al.*, 2007). The classic and much quoted example is that of Romania (Marston & Cleland, 2003). Abortion was legalised in 1957 and, as in so many Eastern European countries where the supply of modern contraceptives was limited (Dorman, 1993), abortion rapidly became the main method of fertility control. Abortion rates in Romania rose from around 30/1000 women or less in 1956 to over 250/1000 women of reproductive age (w.r.a.) by 1965. In 1966 abortion was legally restricted and rates plummeted to around 40/1000 w.r.a in the space of three years. As a result the fertility rate almost doubled. Abortion was further restricted under the Ceausescu regime until 1989 when many of the restrictions were reversed and legal abortion became more accessible again. Maternal mortality and morbidity, which had been low in the 1960s, rose dramatically in the 1980s, probably caused by women resorting to unsafe illegal abortions. An estimated 87% of all maternal deaths in Romania during this period were attributable to unsafe abortions. After the fall of Ceausescu and the improved access to abortion, maternal dcaths declined rapidly.

Fears that legalising abortion will lead to the abandonment of contraception are misplaced. In countries with liberal abortion laws together with easy access to effective contraception regardless of age and marital status – such as the Netherlands – abortion rates are low. In countries where access to effective contraception is limited and where childbirth is a risky business, it makes no sense for abortion to be illegal and nowhere need it be unsafe (Grimes *et al.*, 2006).

References

AbouZahr, C. (1999). Some thoughts on ICPD +5. *Bulletin of the World Health Organization*, 77, 767–70.

All Party Parliamentary Group on Population, Development and Reproductive Health (2007). Return of the population growth factor. Its impact on the millennium development goals. www.appg-popdevrh.org.uk/Publications/Newsletters/Newsletter-14.pdf.

British Medical Association (2007). First trimester abortion. A briefing paper by the BMA's medical ethics committee. www.bma.org.uk.

Bongaarts, J. (2008). Fertility transitions in developing countries: progress or stagnation? *Studies in Family Planning*, 39, 105–10.

Cleland, J., Bernstein, S., Ezeh, A. *et al.* (2006). Family Planning: the unfinished agenda. *Lancet*, 368, 1810–27.

Dorman, S. (1993). More access to contraception? Russian city surveyed. *Population Today*, 21, 5–10.

Frost, J. J., Darroch, J. E. & Remez, C. (2008). Improving contraceptive use in the United States. Issues Brief. *Alan Guttmacher Institute*, 1, 1–8.

Glasier, A. (2002). Contraception – past and future. *Nature Medicine*, 4, S3–6.

Grady, W. R., Bill, J. O. G. & Klepinger, D. H. (2002). Contraceptive method switching in the United States. *Perspectives in Sexual and Reproductive Health*, 34,135–45.

Grimes, D. A., Benson, J., Singh, S. *et al.* (2006). Unsafe abortion: the preventable pandemic. *Lancet*, 368, 1908–19.

Horton, J. (2008). Women on the pill urged to try newer contraceptives. *The Herald*. November 17th. http://www.theherald.co.uk/news/health/.

Horton, R. (2006). Reviving reproductive health. *Lancet*, 368, 1549.

Ikpeze, O. C. (2000). Patterns of morbidity and mortality following illegal termination of pregnancy at Nnewi, Nigeria. *Journal of Obstetrics and Gynecology*, 20, 55–7.

Leathard, A. (1980). *The Fight for Farnily Planning*. London: MacMillan Press Ltd.

Marston, C. & Cleland, J. (2003). Relationships between contraception and abortion: A review of the evidence. *International Familv Planning Perspectives*, 29, 6–13.

National Institute for Health and Clinical Excellence (2005). Long-acting reversible contraception. www.nice.org.uk.

Parkes, A. (1985). *Off-beat Biologist. The Autobiography of Alan S. Parkes.* Cambridge: The Galton Foundation.

Pincus, G. (1965). *The Control of Fertilitv*. New York, NY: Academic Press.

Population Reference Bureau (2004). Transitions in world population. *Population Bulletin*, 59, 1–40. Washington, DC: Population Reference Bureau.

Population Reference Bureau (2005). World population data sheet. www.prb.org.

Potter, L., Oakley, D., de Leon-Wong, E. & Canaman, R. (1996). Measuring compliance among oral contraceptive users. *Family Planning Perspectives*, 28,154–8.

Raufu, A. (2002). Unsafe abortions cause 20,000 deaths a year in Nigeria. *British Medical Journal*, 325, 988.

Rosenberg, M. & Waugh, M. S. (1999). Causes and consequences of oral contraceptive non-compliance. *American Journal of Obstetrics and Gynecology*, 180, 276–9.

Santelli, J. S., Duberstein Lindberg, L., Finer, L. B. & Singh, S. (2007). Explaining recent declines in adolescent pregnancy in the United States: the contribution of abstinence and improved contraceptive use. *American Journal of Public Health*, 97, 1–7.

Sedgh, G., Henshaw, S., Singh, S., Ahman, E. & Shah, I. H. (2007). Induced abortion: estimated rates and trends worldwide. *Lancet*, 370,1295–7.

Singh, S. (2006). Hospital admissions resulting from unsafe abortion: estimates from 13 developing countries. *Lancet*, 368, 1887–92.

Stern Review (2006). The economics of climate change. www.hm-treasury.gov.uk.

Taylor, T., Keyse, L. & Bryant, A. (2006). Contraception and sexual health 2005/06. *Omnibus Survey Report No 30*. London: Office for National Statistics.

Trussell, J. (1998). Contraceptive efficacy. In *Contraceptive Technology*, ed. R. Hatcher, J. Trussell J., F. Stewart *et al.* 17th edn. New York, NY: Ardent Media.

United Nations. (1995). *Report of the International Conference on Population and Development*. Cairo 5–13 September, 1994. New York, NY: United Nations – Sales No.95.XIIL.18.

United Nations (2001). *Road Map towards the Implementation of the United Nations Millenium Declaration*. New York, NY: United Nations: A/56/326.

United Nations.(2005). *World Summit Outcome*. New York, NY: United Nations.

Warriner, I. K., Meirik, O., Hoffman, M. *et al.* (2006). Rates of complications in first-trimester manual vacuum aspiration abortion done by doctors and mid-level providers in South Africa and Vietnam: a randomized controlled equivalence trial. *Lancet*, 368, 1965–72.

12 *Global sexual and reproductive health: responding to the needs of adolescents*

MOLLY SECOR-TURNER, LINDA H. BEARINGER AND RENEE E. SIEVING

Introduction

The health of adolescents has a significant impact on global health. Today, there are more than 1.5 billion adolescents and young adults aged 10 to 25 worldwide (UNFPA, 2003). Representing a large proportion of the global population, the health of these young people will greatly affect the overall health of the world as they become adults and begin families of their own. Adolescence, the transition from childhood to the roles and responsibilities of adulthood, critically sets the life-course (McIntyre *et al.*, 2002). Among the many challenges to the health and well-being of adolescents during this second decade of life – poverty, violence, and communicable diseases, to name a few – sexual and reproductive health issues are paramount. And unsafe sex continues to be a key risk factor for death and illness among adolescents, worldwide. Indeed, in developing countries risky sexual behavior is ranked higher than unsafe water and poor sanitation, and is second only to being underweight (Glasier *et al.*, 2006).

Primary negative outcomes associated with unsafe sex during adolescence are early pregnancy and sexually transmitted infections (STIs), including HIV/AIDS, all of which have the potential to affect the entire life-course of adolescents. At the same time, prevention of these negative outcomes during adolescence has the potential to change the lives of many adolescents, and in turn, adults (Bearinger *et al.*, 2007). Yet, a host of social, cultural, and religious influences challenges the capacity and the likelihood of success in lowering sexual and reproductive risk for young people; norms related to expectations about sexual behavior, age of marriage, and fertility confound promotion, prevention, and treatment strategies, as well.

Sexual and reproductive health issues for young women are especially challenged by social norms as they assume responsibilities related to marriage and childbearing. For example, although there has been a recent global trend in

Reproduction and Adaptation, eds. C. G. Nicholas Mascie-Taylor and Lyliane Rosetta. Published by Cambridge University Press.

delaying marriage and childbearing, early marriage and childbearing continue to be a reality in many developing nations, especially among young women living in poverty (Alan Guttmacher Institute, 1998; Singh & Samara, 1996; World Health Organization, 2004a). Additionally, some girls in the developing world are pressured to become sex workers in order to financially contribute to family economies (McIntyre *et al.*, 2002), which, in turn, substantially increases risk for negative outcomes, including consequences of violence (Bearinger *et al.*, 2007).

Sexual and reproductive health issues are entwined with other contemporary global challenges including poverty, environmental conditions, and health (Glasier *et al.*, 2006). Economically, decreasing fertility may lower poverty rates, improve economic conditions, and increase women's participation in the work force. Further, family planning contributes to increased environmental sustainability by decreasing the overall global population. And family planning contributes to the improved survival and overall health of mothers and children (Glasier *et al.*, 2006). In sum, the benefits of access to and use of family planning are clear. Yet, despite the relatively low cost and relative ease of preventing sexually transmitted infections and early pregnancy, adolescents continue to be affected disproportionately by negative sexual health outcomes, such as unplanned pregnancy and high rates of HIV/AIDS (Bearinger *et al.*, 2007).

Addressing the sexual and reproductive health needs of adolescents requires consideration of the unique needs of adolescents (Bearinger *et al.*, 2007). Adolescents are developmentally distinct from both children and adults, and have different physical, cognitive and social needs and abilities (Bearinger *et al.*, 2007). These distinctions often create a gap in delivery of health services to adolescents, in particular sexual and reproductive health services. If teenaged patients are viewed as children, sexual and reproductive health issues may not be addressed because they are not seen as within the purview of pediatric services (Bearinger *et al.*, 2007). For older adolescents, sexual and reproductive health issues may not be addressed unless within the confines of marriage and with a focus on reproductive health needs (Senderowitz, 1999). This gap in sexual and reproductive health service provision continues to widen owing to increasing delays in marriage and childbearing that create a greater need for services as rates of premarital sex increase in developed countries in response to these delays (Bearinger *et al.*, 2007; Wellings *et al.*, 2006).

Adolescent sexual and reproductive health issues

Adolescent sexual and reproductive behavior is influenced by a myriad of factors related to individual behavior, as well as the social and environmental contexts in which young people live. Variations in sexual behavior exist by gender, race, ethnicity, geography, and socioeconomic status and are influenced by community values and traditions (Marston & King, 2006). Regardless of regional, country, and community variations, five key factors most directly influence sexual and reproductive health outcomes for adolescents worldwide: 1) sexual risk behaviors, such as age at sexual debut, number of sexual partners, and condom use, 2) use of contraception, 3) early pregnancy, 4) abortion and 5) burden of sexually transmitted infections including HIV/AIDS (Bearinger *et al.*, 2007; Glasier & Gulmezoglu, 2006).

Sexual risk behaviors

Sexual debut. Two aspects of sexual debut determine its impact: the age of the young persons, and the context in which sexual debut occurs. The diverse norms around marriage relate to both determinants (Bearinger *et al.*, 2007). For example, in Northern African and parts of Asia, sexual debut most often occurs within marriage (Singh *et al.*, 2000). However, delay in marriage, a global phenomenon, has created a context in which sexual debut is increasingly common outside of the context of marriage (Blanc & Way, 1998). Gender differences also contribute to the context in which first sexual intercourse experience takes place. For boys, first sexual experiences are most often non-marital. In contrast, for a large proportion of girls, many in the developing world, sexual debut occurs within marriage (Singh *et al.*, 2000). While sexual experience during adolescence occurs at similar rates among boys and girls in developed countries, differences by gender are more extreme in developing regions of the world. For example, 56% of females and 59% of males ages 15–19 report sexual experience in Australia (de Visser *et al.*, 2003), compared to 9% of females and 38% of males of the same age in Guatemala (Centers for Disease Control and Prevention, 2005). These variations may reflect societal double standards that encourage males to engage in sexual relationships outside of marriage but stigmatize young women who engage in premarital or extramarital sexual relationships (Glasier & Gulmezoglu, 2006).

Sexual activity during adolescence, both within and outside the context of marriage, creates higher risk for pregnancy and STIs among adolescents. Worldwide, one-third or more adolescent girls and around 40% of males are sexually experienced outside of marriage, with smaller percentages in the Philippines for both genders; eastern Europe, and Latin America for girls; and Nigeria and Rwanda for boys (Bearinger *et al.*, 2007).

Number of sexual partners. The risk of acquiring a sexually transmitted infection, including HIV/AIDS, is greatly increased when adolescents have more than one sexual partner. The proportion of sexually experienced adolescents with two or more sexual partners in the past 12 months varies by gender and by country. Worldwide, the percentage of sexually experienced males who report more than one sexual partner in the last 12 months range from a low of 4% in Rwanda to a high of 58% in Brazil (Bankole *et al.*, 2004; Singh & Darroch, 2000). For female adolescents, the range is 2% in Malawi and Rwanda to 33% in Canada (Bankole *et al.*, 2004; Maticka-Tyndale *et al.*, 2001). The prevalence of multiple sexual partners is higher among males (28% on average) than females (11% on average) in all countries for which data are available (15 countries) (Bearinger *et al.*, 2007). In developing countries, such as Kenya, extreme differences are reported (6% of females compared to 44% of males) (Bankole *et al.*, 2004), while in more developed countries, such as Canada, differences are less drastic (33% of females compared to 41% of males) (Maticka-Tyndale *et al.*, 2001). Again, differences in number of partners may reflect regional variations in social and cultural expectations regarding sexual behavior for men and women in which men are encouraged to prove their masculinity while moral restraint and chastity is expected of women (Alan Guttmacher Institute, 1998; Wellings *et al.*, 2006).

Condom use. Consistent condom use is one of the most effective ways to prevent sexually transmitted infections, including HIV/AIDS, as well as contributing to pregnancy prevention (Alan Guttmacher Institute, 1998). Despite the effectiveness of condoms in preventing negative sexual and reproductive health outcomes, condom use remains relatively low, especially in developing nations, most notably sub-Saharan Africa. In response to the high burden of HIV/AIDS infection in this region, condom use has increased in recent years, but remains at levels too low to combat the spread of HIV/AIDS and other STIs in these countries (Bearinger *et al.*, 2007). Reported condom use at last sex among sexually experienced adolescent females ranges from 13% in Mali to 50% in Uganda and from 21% in Mali to 58% in Zimbabwe for adolescent males (Bankole *et al.*, 2004). In other parts of the world, condom use is comparatively higher. For example, in Latin America adolescents in El Salvador and Honduras report condom use rates greater than 50% for both males and females. In Australia and the United States, condom use at last sex rates are also greater than 50%, with even higher rates among males (54% females vs. 80% males in Australia; 54% females vs. 71% males in the US) (Abma *et al.*, 2004; de Visser *et al.*, 2003). In Eastern Europe, adolescents' use of condoms varies greatly. In Azerbaijan and Georgia, only 2% of girls reported condom use at last sex compared to approximately 50% of adolescent girls in Romania and Ukraine (Centers for Disease Control and Prevention, 2003).

Use of contraception

Contraceptive use among adolescents around the world has important implications for pregnancy prevention and planning to prevent negative sexual health outcomes. Among others, access to contraception, societal attitudes regarding fertility and timing of childbearing, and involvement of partners in decision-making regarding contraception must be considered. Medical contraceptive methods include contraceptive pills, injectables, implants, and intrauterine devices (Bearinger *et al.*, 2007). Among developed countries, sexually experienced adolescent female contraceptive use rates at last intercourse are highest among France, Canada, and the UK (50%, 64%, and 69% respectively) and slightly lower in the United States (42%) (Darroch *et al.*, 2001). In developing nations, contraceptive use varies by age. Contraceptive use among adolescent girls is in general much lower than among adult women. In sub-Saharan Africa, sexually experienced girls, aged 15–19, are infrequent contraceptive users (4% in Benin, 5% in Zimbabwe, 8% in Uganda, 11% in Kenya, and 12% in Mali) (Bearinger *et al.*, 2007). In Latin American and the Caribbean, medical contraceptive use is slightly higher (41% in Brazil and 30% in Nicaragua) (Bearinger *et al.*, 2007).

Several barriers may prevent adolescents from using medical contraceptive services. First, adolescents must have knowledge about modern methods of contraception, as well as access to these methods. Second, adolescent girls may be less likely to access services if they fear stigmatization related to their sexual behavior or if services are perceived as unfriendly to young people (Bearinger *et al.*, 2007; Glasier *et al.*, 2006). Finally, in some countries, use of contraceptives requires parental or husband approval (Dehne & Riedner, 2005).

Early pregnancy and birth

Worldwide adolescent pregnancy and birth rates vary by geography, as well as by socioeconomic status. For most countries, pregnancy and birth rates are represented by reported birth rates (Bearinger *et al.*, 2007). In one recent study comparing birth rates by age 18, among 12 developing countries, birth rates were on average three times higher among girls in the poorest quartile compared to those in the richest quartile (Rani & Lule, 2004). In Asia, sub-Saharan Africa, and Latin America early pregnancy rates were similar (44%, 48%, and 34% respectively) and women in the poorest quartiles experienced higher than average rates of early childbearing (53%, 59%, and 53% respectively) (Rani & Lule, 2004). Geographical variations in birth rates among adolescents vary within and between regions and tend to follow similar patterns based on development status. For example, in sub-Saharan Africa, the average birth rate is 143 per 1,000 girls aged 15–19. Rates range from 37 in Mauritius to 229 in Guinea (Bearinger *et al.*, 2007).

Abortion

Adolescents experience both spontaneous and induced abortions. Rates of spontaneous abortion are most likely similar between populations and across geography (Bearinger *et al.*, 2007); rates of induced abortion vary significantly between and within countries, primarily owing to the confluence of economic, legal, moral, and religious factors as they affect abortion decision-making (Bankole *et al.*, 1999). The accuracy of reported rates varies as well, largely in relation to differences in abortion laws. However, overall, a common pattern exists: the oldest and the youngest aged women have the highest abortion rates (Bankole *et al.*, 1999).

Even among countries with commonalities, e.g. developed countries with legalized abortion, rates among girls aged 15–19 vary. For example, in Bulgaria, the Russian Federation, and the USA the incidence of abortion ranges from 29–44 abortions per 1,000 girls per year. In Australia, Canada, and New Zealand, rates are slightly lower at 20–28 per 1,000. The lowest rates of adolescent abortion occur in Germany, Italy, and Japan (<10 per 1,000) (Bearinger *et al.*, 2007).

Globally, negative health outcomes related to unsafe induced abortions remain one of the most preventable and underestimated issues affecting the sexual and reproductive health of adolescents worldwide (Glasier *et al.*, 2006). Unsafe abortions include those conducted by unskilled providers, using improper techniques, and/or in unsanitary conditions (World Health Organization, 2004b). Each year 19 million unsafe abortions occur resulting in 68,000 deaths. Ninety-seven percent of these deaths occur in developing countries; half of them in Africa. One in four of these deaths are adolescents, resulting in the loss of 17,000 adolescent girls' lives per year (Glasier *et al.*, 2006). The percentage of unsafe abortion differs by age. In developing regions, adolescents account for 12% of all unsafe abortions, in Africa 25%, Asia 8%, and 15% in Latin American and the Caribbean (World Health Organization, 2004b). Recent indications suggest that the incidence and resulting mortality of unsafe abortion are rising among urban unmarried adolescent women. As has always been the case, risks of unsafe abortion and associated mortality are greatest where abortion is illegal and access to family planning services are inadequate (World Health Organization, 2004b).

Sexually transmitted infections and HIV/AIDS

Worldwide, STIs occur in the greatest proportions among young people less than age 25. For some STIs, one-fifth to more than half of all cases occur in young people (Dehne & Riedner, 2005; Panchaud *et al.*, 2000; Weinstock *et al.*, 2004). It is estimated that each year 340 million new cases of curable STIs occur including syphilis, gonorrhea, chlamydia, and trichomoniasis. The largest number of these occur in South and South-east Asia, followed by sub-Saharan Africa, Latin America, and the Caribbean (Bearinger *et al.*, 2007). Surveillance of STIs is limited by inadequate surveillance mechanisms and underfunding for such programs (World Health Organization, 2006). Reported rates are often higher among females than males, as they are more likely to be screened. Overall, rates of STI infection appear to be on the rise in most countries, both developed and developing (World Health Organization, 2006). Each day 6,000 young people are estimated to be infected with HIV worldwide. There are an estimated 10 million youth aged 15–24 living with HIV in the world, the largest percentage living in sub-Saharan Africa (62%), followed by Asia (22%), Latin American and the Caribbean (7%), Eastern Europe and Central Asia (6%), high-income countries (2%), and North African and the Middle East (1%) (UNAIDS, 2004). Exposure to HIV and transmission occurs in multiple ways. In sub-Saharan African, where the majority of adolescents living with HIV live, heterosexual transmission is the most common (UNAIDS, 2004). In addition, HIV disproportionately affects girls, comprising about 75% of all adolescent cases in sub-Saharan Africa. For example, in South Africa rates are 20 to 10 (girls to boys), and as high as 45 to 10 in Kenya, and Mali (UNAIDS, 2004). In other areas of the world, additional modes of transmission contribute to the spread of HIV/AIDS among adolescents. In Central Asia and Eastern Europe, injectable drug use has contributed to the rapid rise of new cases among adolescents, as well as, to a lesser extent, unsafe sex. The transmission of HIV occurs in Latin America and the Caribbean predominantly through heterosexual sexual intercourse, as well as among men who have sex with men. In North American and Western Europe, sexual activity is the main mode of HIV transmission (UNAIDS, 2004).

Addressing the global sexual and reproductive health needs of adolescents

One important outcome of the 1994 International Conference on Population and Development in Cairo was the goal set forth to achieve universal access to safe, affordable, and effective reproductive health care services, including those for young people that promote gender-related issues by the year 2015 (Langer, 2006). Despite this goal and the support of nations involved in creating it, policies and interventions contributing to this goal have largely been met with conservative responses. Decreased funding has severely stressed the capacity to provide needed services, especially to adolescents (Langer, 2006). It has also led to the limiting of prevention strategies, focused now more than before on abstinence and faithfulness with concomitant disregard for broader social and environmental issues (Glasier & Gulmezoglu, 2006). Although the global climate and country-level politics, particularly pronounced in the US during the Bush administration, threaten the provision of evidence-based efficacious strategies for protecting the sexual and reproductive health of adolescents, recent signs of changing sentiments suggest that resources for family planning may increase.

Nonetheless, adolescents need access to appropriate information regarding sexual and reproductive health, access to sexual and reproductive health counseling and services, and opportunities to develop decision-making and interpersonal skills related to sexual and reproductive health (UNFPA, 2003). Worldwide, effective strategies need three components: access to health services, education, and youth development. Specifically, adolescents must be able to access clinical services that provide quality access to sexual and reproductive health care. Sex education programs need to be evidence based and developmentally appropriate. Adolescents also need access to youth development programs to enhance their life skills, foster connections to supportive adults, and provide educational and economic opportunities (Bearinger *et al.*, 2007). In sum, no single, general approach will be effective anywhere; no set of approaches will be everywhere, i.e. any multi-pronged intervention must be tailored to local needs, culture, and the context in which adolescents live (Bearinger *et al.*, 2007; Wellings *et al.*, 2006).

Accessing adolescent sexual and reproductive health services

Universally, adolescents must have access to high-quality, youth-friendly sexual and reproductive health care services. Even if adolescents are armed with information and skills to navigate decision-making about sexual behavior and reproduction, they must have access to commodities and services in order to adequately protect themselves (UNFPA, 2003).

A variety of factors result in barriers for adolescents in accessing high-quality sexual and reproductive health care services. These barriers impede the goals of creating services that are available, accessible, acceptable, and equitable – criteria for judging the adequacy of services as set forth by the

World Health Organization (Tylee *et al.*, 2007). Social norms within a culture may be one of the most powerful influences on adolescents' access to services. Social norms encompass morals, taboos, laws, and religious beliefs, all of which may impede the ease and comfort with which adolescents seek sexual and reproductive health care services (Wellings *et al.*, 2006). Identified barriers for adolescents seeking family planning resources and services include low societal status, strict moral codes about gender role behaviors, legal barriers, financial barriers, attitudinal barriers, and structural barriers (Hock-Long *et al.*, 2003; Senderowitz, 1999; UNFPA, 2003). In accessing STI services, adolescents encounter personal barriers such as lack of knowledge about STIs and symptoms, not taking STIs seriously, and silence about STI symptoms, related to fear or embarrassment (Brindis, 2002; Dehne & Riedner, 2005; Fortenberry, 1997). Structural barriers include geographical locations of clinics, prohibitive operating clinic hours, high cost of clinical services, and legal issues (Dehne & Riedner, 2005; Glasier *et al.*, 2006; Tylee *et al.*, 2007).

One pertinent legal issue in many developing countries is the ability to provide consent for services (Hock-Long *et al.*, 2003). In some countries, accessing sexual or reproductive health services requires parental consent, or consent of a husband. Plus, laws, which widely vary by region and country, define minimum age at which a young person can consent to sex (Dehne & Riedner, 2005). Further, seeking treatment at a minor age might result in legal ramifications regarding sexual experience prior to age of consent. Finally, many adolescents report negative provider attitudes that prohibit them from accessing clinic services. Young people may be ignored, told off or lectured to, hassled or stigmatized for attending clinics (Dehne & Riedner, 2005; Hock-Long *et al.*, 2003; Senderowitz, 1999; UNFPA, 2003).

Typically, there are three access points for sexual and reproductive health care services: facility-based services, pharmacies, and community-based services (Cleland *et al.*, 2006). Facility-based services act as the primary method of delivery for surgical and clinical methods of family planning. Facility-based services provide 80% of all family planning services in developing countries. Pharmacies, shops, and bazaars provide quick, anonymous access to contraception options, including condoms, but do not provide clinical services, such as STI testing or treatment. Community-based services are delivered by community health workers and provide access to address issues related to geography and social norms (Cleland *et al.*, 2006).

Facility-based services are most likely to be provided in clinical settings. Comprehensive clinical services need to address issues of prevention, diagnosis, and treatment of STIs and HIV, prevention of pregnancy, and follow-up care (Bearinger *et al.*, 2007; Senderowitz, 1999). Furthermore, these clinical services should be accessible, ensure confidentiality and privacy, and be free or affordable to adolescents (Ringheim, 2007). Finally, it is imperative these services are provided in a youth-friendly manner.

Youth-friendly services "have policies and attributes that attract youth to the facility or program, provide a comfortable and appropriate setting for serving youth, meet the needs of young people, and are able to retain their youth clientele for follow-up and repeat visit" (Senderowitz, 1999). Youth-friendly services are accessible, acceptable, and appropriate for adolescents. They provide effective, safe, and affordable services, designed to meet the specific needs of adolescents (McIntyre *et al.*, 2002). Specially trained staff serve as the primary providers in these settings; they are competent and sensitive to adolescents, show respect, honor privacy and confidentiality, involve young people and the communities in which they live, and provide adequate time for interaction with youth clients (McIntyre *et al.*, 2002; Senderowitz, 1999). Youth-friendly health facilities have separate space and special times set aside for youth, have convenient operating hours for youth, convenient locations, adequate space and sufficient privacy, and comfortable surroundings (Senderowitz, 1999). Other important characteristics include easy-to-get appointments or drop-in clients welcomed, short waiting times, affordable fees, wide range of services provided, and educational materials available (National Research Council and Institute of Medicine, 2009; Senderowitz, 1999).

Discussions about access to high-quality, youth-friendly adolescent sexual and reproductive health services cannot take place without discussing the social context in which young women experience sexuality and reproduction. Elements of the social context create disparities in access to sexual and reproductive health services directly, as previously described, but also indirectly as they are closely tied to issues related to the ability of women to take part in the decision-making related to sexual behavior and reproduction. Issues related to gender equality, the empowerment of women, elimination of violence against women, and the ability of women to control their own fertility are fundamental to the success of any sexual and reproductive health programs (Glasier *et al.*, 2006; UN, 1995).

Even with access to contraception, condoms, and educational opportunities, many women throughout the world are left powerless over decisions regarding their own sexual experiences. Women are especially disempowered in the developing world, but experience marginalization along gender lines everywhere (Glasier *et al.*, 2006). Living in social contexts in which women are disempowered is directly related to the increased vulnerability of negative sexual risk behavior outcomes among adolescent girls. For example, unintended pregnancy and unsafe abortion are often associated with experiences of violence and sexual coercion (Glasier *et al.*, 2006). Violence against women, including sexual violence (rape, sexual coercion, and child sexual abuse) is accepted in many parts of the world and creates a social context in which these abuses are inevitable and perpetuated by gender inequalities (Glasier *et al.*, 2006; UN, 1995). Adolescent girls living in these social contexts are unable to make decisions regarding their sexual and reproductive health, such as condom or contraceptive use, resulting in elevated risk for sexually transmitted infections, as well as unwanted pregnancies (Glasier *et al.*, 2006). Experiences of forced or coercive sex are confounded by the use of transactional sex to provide income and basic needs for some adolescent girls. Finally, in most regions of the world, men experience a sense of entitlement to sex within and outside of marriage which occurs in contrast to expectations that women do not engage in sex outside of marriage.

Examining elements of the social context in which women live is critical to promoting the sexual and reproductive health of adolescents worldwide. It is imperative that sexual and reproductive health programs address issues of gender inequality and create contexts in which women are empowered to make independent decisions regarding sexual behavior and reproduction. Unfortunately, following the lead in the US of the Bush administration's conservative approach to adolescent sexual health, globally much emphasis has been placed on following strategies that promote abstinence, faithfulness, and condom use to prevent pregnancy and decrease risk for STIs. Because these strategies do not address the social context surrounding decisions regarding abstinence, faithfulness, and negotiating condom use, they place adolescent girls in many parts of the world at greater risk for STIs and unwanted pregnancy. As disempowered decision-makers regarding their own sexual experiences, many women will not have the ability to make decisions about whether or not sex occurs, who her partner has sex with, or whether or not she will use a condom during sex; these are largely decisions made by men who live in social and political contexts that support their entitlement to the sexuality of women.

Summary

Throughout the world, adolescents face diverse challenges in optimizing their sexual and reproductive health. As adolescents navigate paths to adulthood, their experiences with sexual and reproductive health are critical to successful transition. In both developed and developing countries, young people are challenged to make decisions about and respond to their sexual experiences and outcomes. They face challenges in accessing, affording, accepting, and utilizing contraceptives and abortion services. They are making decisions within social and political contexts that often include rigid expectations and values about sexuality that confine their decision-making capabilities. Supporting the sexual and reproductive health needs of adolescents requires not only comprehensive, youth-friendly services, but also policies that protect and support these services.

Adolescents face many barriers in accessing comprehensive, high-quality sexual and reproductive health services. Access to comprehensive clinical services, including sexual and reproductive health, can help keep adolescents healthy and support them in completing their transition to adulthood in good health (McIntyre *et al.*, 2002). Providing youth-friendly services involves creating health care settings and services that are acceptable, accessible, and affordable for adolescents. Addressing the immediate specific and unique sexual and reproductive health needs of adolescents benefits adolescents now and later in life, as well as future generations. Addressing these needs can reduce current morbidity and mortality related to sexual and reproductive health for adolescents, as well as reduce the burden of disease later in life. Lessons adolescents learn as young people about responsible and safe sexual and reproductive health carry over and provide protection into later life. Finally, providing access to sexual and reproductive health services fulfills the human rights of adolescents and protects human capital (McIntyre *et al.*, 2002).

References

Abma, J., Martinez, G., Mosher, W. & Dawson, B. (2004). Teenagers in the United States: Sexual activity, contraceptive use, and childbearing, 2002. *Vital Health Statistics*, 23, no. 24.

Alan Guttmacher Institute (1998). *Into a New World: Young Women's Sexual and Reproductive Lives*. New York, NY: The Alan Guttmacher Institute.

Bankole, A., Singh, S. & Haas, T. (1999). Characteristics of women who obtain induced abortion: A worldwide review. *International Family Planning Perspectives*, 25, 68–77.

Bankole, A., Singh, S., Woog, V. & Wulf, D. (2004). *Risk and Protection: Youth and HIV/AIDS in Sub-Saharan Africa*. New York, NY: The Alan Guttmacher Institute.

Bearinger, L.H., Sieving, R.E., Ferguson, J. & Sharma, V. (2007). Global perspectives on the sexual and reproductive health of adolescents: patterns, prevention, and potential. *The Lancet*, 369, 1220–1231.

Blanc, A.K. & Way, A.A. (1998). Sexual behavior and contraceptive knowledge and use among adolescents in developing countries. *Studies in Family Planning*, 29, 106–116.

Brindis, C. (2002). Advancing the adolescent reproductive health policy agenda: issues for the coming decade. *Journal of Adolescent Health*, 31, 296–309.

Centers for Disease Control and Prevention. (2005). *Reproductive, maternal, and child health in Central America. Trends and challenges facing women and children.*

El Salvador, Guatemala, Honduras, and Nicaragua. Atlanta, GA: United States Agency for International Development.

Centers for Disease Control and Prevention. (2003). *Reproductive, Maternal and Child Health in Eastern Europe and Eurasia: A Comparative Report*. Atlanta, GA: US Department of Health and Human Services.

Cleland, J., Bernstein, S., Ezeh, A. *et al.* (2006). Family planning: the unfinished agenda. *The Lancet*, 368, 1810–1827.

Darroch, J.E., Singh, S. & Frost, J.J. (2001). Differences in teenage pregnancy rates among five developed countries: the roles of sexual activity and contraceptive use. *Family Planning Perspectives*, 33, 244–250.

Dehne, K.L. & Riedner, G. (2005). *Sexually Transmitted Infections among Adolescents: The Need for Adequate Health services*. Geneva: World Health Organization.

de Visser, R., Smith, A., Rissel, C., Richters, J. & Grulich, A. (2003). Sex in Australia: heterosexual experience and recent heterosexual encounters among a representative sample of adults. *Australia New Zealand Journal of Public Health*, 27, 145–154.

Fortenberry, J.D. (1997) Health-care seeking behaviours related to sexually transmitted diseases among adolescents. *American Journal of Public Health*, 87, 417–420.

Glasier, A. & Gulmezoglu, A.M. (2006). Putting sexual and reproductive health on the agenda. *The Lancet*, 368,1550–1551.

Glasier, A., Gulmezoglu, A.M., Schmid, G.P., Moreno, C.G. & Van Look, P.F. (2006). Sexual and reproductive health: A matter of life and death. *The Lancet*, 368,1595–1607.

Hock-Long, L., Herceg-Baron, R., Cassidy, A.M. & Whittaker, P.G. (2003). Access to adolescent reproductive health services: Financial and structural barriers to care. *Perspectives on Sexual and Reproductive Health*, 35, 144–147.

Langer, A. (2006). Cairo after 12 years: Successes, set backs, and challenges. *The Lancet*, 368, 1552–1554.

Marston, C. & King, E. (2006). Factors that shape young people's sexual behaviour: A systematic review. *The Lancet*, 368, 1581–1600.

Maticka-Tyndale, E., McKay, A. & Barrett, M. (2001). *Teenage sexual and reproductive behavior in developed countries. Country report for Canada*. New York, NY: Alan Guttmacher Institute.

McIntyre, P., Williams, G. & Peattie, S. (2002). *Adolescent-friendly health services: An Agenda for change*. Geneva: World Health Organization.

National Research Council and Institute of Medicine (2009). *Adolescent Health Services: Missing Opportunities*. Washington, DC: The National Academies Press.

Panchaud, C., Singh, S., Feivelson, D. & Darroch, J.E. (2000). Sexually transmitted diseases among adolescents in developed countries. *International Family Planning Perspectives*, 32, 24–32.

Rani, M. & Lule, E. (2004). Exploring the socioeconomic dimension of adolescent reproductive health: A multicountry analysis. *International Family Planning Perspectives*, 30, 110–117.

Ringheim, K. (2007). Ethical and human rights perspectives on providers' obligation to ensure adolescents' rights to privacy. *Studies in Family Planning*, 38, 245–252.

Senderowitz, J. (1999). *Making reproductive health services youth friendly*, Washington, DC: FOCUS on Young Adults.

Singh, S. & Darroch, J.E. (2000). Adolescent pregnancy and childbearing: Levels and trends in developed countries. *International Family Planning Perspectives*, 32, 14–23.

Singh, S. & Samara, R. (1996). Early marriage among women in developing countries. *International Family Planning Perspectives*, 22, 148–157.

Singh, S., Wulf, D., Samara, R. & Cuca, Y.P. (2000). Gender differences in the timing of first intercourse: Data from 14 countries. *International Family Planning Perspectives*, 26, 21–28.

Tylee, A., Haller, D., Churchill, R. & Sanci, L. (2007). Youth-friendly primary-care services: How are we doing and what more needs to be done? *The Lancet*, 369, 1565–1573.

United Nations. (1995). Population and development, I:Programme of Action. Cairo: International Conference on Population and Development.

UNAIDS (Joint United Nations Programme on HIV/AIDS). (2004). *2004 Report on the Global AIDS Epidemic*. Geneva: UNAIDS.

UNFPA (United Nations Population Fund). (2003). *The State of the World Population, 2003. Making One Billion Count: Investing in Adolescents' Health Rights*. New York, NY: UNFPA.

Weinstock, H., Berman, S. & Cates, W., Jr. (2004). Sexually transmitted infections among American youth: Incidence and prevalence estimates. *Perspectives on Sexual and Reproductive Health*, 36: 6–10.

Wellings, K., Collumbien, M., Slaymaker, E. *et al.* (2006). Sexual behaviour in context: A global perspective. *The Lancet*, 368: 1706–1728.

World Health Organization. (2006). *Global Strategy for the Prevention and Control of Sexually Transmitted Infections: 2006–2015*. Geneva: World Health Organization.

World Health Organization. (2004a). *Adolescent Pregnancy: Issues in Adolescent Health and Development*. Geneva: World Health Organization.

World Health Organization. (2004b). *Unsafe Abortion: Global and Regional Estimates of the Incidence of Unsafe Abortion and Associated Mortality in 2000* (4th edn). Geneva: World Health Organization.

13 *Understanding reproductive decisions*

ANDREW HINDE

Introduction

Attempts to understand the decisions people make about whether and when to bear children have been vital to the work of demographers and other population scientists for many decades. Yet, despite a vast amount of attention to the issue, progress towards a clear appreciation of what motivates people to have children, and the processes by which they arrive at decisions about fertility, has been, at best, fitful. In the classical version of demography's grandest theory, that of the demographic transition (Notestein, 1945, 1953), the immediate causes of the decline in fertility were less clearly specified than were the causes of the decline in mortality. Even after more than half a century's debate, there is still no consensus about why fertility falls during industrialisation, with some favouring accounts based on a reduced demand for children (Easterlin & Crimmins, 1985) and others advocating a narrative based on the diffusion of new ideas about the acceptability of birth control (Cleland & Wilson, 1987).

Demographers' imperfect understanding of the ways in which decisions about childbearing are made and the factors which determine them reveals itself in discussions about policies and programmes designed to influence fertility. In relation to those countries where fertility is still well above replacement level and population growth is rapid, a great deal of the debate has, at least until recently, taken place at a fairly crude level, focusing simply on the relative importance of family planning programmes (on the one hand) and social and economic changes (on the other) in initiating and maintaining fertility decline. In late twentieth- and early twenty-first century Europe, by contrast, the discussion has been about policies which might encourage childbearing and thereby increase fertility from its current level to something close to replacement level. Yet, with a few exceptions (one of which will be discussed later in this chapter), this debate steps around the process of making decisions about reproduction,

Reproduction and Adaptation, eds. C. G. Nicholas Mascie-Taylor and Lyliane Rosetta.
Published by Cambridge University Press.

concentrating instead on macro-level causes and effects, such as the relationship between aggregate fertility and the way the welfare state is organised.

In defence of demographers, it is clear that reproductive decisions are complex and influenced by many different factors. Moreover human reproduction is an uncertain process: reproductive behaviour is not entirely subject to reasoned action (Fisher, 2006); and, to the extent that it is, the outcome – at least at an individual level – is unpredictable. In this chapter I try to explain how demographers have tried to come to terms with the multi-faceted nature of reproductive decisions. The starting point is a discussion of two simple models of human fertility which emerged independently from one another in the early 1960s. Though seemingly unrelated, I show that they shared several features in common in their view of reproductive decisions. I then move on to discuss a range of developments during the past half century which have extended and (in some cases) superseded these simple models. The chapter concludes with a discussion of the current position with respect to understanding reproductive decisions, and a few suggestions of ways in which demographers might deepen this understanding in the future.

Two models of fertility

Half a century ago, a young American economist gave a talk entitled 'An economic analysis of fertility' at a conference organised by the National Bureau for Economic Research (Becker, 1960). Although he was not the first economist to turn his attention to explaining fertility levels and differentials, Gary Becker's paper is widely regarded as the most important foundational statement of what was to become the micro-economic theory of fertility. This theory holds that fertility is the outcome of utility-maximising decisions made by individual couples. Couples try to achieve the best they can given their income and their preferences for children, and for consumer goods.

A key element of the micro-economic framework is that children are costly and that there are constraints on the amount of money couples wish to spend on each child (what the economists call child *quality*) – most people, for example, aspire to bring up their children to enjoy a similar material standard of living as themselves. Therefore couples attempt to maximise their *expected lifetime utility*, which is a function of the number of children they have, the expected cost per child, and the amount of goods and services unrelated to their children which they consume. Because of a budget constraint there is necessarily a trade-off between the money spent on children and that spent on other goods and services, and the number of children which leads to the maximum utility for a couple will be determined by the couple's relative preferences for children and these other goods and services. Given these preferences, and some level of child quality (which may also be subject to the couple's preferences although it is, as pointed out above, constrained by the couple's own material standard of living), the solution to the couple's maximisation problem is a *number* of children, and the reproductive decision is conceptualised purely as a decision about the number of children.

Some features of this original formulation of the micro-economic theory of fertility can be emphasised. First, the decision-making unit is the couple, and the assumed context is one of lifetime monogamy. Second, there is an implicit assumption that birth control can be switched on or off by the couple at will. Third, reproductive decisions are made subject to constraints. In the micro-economic theory of fertility these are either financial, or treated as fixed preferences. Finally, the reproductive decision is made at the time of marriage, and seems to be singular (made just once, at a given point in the couple's life course) and binding on the future.

One year after Becker's paper was published, the French demographer Louis Henry published a short paper in the now defunct journal *Eugenics Quarterly* entitled 'Some data on natural fertility' (Henry, 1961). In this paper he made a distinction between what he defined as 'natural fertility' and its complement, 'controlled fertility'. Natural fertility was characterised by couples making no effort to influence their reproduction, the latter being determined by a range of biological and cultural factors beyond their control. In other words, under natural fertility couples made no reproductive decisions at all. Variations in fertility from population to population were determined by factors such as differences in the levels of sterility or sub-fecundity, or cultural norms such as the expected duration of breastfeeding. Extended periods of breastfeeding may well have reduced fertility in populations which practised them, but as they were not engaged in with the intention of inhibiting childbearing they could not be characterised as efforts at fertility control.

Controlled fertility, by contrast, was characterised by deliberate decision-making on the part of couples to limit the size of their families. However, Henry had a very specific idea of what fertility control involved. The key feature was that fertility behaviour was *parity-dependent*, that is, couples would alter their behaviour on the basis of the number of children they already had. Typically, this would involve so-called 'stopping behaviour' by which couples attempted to cease childbearing by using birth control methods once they had produced the number of children they wanted.

Henry's distinction between natural and controlled fertility implies that couples decide about the *number* of their children but not about the timing.

Couples have some idea how many children they want (often called a 'desired family size' in the demographic literature). They begin their reproductive careers without using deliberate birth control, allowing children to come naturally at a rate governed by prevailing cultural norms and other biological factors. Once they have successfully produced their desired number of children they 'switch off' their fertility using a method of birth control. However, in contrast to the original version of the micro-economic theory of fertility, reproductive decisions in Henry's model can be made sequentially. Algorithmically, couples are faced with a series of identical decisions at marriage, and then after the birth of each subsequent child, regarding whether or not they wish to have a(nother) child. There are two possible courses of action at each decision point. If the couple decide to 'carry on' childbearing, then they are faced with the same two choices again after their next birth; if the couple decide to stop, this ends their chain of reproductive decisions.

The possibility of reproductive decisions being sequential is important, and I shall return to it later. However, we should also note that apart from this, Henry's model of childbearing shares many of the simplifications of the economic analysis of fertility proposed by Becker. The decision-making unit is the couple and the assumed context is one of lifetime monogamy. It is assumed that birth control can be switched on by the couple at will after the required number of births. Reproductive decisions are also subject to constraints, though in Henry's framework these constraints are handled as biological or cultural background factors.

Henry's work was profoundly influential in the discipline of demography for two or three decades. Methods were developed to assess the extent to which fertility in a population was 'controlled' (Coale & Trussell, 1974, 1978), which amounted to measuring the prevalence of 'stopping behaviour' based on Henry's model, and studies of fertility decline in various parts of the world made use of the methods. A good deal of this work focused on Asian countries, in which fertility was rapidly declining at the time (see, for example, Knodel, 1977). In general, it seemed that Henry's framework was very useful in the Asian context, in that the fertility decline in Asia seemed to be characterised by 'stopping behaviour' of the kind he envisaged. Childbearing in Asian countries took place overwhelmingly within marriage and polygamy was rare, which reinforced the value of framing the analysis of reproductive decisions around the monogamous couple. For at least two decades after its publication, there was very little questioning of the appropriateness of Henry's model of fertility decisions within the demographic literature (for one of the earliest attempts at a critique, see Knodel (1983)).

The influence of context

In recent years, a class of statistical models known as multilevel models has become popular in demographic and social scientific analysis (Goldstein, 2003). The idea underlying the use of these models is that the factors which determine any social phenomenon act at various different levels. So, for example, the risk that a child will die in infancy depends on some factors specific to that particular child (for example his or her birth weight); some factors specific to his or her family, which will affect also his or her siblings (such as the materials from which the family home is constructed); and some factors specific to the community within which he or she lives, and which will affect all infants in that community (such as distance from a health centre). Multilevel statistical models explicitly recognise this hierarchy of causes and effects, and so arrive at an improved understanding of the determinants of social and demographic phenomena, and a more accurate measurement of their effects.

The development of demographers' understanding of reproductive decisions over the last 50 years can be seen as analogous to the development of a multilevel approach to sociological modelling. In this analogy, the Henry–Becker approach takes the decision-making unit as the couple, and assumes that the factors affecting reproductive decisions are all to be measured at the couple level and that the couple's reproductive decisions are made either at a single point in time, or as a sequence of similar decisions in which the only determinant factor which changes is the number of children the couple already have.

It is now explicitly recognised that couples live within a social context, and this social context will influence their decisions about childbearing. Of course, demographers have for many years undertaken analyses of the factors associated with fertility levels and it is well known that fertility in many societies varies with such factors as parental education (Cleland & Rodríguez, 1988), social class (see, for example, Woods, 1985), and the intensity of religious belief (see, for example, Adsera, 2006). Couples with similar educational backgrounds, or who share a similar culture or religious belief, tend to make similar decisions about reproduction.

However the influence of the social context over reproductive decisions amounts to more than just a statement that the characteristics of the couple are important. There are features of a collective of people, for example norms about what is an appropriate number of children to have, which can affect the reproductive decisions of all members of that collectivity. The decisions made by particular couples about childbearing are influenced by the decisions made by other couples with whom they interact, possibly through 'copying' behaviour, but more probably because the fertility decisions made by the members of a community feed into communal attitudes towards childbearing. Thus two couples, identical in every respect other than the fact that one lives in a population where large families are the norm, and the other lives in a population where highly restricted fertility is common, may make different reproductive decisions because of the influence of the community in which they live (Garrett, *et al.*, 2001).

The influence of context over reproductive decisions is not just to be seen at this societal level, though. For each couple does have their own particular situation, which changes over time and over the couple's life course. Reproductive decisions are not made once and for all at the time of marriage, but are made sequentially and updated as time goes on and new information becomes available. In Henry's model of controlled fertility, reproductive decisions were sequential and dependent on parity. But characteristics of the couple other than parity change over the life course, so that successive decisions are made in the light of new information and a changing social and economic situation. Although markers such as ethnicity remain the same, and other characteristics such as the highest level of education attained change rather rarely over the reproductive age range, other aspects of a couple's social and economic situation may undergo great transformations resulting from such events as geographical migration, changes in occupation, or the death of close family members.

In the next section I consider the impact on decisions about reproduction of, first, the wider society; and, second, the individual couple's situational dynamic.

Society and culture

Following the initial exposition of the economic theory of fertility, it did not take long for the importance of social norms for reproductive decisions to be pointed out. Gary Becker had hardly sat down from delivering his paper when James Duesenberry (another economist whose approach was, by his own admission, 'more sociological'), was on his feet explaining that while 'economics is all about how people make choices, sociology is about why they don't have any choices to make' (Duesenberry, 1960, p. 233). Demographers have long recognised that sociological norms and values constrain individual choice, a particularly important example being the emergence of the 'two child norm' in industrialised countries in the twentieth century, especially the United States and Britain (David & Sanderson, 1987).

In a well-known critique of the micro-economic theory of fertility, Judith Blake stressed that children were not like other consumer goods in that there are severe 'social and biological constraints surrounding their acquisition and their "use"'; and parents are producers as well as consumers of children (Blake, 1968, p. 17). While some of Blake's observations reflect the time she was writing, such as her assertion that 'even among the higher income groups, family-size ideals are large enough to insure substantial population growth' (Blake, 1968, p. 25), her insistence on the importance of the social context in constraining the range of possible outcomes of the reproductive decision-making process is of enduring importance.

Demographers' awareness of the role played by macro-level cultural factors was heightened in the late 1970s and 1980s by the advent of new sources of data about fertility from the World Fertility Survey and the Demographic and Health Surveys. These took nationally representative samples of several thousand women from countries all over the world and collected data on reproductive histories (that is, the dates of birth of each woman's children, together with other relevant dates, such as her date of birth and age at marriage). They permitted the extension of rigorous analyses of fertility change to all areas of the world. Particular attention was focused on Africa, the continent for which data on fertility had previously been most scanty. Until the late 1980s there was little evidence of any decline in fertility south of the Sahara. However, between then and the mid-1990s various studies detected the beginnings of what was believed might be a sustained fall in fertility in certain countries, for example Kenya, Botswana, and Zimbabwe. Initial analyses of the beginnings of fertility transition in African countries suggested that it did not fit the Henry model. Africa had long been characterised by a very low natural fertility regime involving lengthy periods of breastfeeding and a substantial amount of sub-fecundity and sterility leading to long intervals between births (Lesthaeghe & Page, 1981). What was surprising, however, was that the decline in fertility in Africa seemed to be associated with *extended birth spacing*, rather than with 'stopping' behaviour. It was as if Africans, accustomed by tradition to spacing births widely, effected fertility transition by simply augmenting and intensifying their existing practice.

This observation led to the idea of 'African exceptionalism' – that Africa was to experience a new kind of spacing-led fertility transition – gaining momentum within the demographic literature during the early 1990s (Caldwell, *et al.*, 1992). If correct, then it implies that even if Africans' long birth spacing is some kind of exogenous cultural 'given', it does not just affect the level of natural fertility but, once decisions about restricting fertility begin to be made, it impinges upon these decisions by affecting the manner in which couples try to achieve lower fertility outcomes.

Anthropological work in Africa has demonstrated that long birth intervals have been considered desirable in African societies not so much because they lead to fewer children being born but because they allow mothers to recover their health and strength between births, and thus slow down the ageing process (Bledsoe, 2002). This is seen as beneficial both for them and for the health and survival prospects of their infants. Decisions relevant to reproduction, therefore, are really decisions about the well-being of women and children. It is worth mentioning that, based on work in rural Gambia, Bledsoe (2002) argues that, although contraception is used to lengthen birth intervals, it is by no means clear that the result is fewer births. For shorter birth intervals increase the risk to women of secondary sterility arising from birth complications, of sub-fecundity deriving from exhaustion, and of maternal mortality. Thus, although long birth intervals may lead to lower fertility in the short term, they increase a woman's chance of living through a long fecund period and thus may maximise the average number of children born per woman over the whole reproductive age range.

Thus the African experience teaches us more than just a lesson about the importance of culture in constraining reproductive decisions. It brings once more into focus the insight that 'reproductive decisions' do not necessarily involve just decisions about the proposed number of children to have. In some societies in the past 'reproductive decisions' took very different forms. In his *Essay on the Principle of Population*, and in subsequent essays, Malthus (1798, 1986, pp. 465–81) viewed the key decision about reproduction to be the decision about whether, and when, to marry. Moreover, subsequent research has suggested that the criticality of this decision was buttressed by a whole set of institutional arrangements surrounding the socialisation of young people, the structure of the household and the organisation of agrarian labour, which constrained the range of possible ages of marriage and thereby constrained reproduction (Hajnal, 1982). This was the basis of Malthus's preventive check by which fertility was restricted and population growth kept under control. Research by historical demographers into reproduction in England during the two centuries before Malthus wrote his *Essay* suggests that his analysis was largely accurate (Wrigley & Schofield, 1981; Wrigley, *et al.*, 1997).

Finally, the relative pervasiveness of the influence of the wider society on the reproductive decisions made by couples varies from culture to culture and over time. At one extreme are societies in which decisions about reproduction are essentially collective, and in which individuals' childbearing is invariably conditioned by the requirements and desires of the wider society. In their book *One Quarter of Humanity*, James Lee and Wang Feng emphasise the collective nature of fertility decision-making in China since at least 1700. As they put it, 'such decision making has almost never been an individual prerogative. Rather it has been a familial or community decision or a national policy' (Lee & Feng, 1999, p. 99). The acquiescence of the Chinese population in the face of what Western observers see as the draconian one-child policy of the late twentieth century is a modern manifestation of the Chinese tradition that decisions about fertility should be made with collective interests in mind. By contrast, in Western Europe, and perhaps most clearly in England, reproductive decisions, including the decisions about marriage which in the Malthusian world of the seventeenth and eighteenth centuries were also decisions about fertility, were largely within the remit of individual couples, to be made with minimal interference from other family members or the wider community.

Community-level constraints on reproductive decisions arise in forms other than social or cultural norms. Discussion of constraints in the recent literature has focused on the practical implementation of birth control and specifically on the range, and reliability, of methods of birth control available to couples wishing to achieve small family sizes. An important development here was the observation that in historical populations, for example that of England and Wales during the early twentieth century, very low fertility (an average of fewer than two children per woman) was achieved largely by the use of *coitus interruptus* (withdrawal) and abstinence, with a limited contribution from condoms and illegal abortion (Szreter, 1996; Fisher 2006). Withdrawal and abstinence have largely been dismissed by demographers of the contemporary world as 'traditional' or 'inefficient', placed in the same category as the wearing of charms around the neck. Of course, this is consistent with the original Becker and Henry models, which implicitly require the efficient and effective 'switching off' of fertility once the desired family size has been reached.

But couples who were relying on withdrawal and abstinence could not 'switch off' fertility in this way. Withdrawal *was* relatively ineffective compared with modern methods such as the pill and the intra-uterine device, and there were likely to be failures even where it was practised assiduously. Similarly, prolonged abstinence was onerous, and occasional lapses were likely. What emerges, especially from the work of Kate Fisher, who interviewed couples who were having their children in inter-war Britain (Fisher, 2006), is that people realised this and adopted a strategy of using these methods of contraception on a more-or-less permanent basis, allowing the few failures to produce the small family they wanted. In other words, these couples were reducing their fecundity, or 'turning their fertility down' on a long-term basis, rather than switching it on and off. The result was fertility control achieved by long spaces between births, not by the 'stopping' behaviour inherent in Louis Henry's model of controlled fertility.

Constraints because of the availability of birth control methods make themselves felt in modern societies too. In Brazil there was a rapid rise in the proportion of women who were sterilised, from 29 per cent in 1986 to 40 per cent in 1996 (Leone & Hinde, 2007, p. 459). There was also a simultaneous rise in the proportion of births taking place through elective Caesarean sections, from 15 per cent in 1971 to 37 per cent in 1996. These two trends are closely related: female sterilisation became popular because other reliable methods were in short supply as a result of legal restrictions, but sterilisation was legally available to women who had had two Caesarean sections, and was free of charge to women if carried out at the same time as a Caesarean section (Leone & Hinde, 2007, p. 462)

The sequential nature of decisions about fertility

If questions about the universality of the Henry framework of natural and controlled fertility were a consequence of the evolution of fertility trends in some poor countries between 1970 and the 1990s, the experience of much richer populations was, at around the same time, causing economists to reconsider and extend the micro-economic theory of fertility. The main impetus for this was the increasing tendency for married women in industrial countries to work outside the home and to have careers of their own.

It was realised early on that the existence of a substantial contribution to the household budget from the earnings of the wife complicated the couple's utility-maximisation problem. The incompatibility of paid work outside the home for the wife and childbearing meant that having children reduced the household budget, through the 'opportunity costs' of childbearing – specifically the wife's foregone earnings. More seriously, however, 'opportunity costs' did not just comprise the earnings lost while a mother was at home looking after her children. For women in occupations with a clear career structure, the early-career interruptions occasioned by childbearing could have long-term consequences manifest as a reduced earning capacity later in life. Technically speaking, in these occupations there are strong positive returns to working experience, and mothers had to forego some of that experience and the concomitant positive returns. A consequence of this was that the *timing* of births mattered as well as their number. Maximising expected lifetime utility entailed making decisions about the impact on the wife's prospective earning capacity of taking 'time out' to have children at different stages in her career (Ní Bhrolcháin, 1986a, 1986b). For many professional women in occupations requiring lengthy periods of training and the acquisition of postgraduate qualifications, it made sense to delay childbearing until qualification was complete. Decisions of this kind have contributed to the marked decrease in birth rates in many Western countries among women in their twenties during the last 25 years or so, and a parallel increase in the birth rate among women in their thirties, especially their late thirties.

The need for a couple to make decisions about childbearing in a context where both partners are potentially economically active reinforces both the sequential nature of fertility decision-making and the fact that even within the utility-maximisation framework couples can be imagined to update their calculations as time goes on and new information is acquired. It is unrealistic to suppose that couples make decisions about the number (still less the timing) of their childbearing only at the start of their reproductive careers. Decisions are continually reviewed as economic circumstances change. It is possible to incorporate this sequential decision-making into micro-economic models of fertility by making the outcome variable whether or not to try to have a child within a certain time-frame. Effectively what is required is to ask couples to evaluate their expected lifetime utility under two outcomes: the first in which they have a child soon, and the second in which they do not. If the expected lifetime utility is higher under the first outcome, the couple will avoid the use of contraception and attempt to have the child. The model then moves forward in time and the couple perform the same calculation at the next time point, incorporating additional information they may have acquired in the interim including, most notably, the possibility that they will have had another child by then. This model combines the utility-maximisation of the micro-economic model with the repeated decision-making of Henry's model, while acknowledging the reality that factors other than parity change as time goes on, allowing couples to make decisions at any time (rather than just immediately after the birth of a child) and permitting a broader range of fertility outcomes than are allowed within Henry's family limitation framework.

The incorporation of sequential decision-making, and especially the need to take account of the economic consequences of current parity, presents another challenge to the micro-economic model. The assumption that preferences are fixed becomes untenable. Preferences relate to the relative value a couple places on material possessions, the non-material satisfaction derived from having and raising children, and so on. In the original formulation of the micro-economic theory of fertility these were considered to be immutable, an assumption which was essential when the fertility decision-making was viewed as a once-and-for-all event made at the beginning of a couple's reproductive history. However as soon as decision-making becomes sequential, the possibility ought to be admitted that tastes may change over time; moreover the event most likely to trigger a change in preferences is the birth of a child.

There is some empirical evidence in support of sequential utility-maximising decision-making with respect to childbearing. A well-known example comes from Sweden in the early 1990s. In Sweden at this time, maternity pay was allowed for each child for up to 12 months at 90 per cent of the earnings of the mother before her confinement. This typically meant that for the first child a woman enjoyed substantially higher maternity pay then she did for subsequent children. Before her first child she would typically be working full time; whereas immediately before having her second child, she would often only be working part time – because of the existence of the first child. However, as Jan Hoem points out, '[d]uring the 1970s . . . it became established legal practice for a parent to retain the right to the level of income compensation paid after one birth during parental leave for the next birth, provided the interval between the two births did not exceed the period of statutory leave plus six months' (Hoem, 1990, p. 744). In other words, if a woman had, say, her second child within a certain time period after her first child, she could be entitled to maternity pay for the second child based on her earnings *before her first child was born.* Initially, the interval between the first and the second child had to be very short to take advantage of this benefit (no longer than 12 or 15 months) but this period was extended in 1980 to 24 months, and in 1986 to 30 months.

Clearly, this policy raised the returns to rapid childbearing, at least in the short run. This was particularly true after the extension of the qualifying period to 24 and 30 months, as with a very short qualifying period a couple's biological fecundity was a rather important determinant of the ability to take advantage of the concession. It made economic sense for couples who had recently given birth, and who were already considering having another child, to have that child soon. Certainly this is what seems to have happened, as the fertility rate in Sweden increased substantially in the years immediately following the introduction of the policy. For example, the second-birth rate among women born in 1956 was substantially higher at all durations up to three years since the first birth than it was for women born in 1951 (Hoem, 1990, p. 740). Starting in the late 1970s there was an increase in both second and third birth rates, which was especially marked at durations under three years since the previous birth (Hoem, 1990, pp. 741–2).

Of course, the decision to accelerate childbearing has implications for long-run economic prospects. The couple has to balance the short-term economic gain against the possible long-run opportunity costs of the wife spending a single, quite lengthy period largely outside the labour force, rather than two or more shorter periods between which she returns to work. As Hoem notes, 'many Swedish couples [were] willing to adjust the timing of their childbearing after the first birth, even at a possible long-term cost to the woman's career. This behavioural response suggests that childbearing preferences have priority over the wife's job prospects for many couples at the prevailing level of financial incentives and given current expectations about the family's economic standard of living' (Hoem, 1990, p. 745).

The individual and the couple

The models of fertility decisions developed in the 1960s were based around the monogamous couple as the decision-making unit. However, developments since the 1960s have made it increasingly untenable to see the couple as a permanent, indivisible entity in this way. A rise in individualism and the freedom of the *individual* to choose his or her own lifestyle has unleashed powerful forces which have transformed Western societies so that monogamy, rather than being a cultural norm enforced by draconian social sanctions, has become one of many 'lifestyle choices'. So influential has this change been on the demography of Western countries that some have labelled it and its demographic consequences a 'second demographic transition' (Van de Kaa, 1987; Lesthaeghe, 1995).

In many countries, a large and increasing proportion of births now take place outside a formal marriage, or even outside a long-term monogamous relationship. In England and Wales in 2007, 44 per cent of all births took place outside marriage (Office for National Statistics, 2008, p. 1). In some countries of Latin America, informal unions outnumber formal marriages, and in much of sub-Saharan Africa pre-marital fertility is common. The existence of multiple, fluid partnerships complicates the analysis of reproductive decision-making. Paradoxically, it can increase the constraints surrounding these decisions.

Consider, again, the example of Brazil. Brazilian unions (many of which are not legal marriages) are frequently unstable, with both men and women often having several partners during their lives. At the same time, the social and economic position of women can be precarious. These two features of Brazilian society greatly complicate the calculation which women, in particular, need to make when deciding about their future reproduction. The use of sterilisation as a means of contraception, which, it was suggested earlier, had a certain rationality in the Brazilian context, becomes risky if women believe their current union may end. The precarious economic situation of single women means that finding another partner is important. If a suitable partner is found, however, that partner (and, for that matter, the woman herself) may want a child to 'cement' the new union. If the woman is sterilised, her inability to have this child may jeopardise the new relationship.

Apart from the permanence of the couple, its indivisibility has also been challenged. The internal dynamics of the couple were largely ignored by Becker and Henry. There was, perhaps, an assumption that the couple came to a mutually agreed decision. It was certainly recognised that population-level social and cultural factors could be an important influence on the decisions of the couple, but the view was that they operated to constrain what was still the couple's decision. Even when the value of work outside the home by the wife became incorporated into fertility decisions, the idea that the husband

and wife might have conflicting fertility agendas was hardly acknowledged. The respective employment careers of the husband and wife were two separate elements to be factored into decisions made by the couple as a unit.

Recent demographic work has, however, begun to consider that partners within the couple might have different views on the appropriate number of children; and that reproductive decisions can be influenced by the nature of the relationship between the spouses. Much of this work has been based on large-scale individual-level surveys in developing countries, notably the Demographic and Health Survey programme which started in 1985 and now includes almost 200 surveys in 77 different countries (Demographic and Health Surveys, 2009).

The role of men in reproductive decisions has been especially problematical. There was a time when men were viewed mainly as obstacles to fertility decline, especially in developing countries. Most birth control methods which were being promoted in those countries were female-based (for example the contraceptive pill, tubal ligation, and intra-uterine devices). Unfortunately society in many of these countries was highly patriarchal, with prestige within the extended family or clan being conferred by a combination of age and maleness. In such circumstances men naturally favoured high fertility, since the status of a man in his older years was directly related to the number of descendants he had. Even where women wished to restrict their fertility, family planning programmes found it hard to persuade them to adopt contraceptive methods because of their subservient position within the family, and the objections, perceived or real, of their husbands.

This simplistic 'men as problem' position is now being refined. It is now realised that drawing men into the debate about fertility control, and especially improving communication between spouses about family planning and related matters, is likely to lead to couples making decisions which better reflect their mutual wishes and (it would appear) involve the adoption of birth control with the husbands' blessing (see, for example Dahal *et al.*, 2008). The notion that birth control is largely a matter for women developed as a response to the era of the pill, intra-uterine device, and (more recently) injectable contraception. However, even in this era when most efficient, reversible, methods of birth control are female, such methods are likely to be practised most effectively when spouses and partners are actively supporting their wives.

In some contexts, the involvement of men in both the formulation and the implementation of reproductive decisions seems to have been dominant. Fisher (2006), in her study of working-class couples in Britain in the 1920s and 1930s, found that for many couples, birth control was largely a matter for the husband. Of course, when it is considered that the methods available consisted of abstinence, *coitus interruptus*, and the condom, this becomes unsurprising.

Conclusion

In this chapter I have tried to show how demographers' understanding of reproductive decisions has progressed from the simple models of the 1960s. The most important developments have been the recognition of the range and variety of higher-level (i.e. societal) constraints; the realisation that in certain cultural contexts decisions relevant to reproduction may be made indirectly; the extension of models of reproductive decisions to incorporate sequentiality and situational changes over the life course; and the acknowledgement that couples are neither permanent nor indivisible.

However, these developments over the last half century have not yet led to new models of reproductive decision-making which have the simplicity and power of the old. The inadequacies of the old models are acknowledged, but they have been replaced so far not by new models but by partial explanations of reproductive decisions in specific populations. We now have a situation in which there is no universal theory of fertility; indeed with only a little exaggeration it could be asserted that we are presently in a state in which theory has been abandoned and attempts to understand reproductive decisions are focused simply on detailed analysis of the contexts in which specific individuals find themselves.

This may be just the stage we are at, and in due course better-developed, new unified models of reproductive decisions might emerge. In conclusion to this chapter I shall offer some suggestions as to the likely nature of such models. I think the starting point is that most of the context-specific accounts of fertility behaviour which have been written in recent years eventually conclude that some form of reasoned action is at work. People do behave rationally with respect to childbearing, and when they appear not to, this is because we do not understand completely enough the situation in which they find themselves. Fisher & Szreter (2003) argue that British working-class couples of the 1920s and 1930s seem not to have expressed a desire for a specific number of children, but this does not seem to me to mean that they abandoned the idea of reproductive decisions at all. Indeed, given their situation, what they were doing was sensible, reasoned, and purposeful. They were faced with a culture in which small families were becoming the norm, and a social and economic environment in which the opportunities for expenditure on consumer goods other than children were increasing, and in which expectations of a material standard of living were rising rapidly. Therefore a 'smallish' family of, perhaps, two children was seen as the ideal. There was little point having fertility intentions which were more precise than this when the main means of birth control considered usable were the drastic one of abstinence and the inefficient one of withdrawal (the available appliance methods were considered too intrusive and inconvenient). The strategy adopted to achieve a 'smallish' family was to 'turn fecundity down' by infrequent sexual intercourse (but not complete abstinence) and the use – most of the time – of withdrawal in the full knowledge that these methods were imperfect and that eventually they would fail, but in the hope that they would fail only rarely during the couple's reproductive lifetime. Of course, if for a particular couple, failure happened too often, then it was always possible to resort to more effective (but more costly) methods, such as condoms or extended periods of abstinence. On the other hand, perhaps a more pervasive risk was that the employment of withdrawal and partial abstinence would be more successful than anticipated, and that fewer than two children would result. This could explain the relatively large number of one-child families in Britain in this period.

Therefore I think that in order to develop new models of reproductive decisions, it will be important to understand fully the nature of, and constraints upon, rational decision-making with respect to fertility. With this in mind, it is a shame that in European countries (though less so in the United States) there is a divide between demographers and economists. For the idea and consequences of rational decision-making have been much more fully worked out in the discipline of economics than they have in other social scientific disciplines, and recent developments have involved the incorporation into economic thought of challenges to rationality posed from outside the discipline (see, for example, Kahneman & Tversky, 2000). Few demographers have a formal training in economics: consequently most find it difficult to read the economics literature (and unfortunately economists do not make it easy for them). However, I think that future progress will involve demographers increasing their engagement with the ideas about decision-making which have been developed within, or have been assimilated into, the economics literature, including ideas about decision-making under uncertainty and the relative valuation by individuals of present and future benefits. In engaging with this body of theory, demographers will bring to bear their knowledge of the societial and technical constraints acting upon people, and about the context within which individuals operate. The result is likely to be a theory of fertility based on reasoned action, but one which cannot easily be written down in algebraic terms and 'solved' using conventional algorithms. It will be more fuzzy than this, but more complex and, I hope, more realistic.

Acknowledgements

I should like to thank Melanie Frost for reading and commenting on a draft of this chapter, and the other participants at the Parkes Foundation seminar in

Cambridge in September 2007 at which these ideas were first presented for their encouraging comments.

References

Adsera, A. (2006). Marital fertility and religion in Spain, 1985 and 1999. *Population Studies*, 60, 205–21.

Becker, G. (1960). An economic analysis of fertility. In *Demographic and Economic Change in Developed Countries, Conference of the Universities-National Bureau Committee for Economic Research, a Report of the National Bureau of Economic Research.* Princeton, NJ: Princeton University Press, pp. 209–31.

Blake, J. (1968). Are babies consumer durables? A critique of the economic theory of reproductive motivation. *Population Studies*, 22, 5–25.

Bledsoe, C.H. (2002). *Contingent Lives: Fertility, Time, and Aging in West Africa* (with contributions by F. Banja). Chicago, IL: University of Chicago Press.

Caldwell, J.C., Orubuloye, I.O. & Caldwell, P. (1992). Fertility decline in Africa: a new type of transition? *Population and Development Review*, 18, 211–42.

Cleland, J. & Rodríguez, G. (1988). The effect of parental education on marital fertility in developing countries. *Population Studies*, 42, 419–42.

Cleland, J. & Wilson, C. (1987). Demand theories of the fertility transition: an iconoclastic view. *Population Studies*, 41, 5–30.

Coale, A. & Trussell, J. (1974). Model fertility schedules: variations in the age structure of childbearing in human populations. *Population Index*, 40, 185–258.

Coale, A. & Trussell, J. (1978). Technical note: finding the two parameters that specify a model schedule of marital fertility. *Population Index*, 44, 203–13.

Dahal, G.P., Padmadas, S.S. & Hinde, P.R.A. (2008). Fertility-limiting behavior and contraceptive choice among men in Nepal. *International Family Planning Perspectives*, 34, 6–14.

David, P.A. & Sanderson, W. (1987). The emergence of a two-child norm among American birth controllers. *Population and Development Review*, 13, 1–41.

Demographic and Health Surveys (2009). http://www.measuredhs.com/aboutsurveys/search/search_survey_main.cfm?SrvyTp=country.

Duesenberry, J.S. (1960). Comment. In *Demographic and Economic Change in Developed Countries, Conference of the Universities–National Bureau Committee for Economic Research, a Report of the National Bureau of Economic Research.* Princeton, NJ: Princeton University Press, pp. 231–4.

Easterlin, R.A. & Crimmins, E. (1985). *The Fertility Revolution: A Supply – Demand Analysis.* Chicago, IL: University of Chicago Press.

Fisher, K. (2006). *Birth Control, Sex and Marriage in Britain, 1918–1960.* Oxford: Oxford University Press.

Fisher, K. & Szreter, K. (2003). 'They prefer withdrawal': the choice of birth control in Britain, 1918–1950. *Journal of Interdisciplinary History*, 34, 263–91.

Garrett, E., Reid, A., Schürer, K. & Szreter, S. (2001). *Changing Family Size in England and Wales: Place, Class and Demography, 1891–1911.* Cambridge: Cambridge University Press.

Goldstein, H. (2003). *Multilevel Statistical Models*, 3rd edn. London: Arnold.

Hajnal, J. (1982). Two kinds of preindustrial household formation system. *Population and Development Review*, 8, 449–94.

Henry, L. (1961). Some data on natural fertility. *Eugenics Quarterly*, 8, 81–91.

Hoem, J. (1990). Social policy and recent fertility change in Sweden. *Population and Development Review*, 16, 735–48.

Kahneman, D. & Tversky, A., eds. (2000). *Choices, Values and Frames*. Cambridge: Cambridge University Press.

Knodel, J. (1977). Family limitation and the fertility transition: evidence from the age patterns of fertility in Europe and Asia. *Population Studies*, 31, 219–49.

Knodel, J. (1983). Natural fertility: age patterns, levels and trends. In *Determinants of Fertility in Developing Countries:* Vol. 1, *Supply and Demand for Children*, ed. R.A. Bulatao and R.D. Lee. New York, NY: Academic Press, pp. 61–102.

Lee, J. & Feng, W. (1999). *One Quarter of Humanity: Malthusian Mythology and Chinese Realities*. Cambridge, MA: Harvard University Press.

Leone, T. & Hinde, A. (2007). Sterilization and union instability in Brazil. *Journal of Biosocial Science*, 37, 459–69.

Lesthaeghe, R. (1995). The second demographic transition in Western countries: an interpretation. In *Gender and Family Change in Industrialized Countries*, ed. K.O. Mason and A.-M. Jensen. Oxford: Clarendon Press, pp. 17–62.

Lesthaeghe, R. & Page, H., eds. (1981). *Child Spacing in Tropical Africa: Traditions and Change*. London: Academic Press.

Malthus, T.R. (1798). *An Essay on the Principle of Population, as it Affects the Future Improvement of Society, with Remarks on the Speculations of Mr. Godwin, M. Condorcet and Other Writers*. London: J. Johnson.

Malthus, T.R. (1986) (First published 1798). *An Essay on the Principle of Population: the Sixth Edition (1826) with Variant Readings from the Second Edition (1803) in The Works of Thomas Robert Malthus*, ed. E.A. Wrigley and D. Souden, vol. III. London: Pickering.

Ní Bhrolcháin, M. (1986a). Women's paid work and the timing of births: longitudinal evidence. *European Journal of Population*, 2, 43–70.

Ní Bhrolcháin, M. (1986b). The interpretation and role of work-associated accelerated childbearing in post-war Britain. *European Journal of Population*, 2, 135–54.

Notestein, F.W. (1945). Population – the long view. In *Food for the World*, ed. T.W. Schultz. Chicago, IL: University of Chicago Press, pp. 36–57.

Notestein, F.W. (1953). Economic problems of population change. In *Proceedings of the Eighth International Conference of Agricultural Economists*. London: Oxford University Press, pp. 13–31.

Office for National Statistics (2008). *Birth Statistics* (series FM1 no. 36). London: Office for National Statistics. http://www.statistics.gov.uk/downloads/theme_population/FM1_36/FM1-No36.pdf.

Szreter, S. (1996). *Fertility, Class and Gender in Britain 1860–1940*. Cambridge: Cambridge University Press.

Van de Kaa, D. (1987). Europe's second demographic transition. *Population Bulletin*, 42, 1–59.

Woods, R. (1985). The fertility transition in nineteenth-century England and Wales: a social class model. *Tijdschrift voor Economische en Sociale Geografie*, 76, 180–91.

Wrigley, E.A., Davies, R.S., Oeppen, J.E. & Schofield, R.S. (1997). *English Population History from Family Reconstitution 1580–1837*. Cambridge: Cambridge University Press.

Wrigley, E.A. & Schofield, R.S. (1981). *The Population History of England 1541–1871: A Reconstruction*. London: Edward Arnold.

Index

Printed in the United States
by Baker & Taylor Publisher Services

Printed in the United States
by Baker & Taylor Publisher Services